# AIRCRAFT SHEET METAL WORK

FUSELAGE SECTIONS OF FLYING FORTRESSES ON ASSEMBLY FLOOR WHERE WIRING AND INSTRUMENTS ARE INSTALLED
*Courtesy of Boeing Aircraft Co., Seattle, Wash.*

# *Aircraft* SHEET METAL WORK

## HOW TO DO

Blueprint Reading, Template Layout, Patterns for Bends, Riveting, Soldering, Brazing, Welding, Drop Hammer Work

By CLARENCE ALLEN LeMASTER

*Member United Air Lines Supervisory Staff*

Originally published in 1944

ISBN: 978-1-940001-34-0

The Aviation Collection
by
Sportsman's Vintage Press
2015

# Preface

THIS BOOK is planned to serve as a basic course of instruction for apprentices and other students of aircraft sheet metal work and as a refresher for mechanics who are more or less experienced in the work of this trade. The general contents of the book were inspired by, and are based on, the author's many years of experience as a mechanic in the trade and as a teacher of apprentices. Both experiences, especially the latter, gave the author ample opportunity to observe, understand, and devise means of overcoming the learning difficulties encountered by beginners, and to build up a course suitable in approach and presentation to meet the needs of apprentices or students and experienced workers alike.

The general plan of the book is quite different from that of the usual textbook in aircraft sheet metal work, a plan which the author formulated to teach beginners and experienced workers how to carry on the various processes and operations required of a finished sheet metal mechanic in the aviation field.

The first chapters in the book have to do with safety rules, personal and shop-furnished tools, and blueprint reading. The author considers these features of first importance, especially to beginners. The chapter on tools not only includes a description of personal tools, but suggests a method of gradually accumulating them in pace with expanding knowledge; this chapter is especially valuable because of its instruction in how to use the tools. The author feels that the ability to read blueprints ranks first in importance and has explained how to do this by a method entirely independent of mechanical drawing and other subjects which are not required of sheet metal workers.

The remaining chapters progress from simple to more complicated processes and operations. Emphasis is always placed on how to do the work—how to rivet, how to weld, how to use the drop hammer, how to figure bend allowances, how to do all the things the sheet metal man must be able to do before he can be classed a full-fledged mechanic. The author has an ever-growing conviction that begin-

# PREFACE

ners must actually do sheet metal work as they study it. For that reason he has included many simple yet practical projects which can be done using few and ordinary tools and but little material. Projects are at the end of those chapters which deal with processes; and they are progressive and in line with the chapter content.

Many instructive pictures showing actual operations in the shop, plus an even greater number of drawings, are used throughout the book to illustrate typical sheet metal work and the principles explained. Questions and answers are supplied to help the reader test his accumulating knowledge. Questions without answers are also presented for test purposes.

Extensive explanations relative to the materials a sheet metal worker encounters, such as aluminum and steel, have been included so that he will know their properties, especially in regard to working, heat-treating, strengths, and values.

C. A. LeMaster

Cheyenne, Wyoming
October 1943

*The material in this volume is also available in the Aviation Cyclopedia.*

# Acknowledgments

THE AUTHOR gratefully acknowledges the cooperation of the following aircraft and tool companies and other organizations which have aided him in giving this book the scope he desired:

>    Aircraft Tools, Inc., Los Angeles, Cal.
>    Aluminum Co. of America, Pittsburgh, Pa.
>    American Standards Association
>    Behr-Manning Corp., Troy, N. Y.
>    Bell Aircraft Corp., Buffalo, N. Y.
>    Boeing Aircraft Co., Seattle, Wash.
>    Chambersburg Engineering Co., Chambersburg, Pa.
>    Chicago Pneumatic Tool Co., New York, N. Y.
>    Civil Aeronautics Administration, Washington, D. C.
>    Cleveland Pneumatic Tool Co., Cleveland, Ohio
>    Cleveland Twist Drill Co., Cleveland, Ohio
>    Continental Machine Specialties, Inc., Minneapolis, Minn.
>    The Dill Manufacturing Co., Cleveland, Ohio
>    Douglas Aircraft Co., Inc., Santa Monica, Cal.
>    E. I. du Pont de Nemours and Co., Inc., Wilmington, Del.
>    B. F. Goodrich Co., Akron, Ohio
>    Independent Pneumatic Tool Co., Chicago, Ill.
>    Lockheed Aircraft Corp., Burbank, Cal.
>    The Glenn L. Martin Co., Baltimore, Md.
>    The Nicholson File Co., Providence, R. I.
>    Society of Automotive Engineers, Inc., Twenty Nine West
>       Thirty Ninth Street, New York, N. Y.
>    The L. S. Starrett Co., Athol, Mass.
>    United Air Lines, Chicago, Ill., and Cheyenne, Wyo.
>    Vought-Sikorsky Aircraft Division, United Aircraft Corp.,
>       East Hartford, Conn.
>    J. Wiss and Sons Co., Newark, N. J.

Care has been exercised throughout the book to acknowledge all cooperation as it occurs. Special acknowledgment is due A. E. Burke, Head of American Technical Society's drafting department, for his work on the illustrations.

Above. Sheet metal for cowling parts of a Flying Fortress is being drawn and stretched over a die in a hydraulic stretch press.
*Courtesy of Boeing Aircraft Co., Seattle, Wash.*

Lower Left. Workman aligning sheet metal nose rib for a Martin bomber on steel jigs preparatory to drilling operation. Lower Right. Skin has been placed over nose ribs and drill jig lowered onto sheet metal. Worker is drilling rivet holes using drill guides in jig.
*Courtesy of The Glenn L. Martin Co., Baltimore, Md.*

# Contents

| | |
|---|---|
| Hints for Safety and Production | 1 |
| Tools | 5 |
| Files and Their Uses | 29 |
| Blueprint Reading | 37 |
| Measuring and Measuring Tools | 77 |
| Template Layout and Bench Work | 87 |
| Pattern Development for Bends | 111 |
| Rivets and Riveting | 133 |
| Skin Fitting, General Fabrication | 189 |
| Soldering, Brazing, and Welding | 195 |
| Use of Drop Hammer | 241 |
| Assembly, Repairs, Techniques, Projects | 271 |
| Aluminum and Related Metals | 317 |
| Steel in Aircraft Construction | 353 |
| Index | 371 |

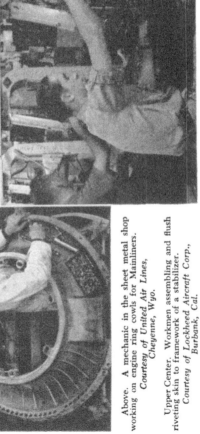

Above. A mechanic in the sheet metal shop working on engine ring cowls for Mainliners.
*Courtesy of United Air Lines, Cheyenne, Wyo.*

Upper Center. Workmen assembling and flush riveting skin to framework of a stabilizer.
*Courtesy of Lockheed Aircraft Corp., Burbank, Cal.*

Above. This hydraulic press has a pressure capacity of 5,000 tons and uses rubber mats which act as female dies in forming sheet metal.

Left. Drill presses in operation. Man in foreground is working with small drill die in the fabrication of small part.
*Courtesy of Douglas Aircraft Co., Inc., Santa Monica, Cal.*

# AIRCRAFT SHEET METAL WORK

*Preliminary Chapter*

## Helpful Hints for Safety and Production

We begin this book by saying that the aircraft industry is faced with tremendous and expanding production schedules. To maintain such schedules, every worker must do his part. Here in this preliminary chapter, we submit a few suggestions which, if followed, will aid in this production program. The first ten of these are safety rules, since the disregard of safety is one of the worst enemies of production; the other twelve are suggestions for making you more efficient.

Good luck to you!

1. Management is deeply concerned with Safety. Co-operate!

2. No matter where you work on airplanes, in a factory, for Army or Navy, in a repair base, or for a private owner, NEVER use electric tools, welding apparatus, etc., before making sure the plane is grounded by approved grounding equipment. Ordinarily, the friction of air over the various surfaces of the plane in flight tends to generate a static charge, which should be reduced to ground potential as soon as possible after landing.

3. Abrasion stripping, made from cotton fabric for use between removable fillets which screw on the plane, is sometimes soaked in soya bean oil. Care must be taken in storing these strips, because of the tendency to spontaneous combustion. In general, keep an eye out for any kind of oily rags; they constitute a fire hazard.

4. Always use soapy water and look for bubbles in a search for a leak in pressurized gas tank, oil tank, natural gas line, or welding manifold or hose. NEVER use a match, torch, etc., in testing for leaks.

5. NEVER use oil or grease on or around oxygen tanks or gages.

6. ALWAYS wear goggles when welding.

7. Guard your eyes not only from flying particles or objects,

and the glare of visible light, but also from the potentially dangerous ultraviolet and infrared rays. Do not watch arc welding operations, and ask a doctor or your optometrist about symptoms of eyestrain.

8. Watch the squaring shears. They might be called the guillotine of the sheet metal shop. They will cut off fingers as easily as you can cut cheese.

9. In hot climates or hot weather, or if you are working around drop hammers or heat-treating tanks, the use of salt tablets with your drinking water will help prevent heat fatigue.

10. Horseplay is always dangerous, and with an airhose it has been known to be fatal. NEVER fool with an air hose. It is an essential piece of equipment, but it is also a dangerous weapon. A recent item in the *National Safety News* describes a freak accident to a man who, at the close of a hot day, attempted to clean fine wood dust off his clothes with the air hose.

The following suggestions may help you make your shop work smoother and more effective.

1. Never be afraid to go to the foreman or instructor with any problem. Think your question over carefully; then if you are still in doubt as to procedure, he will be glad to help you.

2. Co-operate, always, in solving problems of work simplification or increased efficiency.

3. Do not wait for the foreman or instructor to discover you are out of work. When you finish a given job, keep busy at something until you have a chance to report that you have finished the job you were working on. A successful mechanic must contain his own self-starter.

4. If you are an apprentice or mechanic, you are being paid for the time you waste. And wasting time does not necessarily mean loafing on the job. Some very hard workers accomplish less than others who do not seem to work so hard. Plan your work. Be sure you understand what you are to do, then do it efficiently.

5. Make every job as perfect as you can, and NEVER conceal a mistake.

6. If another shift is to carry on, do not leave any job unfinished unless someone knows how to follow it through, or unless your foreman is informed.

## HINTS ON SAFETY AND PRODUCTION

7. You will not be expected to memorize charts on steel or other material specifications. But you are expected to be familiar enough with them so that you will know how and when to refer to them when you have need of such information.

8. Do not walk on wings, center sections, fuselage, or stabilizers without first protecting the surface with wing walks, pads, or suitable protective measures. Inspect the soles of your shoes for protruding nails and for imbedded chips or shavings that might scratch the surface of the skin. Watch where you step, and be sure the structure is sufficiently strong to hold your weight before you apply it.

9. When drilling hard metals, never use a pilot hole over half the diameter of the final hole, as it will cause the lips to burn at the outer edges, and consequently ruin the drill.

10. Watch the tools. If you lose them, leave them in a wing tip, center section, or under the floor boards of an airplane they are apt to be returned with some stern advice.

11. A Master Mechanic takes pride in his work, takes pride in his tools. He is careful with his own as well as the company's tools.

12. DO NOT WASTE MATERIALS.

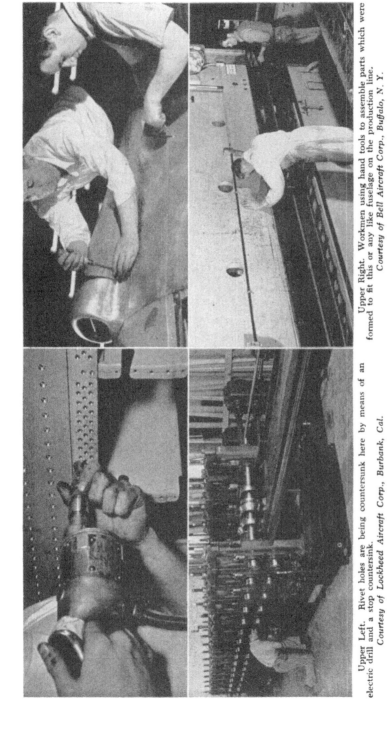

Upper Left. Rivet holes are being countersunk here by means of an electric drill and a stop countersink.
*Courtesy of Lockheed Aircraft Corp., Burbank, Cal.*

Lower Left. The cold rollers shown here are used for the fabrication of corrugated sections.
*Courtesy of Douglas Aircraft Co., Inc., Santa Monica, Cal.*

Upper Right. Workmen using hand tools to assemble parts which were formed to fit this or any like fuselage on the production line.
*Courtesy of Bell Aircraft Corp., Buffalo, N. Y.*

Lower Right. These two workmen are adjusting a piece of sheet metal prior to forming in a huge power brake.
*Courtesy of Douglas Aircraft Co., Inc., Santa Monica, Cal.*

*Chapter I*

# Tools

## PERSONAL TOOL REQUIREMENTS

It is essential that a mechanic possess his own kit of tools. Potentially good craftsmen handicap themselves by not having the proper assortment of tools. Borrowing from a fellow worker is not considered good shop ethics, for, besides loss of independence and prestige, it is bound to cause unnecessary conversation and may lead to ill feeling and shop friction. A complete set of tools of highest quality is one investment a mechanic will never regret.

However, a complete set of tools is too expensive to be purchased at one time by the large majority of students and apprentices. Tool requirements, therefore, are listed in the order in which the tools should be bought. Under ordinary circumstances, an employed mechanic should be able to complete the list in two to three years. Because of the importance of having a machinists' tool box or kit for proper care of the fine tools used at bench work, this should be purchased first. A bag-shaped kit with folding trays is convenient for carrying the tools to different jobs.

In the list here given, the first fourteen are the tools usually required by student or apprentice:

1. Flexible scale, graduated in 64ths, 6"
2. Flexible rule, 6'
3. Scriber or scratch awl
4. Dividers, toolmakers' spring, 4"
5. Combination square, 12"
6. Gas pliers, slip joint, 6"
7. Diagonal cutters, 6"
8. Center punch
9. Hammer, ball peen, 6 oz.
10. Tin-snips, Wiss No. 18
11. Hack-saw frame
12. Screw driver, medium size
13. Jackknife
14. Small cold chisels, one set
15. Airplane snips, one pair of left and one pair of right
16. Crescent wrench, 6"
17. Crescent wrench, 10"
18. Open-end wrenches, sizes 3/8", 7/16", 1/2", 9/16"
19. Spin tight, or midget socket wrenches, one set
20. Mallets (rawhide, pyralin, or fiber)
21. Beveled edge layout scale, graduated in 64ths, 18"
22. Hermaphrodite calipers
23. Flashlight
24. Prick punch
25. Screw driver, large size
26. Shorty screw driver
27. Riveting hammer
28. Setting hammer

29. Drive pin punches, one set, 1/16", 3/32", 1/8", 5/32", 3/16", 1/4"
30. C clamps, small
31. Long nose pliers
32. Smoothing hammer
33. Slip nut dividers, 8"
34. Drill box for No. drills
35. Micrometer, 1" (1/2" tubing micrometer optional)
36. Trammel points, one pair
50. Offset screw driver
51. End snippers
52. Ding hammer
53. Toe dolly
54. Gage, drill, tap and steel wire
55. Calipers, 6", inside and outside
56. Plumb bob
57. Steel tape, 50'
58. Universal surface gage, with 9" spindle
59. Drill blocks and clamps

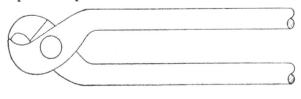

*(A)* FENDER PLIERS

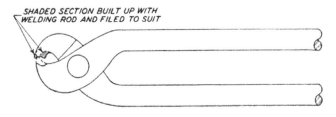

*(B)* WIRING TONGS

Fig. 1. How Fender Pliers Are Made into Wiring Tongs

37. Protractor head, to fit combination set
38. Yankee automatic screw driver
39. Yankee hand drill
40. Automatic center punch
41. Saw frame, jewelers
42. Tap wrench, T-handle ratchet
43. Lining-up punches, two, medium size
44. Fraction drill case
45. Oil stone
46. Cross peen hammer, 8 oz.
47. Thickness gage, .0015" to .015", in case
48. Pocket slide calipers, 3" Starrett
49. Oil can

60. Fender pliers for making wiring tongs, Fig. 1. (Wiring tongs are sometimes made by reworking the jaws of pincers or nippers, but the fender pliers are preferable because they are designed to give the greater leverage needed for wiring heavier gage metals.) Build up the upper jaw with welding rod and file to a radius which will admit several sizes of rolled edges. Work the lower jaw so that it will set the metal tightly against the wire. File and polish the jaws to desired shape by trial and error.

*Note:* Drills, hack-saw blades, electric drills, soldering irons, air hammers, etc., are usually furnished by toolrooms.

# TOOLS 7

It should be understood, of course, that this list is a general one and does not include all the tools a mechanic will accumulate. Any special tool for which you have considerable need should be added to the collection.

It pays to invest in tools of good quality, as their manufacturers usually guarantee them against breakage. Keep your tools sharp, free from rust, and in order in your tool kit.

**SCRIBER AND DIVIDERS.** The scriber and dividers should be kept sharp, with tapered points. When sharpening is needed, place in a vise and use strip emery cloth on the points, the way

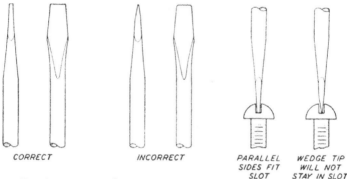

Fig. 2. Correct and Incorrect Way to Sharpen a Screw Driver by Grinding

a bootblack polishes shoes. If the point becomes too thick, grind on a fine wheel, then finish with the emery cloth.

The use of a scriber is confined to marking when laying out patterns, and marking around patterns the line to be followed in cutting. Never scribe a line for reference or to denote bend locations, except on patterns. The dividers are used for scribing radii and transferring dimensions from the scale to the pattern in layout. You will find more about the scriber in Chapter IV, where measuring tools are considered.

**FILES.** Files will be discussed in Chapter II.

**SCREW DRIVERS.** If a screw driver becomes worn, either replace it with a new one or grind it correctly; see Fig. 2. Besides his assortment of conventional screw drivers, a mechanic should have one each to fit the Phillips screw and the Reed and Prince screw.

# AIRCRAFT SHEET METAL WORK

**TWIST DRILL.** The parts of a twist drill, as shown in Fig. 3, consist of shank, body, lip or cutting edge, margin, flute, chisel point, or dead center, and tang. Body clearance is built into the body of the drill and requires no attention, but lip clearance must be ground each time the drill is sharpened.

Always tighten the drill in the chuck sufficiently to prevent slipping on the shank, as this will scrape off the identification mark, as well as allow the drill to wobble in the hole. There are proper

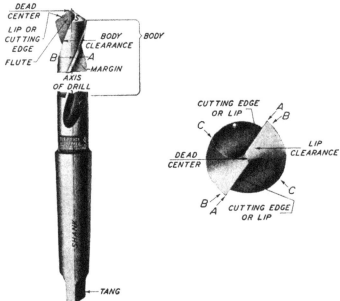

Fig. 3. Parts of Twist Drill. A-B Is Margin; B-C Is Body Clearance
*Courtesy of the Cleveland Twist Drill Company, Cleveland, Ohio*

speeds and feeds for drilling with either carbon-steel drills or high-speed drills.

**Grinding of Twist Drills.** For all general purposes, the cutting edge of a twist drill should include an angle of 118°, and 12° to 15° lip clearance. An angle of chisel point 125° to 135° is best suited for drilling aluminum alloys, iron, and mild steel. The cutting edges must have a proper and uniform angle with the longitudinal axis of the drill, so that they are of exactly equal length, with the proper lip clearance; see Fig. 4. In Fig. 4, A indicates a drill that has been ground incorrectly. The angles of the lips are equal, but their lengths are different. Here the hole is much larger than the drill. Such action is very hard on the drill. The drill shown at B is ground correctly, with the two lips of the same length and at the same angle to the axis of the drill. C, in this same figure, illustrates a drill sharpened for cutting large holes in thin sheet metal, up to .064" in thickness. A drill sharpened in this manner cuts a smooth round hole and avoids chattering. It is used in drill sizes above ¼".

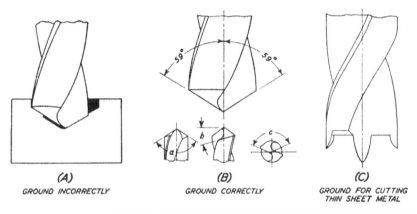

(A) GROUND INCORRECTLY    (B) GROUND CORRECTLY    (C) GROUND FOR CUTTING THIN SHEET METAL

| Material Used on | a | b | c |
| --- | --- | --- | --- |
| Average class of work | 118° | 12° to 15° | 135° |
| Heat-treated steels, drop forgings, Brinell hardness 250 | 125° | 12° | 135° |
| Steel rails 7% to 13% manganese and hard material | 150° | 10° | 135° |
| Cast iron—soft | 90° | 12° | 135° |
| Brass | 118° | 12° | 135° |
| Wood, hard rubber, Bakelite, and fiber | 60° | 12° | 135° |
| Copper | 100° | 12° | 135° |

Fig. 4. Cutting Edges on Twist Drill. All Twist Drills Are Right-Hand (Cut Clockwise) for General Use. Left-Hand Drills Are Used in Gang Drills

In the larger shops there are drill-grinding machines in the toolrooms where the correct point can be ground much more cheaply than if you try to do the job yourself. Nevertheless, every mechanic should learn the knack of sharpening his own drills, since he may not always be working in a shop that maintains its own drill-grinding machine.

Cutting lips that are not of uniform length result in wear on the margin of one cutting lip, in addition to drilling the hole oversize. Insufficient lip clearance will prevent the cutting edge from entering the metal, and this will result in burning the heel of the drill. Running the drill too fast will break down the outer corners of the cutting lips and cause considerable wear on the margin of the drill.

No cutting lubricants are used with high-speed drills in drilling small holes in aluminum, aluminum alloys, and mild steel. Lubricants are used to keep the tools cool at fast cutting speeds and to reduce wear on the cutting edges. Lard oil when used in its pure state is highly effective on alloy steel or mild steel and on cast iron. Saponified oil of a mineral base mixed with water is widely used.

**RIVET SETS AND RIVET SNAPS.** Most mechanics prefer to own their own rivet sets. In the factory, the riveting is done to a large extent by specialists, but a master mechanic must be able to drive rivets just as accurately as these men. It is therefore essential to have the best tools with which to do the job. There are not many sets on the market whose performance will satisfy the rigid workmanship requirements of the aircraft industry. Fig. 5, therefore, gives the specifications for those who may wish to make their rivet sets or have them made by someone who has access to a turning lathe. Thousands of sets have been made to these dimensions and have proved very popular.

Generally speaking, squeeze riveters are preferable to other types for driving rivets in aluminum. Their use assures properly upset shanks and well-centered heads. In the assembly and actual fabrication and repair of an airplane, however, it is often impossible to use any kind of a squeeze riveter, and the work must be done with air gun and rivet sets. Individually owned sets, naturally, are not subject to as hard wear as those in general use. This is one reason why a mechanic prefers to buy or make his own; the convenience of not having to check rivet sets out of the tool crib, thus depleting his supply of tool checks, is another consideration.

In making these rivet sets, the material used must be a high-grade carbon tool steel, such as Bethlehem 71, Braeburn Extra Annealed (carbon tool steel), Ceswic Special Annealed (carbon tool steel), or any tool steel recommended for chisel stock. The cup should be slightly wider and shallower than the manufactured head of the rivet, so that the initial contact will be at the end of the head,

# TOOLS 11

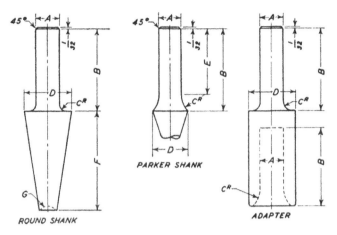

|   |              | A    | B       | C     | D     | E      |
|---|--------------|------|---------|-------|-------|--------|
| I | Parker Shank | .401 | ......  | .500  | 3/4   | 1.164  |
| II | Parker Shank | .497 | ......  | .75   | 27/32 | 1.4175 |
| III | Parker Shank | .498 | ......  | .75   | 27/32 | 1.105  |
| IV | Round Shank | .401 | 1.4062  | 1/16  | 5/8   | ..... |
| V | Round Shank | .497 | 1.75    | .0937 | 13/16 | ..... |
| VI | Round Shank | .401 | 1 13/32 | 5/64  | 11/16 | ..... |
| VII | Round Shank | .498 | 1 7/16  | 1/16  | 27/32 | ..... |

Rivet Set for Thor Riveter 03-04-00G-00L-00B-3S-4S: Use Dimensions I or IV.

Rivet Set for Thor Riveter OP-OG-OL-OGR-OPR-OLR-MM-NN: Use Dimensions II or V.

Rivet Set for Cleco Style G, Use Dimensions VI; Style GP, I; Style JP, III; Style J, VII.

Rivet Set for Chicago Pneumatic CP-215 Simplate, UU-Ring Valve, O-Boyer, U-Boyer, OO-Boyer, UU-Boyer: Use Dimensions VI. For CP-204 One-Shot, CP-6 Aero Slow-Hitting: Use Dimensions III. For CP-3 Aero Slow-Hitting: Use Dimensions I.

Rivet Set for Ingersoll-Rand AV-1, AV-2, AV-3 (Short-Stroke) and AV-11, AV-12, AV-13 (Long-Stroke): Use Dimensions I or VI.

Fig. 5. Rivet Sets or Snaps

directly in line with the shank as shown in the top center illustration in Fig. 6. This will prevent the shank from being driven up into the head and will facilitate uniform upsetting throughout the length of the shank.

After the rivet snaps have been turned in the lathe, they must be heat-treated to develop their strength and wear resistance. Hardening temperatures are indicated by the steel manufacturers. The recommended temperature should be adhered to unless the user is willing to experiment by judging colors to find a temperature that will suit the particular requirements. If no hardening furnace is available, the oxyacetylene welding torch must be utilized. Heat the steel to a full red heat or above the critical, then quickly plunge it into a bath of clean cold water or oil. Move it backward

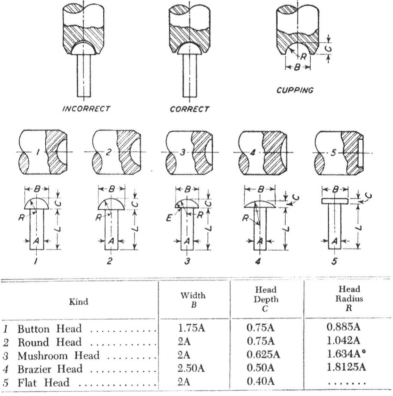

| Kind | Width B | Head Depth C | Head Radius R |
|---|---|---|---|
| 1 Button Head | 1.75A | 0.75A | 0.885A |
| 2 Round Head | 2A | 0.75A | 1.042A |
| 3 Mushroom Head | 2A | 0.625A | 1.634A* |
| 4 Brazier Head | 2.50A | 0.50A | 1.8125A |
| 5 Flat Head | 2A | 0.40A | ....... |

\* Edge radius "E" for mushroom head is 0.50A.

Fig. 6. Rivet Set or Snap Data

and forward through the liquid, thus hastening the cooling by continued contact with cold water or oil and preventing the formation of an insulating layer of steam around the tool.

As soon as the hardening is complete, the tempering process should be begun. In this operation, a piece of emery cloth is used

# TOOLS 13

to remove scale and give the entire tool a polished surface. Apply the heat in the center of the length of the set, which must be drawn to a softer temper, especially at the root of the shank, to prevent breakage in use. The end of the shank and the cupping especially require the proper hardness, and the tool should be quenched when the color at these points is changing from purple to dark blue.

**AIRCRAFT SNIPS (LEFT AND RIGHT).** Snips of the type shown in Fig. 7 are indispensable to the aircraft mechanic. A set

Fig. 7. Aircraft Snips (Left and Right)
*Courtesy of J. Wiss and Sons Co.,
Newark, N. J.*

of aircraft snips consists of two pair, one for cutting to the left and one for cutting to the right. They are used as a major tool for cutting out circles and squares, also intricate designs and irregular patterns, on stainless steel, aluminum alloys, tubing ends, exhaust manifolds, and the like. The handles, which are of pressed steel, have a compound action. The snips are made with serrations which prevent the blades from slipping out when starting a cut from a punched hole or in close quarters. They are pocket size.

**SHEARS.** The hand shears or "snip" does not cut hard or heat-treated metal, but parts it in a shearing action. Try cutting a piece of such metal, and you will actually see the metal parting before there is any cutting action. For this reason an oil stone should never be used on a pair of snips. Keep the jaws polished and free from rust, and when the cutting edges need sharpening, disassemble the jaws by removing the bolt and grind across the edge at a slight bevel on a medium fine (½"x6") wheel. Usually one sweep across the face of the wheel will do a good job. The fine grinder marks left by the wheel are considered desirable in a pair of snips, as shears sharpened in this way are not so likely to slip back out of the work on heavy cutting.

On a job where there is much heavy cutting and the mechanic does not have a powerful forearm, the work can be done much more easily by resting the bottom handle of the shears on the bench and bearing down with the body, also.

Do not cut wire, rod, or anything that may spring the jaws. Canvas will buff the edges of a sharp pair of snips rapidly.

## SHOP-FURNISHED TOOLS

**SQUARING SHEARS.** Sheets are fed into the machine from either front or rear, depending on which is more convenient. When the sheet enters from the rear, the front bed gages are used, and, if the piece to be cut is larger than the bed in the machine, the gage is carried out on the front arms provided for gage extension.

In squaring and trimming small pieces, the material is inserted between the cutting blades from the front of the machine, the work being gaged either by sight to scribed marks, or on the bed with the short or the long bed gage, or with the rear gage, according to the work that is to be sheared. Squaring shears should never be used for material exceeding the maximum thickness specified by the manufacturer, even if the pieces to be cut are narrow.

Well-equipped sheet metal shops also provide other shears, including lever slitting shears, scroll shears, rotary slitting shears, rotary circular shears, ring and circumference shears, and many others, both hand and power operated.

**BENDING BRAKE.** This machine is one of the most indispensable units in the shop. Unfortunately, it is often the most neglected, also, because it is used by everyone.

Adjustments must be set for each different thickness of metal, angle of bend, and radius of bend. As the brake does not have a gage, the sheet or piece to be bent is laid out, previous to forming, with fine pencil marks. The pencil mark used for lining up the sheet by eye with the edge of the folding blade radius must be the distance of the radius out from the beginning point of the bend.

Different sized radii are obtained by building up the upper jaw or nose of the folding blade with different thicknesses of metal, after which the brake is adjusted for clearance for the metal to be bent. If several pieces are to be bent with one setup, straps can be welded to the rear of the radius filler and clamped to the rear section of the clamping bar to hold it firm.

Adjustments are provided by the eccentric connection to maintain a maximum pressure on the material in the machine while in the process of forming, and thus prevent the work slipping from the true line. In order to obtain the correct radius throughout the bend, make sure the bending plate at top of bending leaf is parallel with the bed or lower jaw of the brake. Also, see that clearance for metal thickness is uniform throughout the length of the brake jaws.

For heavy bending there is an angle-iron attachment which fits on the bending leaf and is held by friction clamps or screws. Do not use a steel hammer on work in the brake, or any tool that will damage the jaws which should be kept free from dents.

Knowledge of the direction of grain flow is essential for correct practice in bending. In most cases, the bending operation should be performed crosswise to the grain for maximum strength. However, where strength is required across a long angle or channel, the material, in all cases, should have the grain flow lengthwise regardless of bends.

The box brake has sectional upper jaws of various widths which are removable. By inserting the proper jaws, almost any size can be formed on all four sides; this cannot be done on the cornice brake. The bar folder is used for braking thin material and folding metal over 180° for making seams.

**FORMING MACHINE OR ROLLS.** There is no fixed rule that may be applied for the setting of the rolls to secure any desired circle. The necessary adjustments are best determined through experiment. The sheet metal to be formed is inserted from the front of the machine and passed through the two front gripping rolls, securing uniform pressure of the rolls on the material through adjusting screws so that the sheet will ride through the rolls freely. In forming cylinders of small diameters, the metal should be given sufficient curvature by bending the material upwards as soon as it enters through the gripping rolls. This prevents a flat space, and helps the material strike the rear forming roll in proper position for the shaping of the cylinders.

The slip roll forming machine has the advantage of allowing the formed work to be slipped off the end of the front upper roll.

**BENCH MACHINES.** There are so many kinds of bench machines that it would be impossible to describe them all in this book. Their use in airplane manufacture is somewhat diminishing as newer construction methods are developed. However, there are times when the turning machine, burring machine, setting down machine, crimping and beading machines, and other like equipment are necessary. The punching machine is still as widely used as the lever slitting shear. It is easy to learn to operate these machines, and the student will develop skill as he uses them.

**BENCH STAKES.** It is often necessary to form sheet metal by hand, over anvils of peculiar form known as *stakes.* These fit into holes cut in a bench plate or into a stake holder that is fastened to the bench. A detailed description of each stake need not be given, for in most cases its shape indicates its uses. This is illustrated in Fig. 8, where the following stakes are shown:

(1) beakhorn; (2) candle mold; (3) blowhorn stake; (4) creasing stake with horn; (5) needlecase stake; (6) square stake; (7) creasing stake; (8) double seaming stake; (9) round head stake, and (10) hollow mandrel stake. Also shown are (11) ball peen hammer and (12) setting hammer.

The setting hammer included in Fig. 8 is so named because of the peen which is used for setting metal down in corners, etc. The riveting hammer is similar, except that the peen has a round nose. A mallet is used for flanging and forming metal over flanging blocks; or, rounded to ball shape at one end, it is used with a sandbag to bump metal into concave shape. Mallets are made of rawhide, pyralin, rubber, fiber, or wood. Lignum vitae makes a very good mallet for forming operations, standing up well when turned with ball ends; but because this wood is very expensive, it is not used for flat-faced mallets.

**RIVETERS.** Riveters for use in airplane manufacture and repair are now made in various designs by several pneumatic tool companies. Most airplane manufacturers stock the riveters of a number of firms, usually several types made by each, including high-

# TOOLS

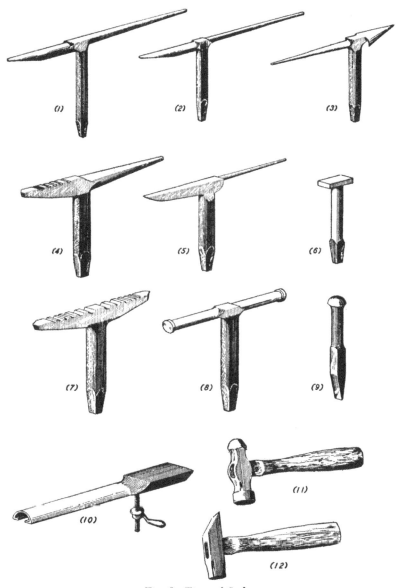

Fig. 8. Tinners' Stakes

speed or *fast-hitting* light riveters, the single-impact or *one-shot* riveters, the compressing or *squeeze* riveters, and, a more recent development, the long-stroke, or *slow-hitting* riveters.

The terms slow hitting and fast hitting are, of course, relative when applied to aircraft work. What might be fast for one job might be slow for another. Generally speaking, riveters ranging up to 2,500 blows per minute might be termed slow hitters, while those delivering 2,500 to 5,000 blows per minute can be called fast hitters.

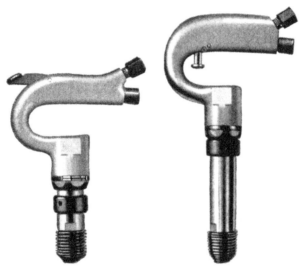

Fig. 9. (Left) "Slow-Hitting" and (Right) "Fast-Hitting" Cleco Riveters for Aircraft Work
*Courtesy of the Cleveland Pneumatic Tool Co., Cleveland, Ohio*

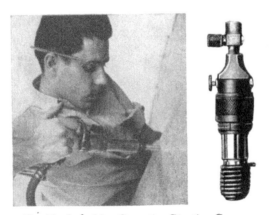

Fig 10. Left, Man Operating Riveting Gun; Right, UU Boyer (Push Button)
*Courtesy of Chicago Pneumatic Tool Co., New York, N. Y.*

TOOLS 19

The size of the rivet, the material involved, its relative hardness—all these enter into the problem, so that it is necessary to determine by practical test the tool best suited to the work.

The important thing is to select a tool sufficiently powerful to drive the rivet with a comparatively small number of blows; this

Fig. 11. 3S and 4S Thor "Slow-Hitting" Riveter with Offset Handle
*Courtesy of Independent Pneumatic Tool Co., Chicago, Ill.*

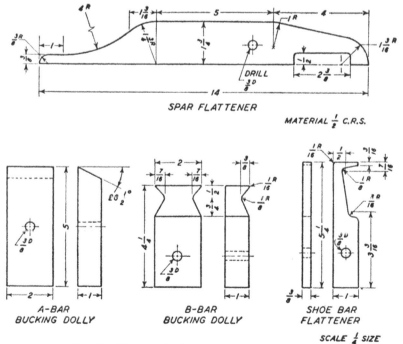

Fig. 12. Riveters' Bucking Bars and Flattening Bars

20                    AIRCRAFT SHEET METAL WORK

makes for well-driven rivets and economy in labor. Fig. 9 shows Cleco riveters of both the slow-hitting and the fast-hitting types. Fig. 10 illustrates a riveting gun in operation, together with a view of the UU Boyer (push button). The slow-hitting riveter shown in Fig. 11 is especially desirable for dimpling operations, counterpunch operations, and driving flush rivets. The slow-hitting hammers strike

Fig. 13. Battery of Doall Contour Saw and File Machines in Aircraft Production

*Courtesy of Continental Machine Specialties, Inc., Minneapolis, Minn.*

fewer blows per minute, but with more force to each blow; the fast-hitting hammers strike a faster but a lighter blow.

**RIVETERS' BUCKING BARS AND FLATTENING BARS.** Such bars are made in endless variety, and every shop is equipped with a great assortment. They are usually ground to fit individual requirements. Fig. 12 shows a small assortment, for the benefit of the mechanic who wishes to make and caseharden them for his individual use. See also Figs. 10 and 11 in Chapter VII. Chapter VII is devoted entirely to rivets and riveting.

# TOOLS

**CONTOUR SAW AND FILE MACHINE.** New procedures for making dies, jigs, fixtures, and for gang-cutting stacks of flat stock have been promoted by the development of the contour saw and file machine, shown in Figs. 13, 14, and 15. This machine is a self-contained, variable-speed, motor-driven unit employing an endless continuous narrow band saw for cutting internal openings or

Fig. 14. Contour Saw and File Machine Used by Northwest Airlines, Inc. Makes Propeller Wrenches, Special Calipers, and Other Parts for Aircraft Maintenance

*Courtesy of Continental Machine Specialties, Inc., Minneapolis, Minn.*

for sawing out fittings in multiple. It is used for cutting many different types of metals, sawing metal thicknesses above three inches to surprisingly close tolerances throughout the cut. It leaves a clean narrow slit in the wake of the cut and does not burn or distort the metal. The saws come in 100' coils of various pitches, and in eleven different widths, from $1/16''$ to $1''$. See Fig. 16 for types of teeth.

The contour sawing machine, one of the newest of machine tools, is already extensively used in airplane sheet metal shops for sawing out parts. For cutting odd-shaped holes, the machine is provided with an automatic electric butt welder, so that the saw may be joined back into a band after one end has been passed through a starting hole in the work. The machines come in different throat capacities with speed ranges from 50' to 1,500' per minute. A built-in air compressor furnishes air to a jet for keeping work and

Fig. 15. Disc Cutting Attachment Clamped to Rigid Saw Guide Post (A centering drill is used to provide center required. Pivoted on this center the work is revolved into saw, thus automatically making true circles. Center may be set any distance from saw, to obtain any desired diameter.)

*Courtesy of Continental Machine Specialties, Inc., Minneapolis, Minn.*

table clear of cuttings. The unit consists of a tiltable table, tachometer for measuring speed of travel of saw, panel light, automatic butt welder, grinder for dressing off welding flush, handwheel that controls variable speed pulley, and power feed with pressure control. Various other attachments are contained in a metal cabinet.

# TOOLS

**HACK SAWS.** The hack-saw frame should be a heavy one, rigidly constructed so it will not be cramped by straining of the blade. It should have an adjustable back to take saws up to 12", which may be set to cut in each of four directions. The pistol grip handle is recommended for all-round sawing, especially for heavy work. However, the horizontal handle makes for easier control in fine, accurate work and is preferred by toolmakers. The depth of

*Pitch of Teeth:* Band saws are cut in 8-10-12-14-18-22-24-32 pitch, or teeth per inch. They are measured thus—

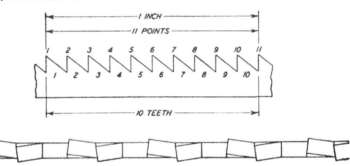

*Raker Tooth.* One straight tooth alternates with two teeth set in opposite directions. The straight tooth clears the kerf of chips. This tooth construction is used exclusively for cutting all iron or steel except sheets, tubing, or angles.

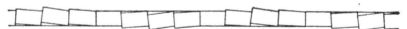

*Wave Tooth.* This is the "group" set, one set of waves to the right and the next set to the left, etc. This gives the smallest possible tooth spacing. For cutting sheets, tubing, etc.

*Straight Tooth.* This is the smallest set, i.e., one tooth set to the right, the next to the left. For brass, copper, plastics, and nonmetals.

Fig. 16. Tooth Construction of Contour Saws
*Courtesy of Continental Machine Specialties, Inc., Minneapolis, Minn.*

the frame from the cutting edge of the blade should not be less than 3".

**Hack-Saw Blades.** Too little attention often is given to the use of hack-saw blades, compared with other tools used in the shop. The cost of one good hack-saw blade is over twenty cents, so using them correctly is not only a matter of efficiency, but also of economy. The blade should be kept tightly enough in the frame to hold it straight and taut. On the other hand, too much tension is apt to break the blade at the pinholes, if the saw is twisted or cramped in use. Short cutting strokes should be avoided; a long steady downward stroke will produce a faster and cleaner cut and will prevent binding. Do not start a cut on corners, or, if this is unavoidable, use a very fine tooth blade with a steady stroke until the teeth cut through the corner into the thicker stock.

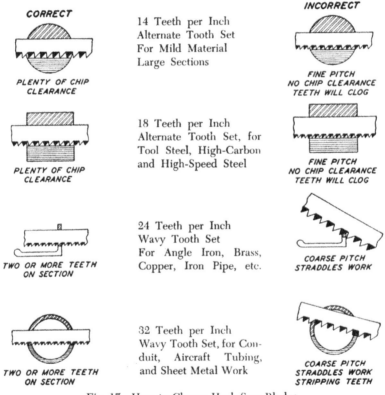

Fig. 17. How to Choose Hack-Saw Blades

In cutting tubing or sheet metal, use a fine wavy-tooth blade, engaging as many teeth as possible. The blade should be held at an angle. Coarse tooth blades used on thin wall tubing would straddle the work, which would rip out the teeth immediately. To prevent this, at least two teeth should be cutting at the very thinnest sections of the material. The larger the surface to be cut, the

# TOOLS

coarser the teeth, Fig. 17. Use of the finer tooth saws on heavy cutting would result in clogging the teeth with chips, making cutting difficult. The coarser tooth blades dig in more deeply, removing larger chips, and there is enough space between the teeth to carry the chips through the cut and drop them on the outside.

Too much weight wears out the saw blade quickly and will cause the saw to run; too little weight causes the saw to slide over the work without cutting, dulling the teeth quickly. The correct speed and feed are governed by practice and feel. As a saw dulls, increase the pressure to keep the blade cutting rather than rubbing.

## QUESTIONS

1. To what angle should a twist drill be ground for general use?
2. Name five tools used in shearing and cutting metal.
3. Name five different types of forming stakes.
4. Distinguish between a cornice brake, box brake, and bar folder.
5. By what other name is a rivet set sometimes called?
6. Why are hack-saw blades made with different size teeth?
7. Should metal-cutting snips have a sharp smooth edge?
8. Name three styles of teeth in use in metal-cutting band-saw blades.
9. Describe a ball peen hammer, riveting hammer, and setting hammer.
10. Where, and why, would you use a slow-hitting riveter? A fast-hitting riveter?

### PROJECT 1.  SHARPENING DRILLS

**Objective**
1. Practice in proper method of sharpening twist drills on electric grinder.
2. Practice in proper method of checking drill angles with drill gage.

**Tools and Equipment**
1. Electric grinder
2. Goggles
3. Twist drills (assorted)
4. Drill gage

**Materials**
°1. Lysol solution and sterile cotton (for sterilizing the goggles)
2. Water

**Procedure**
1. Sterilize the goggles.
2. Select a large twist drill 1/4" to 1/2" that is ground properly and, using the fine wheel, *polish* rather than *grind* the point.
*Note:* This is done in order that you may become familiar with the position and movements involved in grinding a drill properly.
3. Hold the drill in the right hand, between the first finger and thumb, and steady it by resting the fingers against the grinder rest. With the shank

---
° Some shops furnish individual goggles which do not need sterilization.

of the drill grasped in the left hand, line up the cutting edge of the drill with the face of the wheel. The fingers of the right hand turn the drill, while the left hand swings the shank in a downward motion, to control the grinding of the lip clearance. Through this operation the drill rocks or pivots in the fingers of the right hand.

4. Practice grinding drills of various sizes, referring again to the text on grinding twist drills, in this chapter.

5. Check drill angles with a drill gage.

6. Have the instructor inspect your work.

## PROJECT 2. HOW TO OPERATE AND ADJUST A RIVET GUN

**Objective**

1. To develop operating skill by becoming familiar with the "feel" of the riveter

2. To develop the knack of adjusting a riveter

**Tools**

1. Rivet gun
2. Rivet snap

**Material**

Small flat board, about one-half board foot

**Note**

The most important operation in riveting is to "gun the rivet" just the proper amount, i.e., time your shots; drive each rivet with one pull of the trigger or one push of the button. The gun must be adjusted so that the riveting process may be done in the fewest possible blows, though not so few that the timing cannot be judged. Too many blows on a rivet will work-harden it and destroy its physical properties or strength. Rivets of different diameter require a different adjustment of the riveter, or of the air pressure to the riveter.

In most cases two operators are required in riveting, one to operate the riveter and one to buck up the rivet with a bucking bar or dolly. When two operators are working together in a factory on large assemblies such as wings or fuselage, there is usually too much noise for them to hear each other's voice clearly, and signals must be used to denote whether a rivet is driven sufficiently.

If the bucker-up judges the rivet sufficiently driven when the riveter has finished, he taps two sharp blows in the vicinity of the rivet with the bucking iron, thus advising his partner to drive the next rivet. If the rivet is insufficiently driven, he replaces the bucking iron on the rivet with enough force to be felt by the riveter as a single blow, which means that he is to "gun it" again. The riveter and snap are not removed from the rivet head until the signal of two blows is given.

**Caution**

Never depress the trigger on an air riveter, apart from normal operation, unless a set is in it and pressure applied by hand or the set placed against a block of wood. If the plunger is forced against its retainer at full speed, it may result in a damaged retainer which will put the gun out of order.

# TOOLS

**Procedure**

1. Hold the riveter in your hand as though about to rivet.
2. Point riveter down on block of wood with snap inserted, holding it at 90° to the work.
3. Depress the trigger for about half a second, noting the amount of pressure required to keep the snap from jumping about.
4. Repeat and adjust the air pressure until you become familiar with lighter and harder blows. After some experience, you will be able to tell whether the riveter is set to drive a particular size of rivet in the proper length of time, or the minimum amount of blows.

*Note:* Where air regulators are used in the hose line, or at the manifold, your gun may have had its regulator removed and the hole plugged. In this case, the adjusting must be done with the air regulators.

## PROJECT 3. HOW TO RIVET WITH A RIVETER

**Objective**

To develop further the "feel" of the gun and the ability to rivet with it

**Tools**

1. Rivet gun
2. Set
3. Scrap metal with rows of drilled holes 1/8" set up in vise on solid frame of wood, preferably a thick and thin sheet together

**Material**

Rivets

**Exercise**

Pick up eight to sixteen rivets in the left hand and practice feeding them one by one to your finger tips, so you may be inserting one rivet while driving the other.

Learn to time the riveter to avoid repeating.

In depressing the trigger or button, do so as quickly as possible, holding it down to the stop and thus allowing the full volume of air to pass through the valve. This is good riveting practice and is important to timing the operations. Pushing the trigger down slowly, or only part way, prevents the establishment of a fixed rhythm of operations.

**Procedure**

1. Hold the riveter in the right hand.
2. Put rivet in the hole with the left hand.
3. Put set on rivet.
4. When you feel the bucker rap the rivet, press the trigger, at the same time holding enough pressure on the rivet to prevent the set from jumping off and damaging the surrounding metal.
5. As you are doing step 4, be inserting another rivet in the next hole with your left hand. You must keep your fingers on it, too, or the bucking tool will knock it out before you have a chance to get the set over the rivet head.

*Note:* Do not pull the trigger before you have applied pressure on the rivet. Do not hurry; speed will be acquired with practice.

## PROJECT 4. TOOL IDENTIFICATION

**Objective**

    Learning to etch tools for identification

**Tool**

    Scriber

**Materials**

    Medicine dropper
    Beeswax
    Etching solution. (The etching solution may be muriatic acid, 3 parts by volume, and nitric acid, one part by volume. Or, dissolve in 150 parts of vinegar, 30 parts sulphate of copper, 8 parts alum, 11 parts salt. Add a few drops of nitric acid, depending on how long the mixture is to act.)

**Note**

    When a new tool is purchased, for its protection it should be marked with the owner's initials and name. Since tools of good quality are too hard to punch letters in, another means of marking must be used—etching.

**Procedure**

    1. Clean the surface of the tool and apply a thin film of beeswax. This can be done easily by warming the tool just to the point where the wax will melt and flow evenly over the surface, then letting it cool.

    2. When the wax is cold, write your name and any desired data with the scriber.

    3. Using a medicine dropper, apply a few drops of etching acid on the surfaces exposed by the scriber. Care should be taken not to allow the etching solution to run onto any uncoated or unprotected metal surface. Wax may be removed when the etching process is complete.

*Chapter II*

# Files and Their Uses

The earliest tools of mankind were abrading tools made from rough stones. Iron files were introduced in very early times. The file is still one of the most universally used tools, and also one of the most neglected and unappreciated, that man has at his service. Few mechanics know how to get the utmost service from a file through proper application and care. Expert hand filing requires much practice in order to file smoothly without making an untrue surface. While they are designed primarily for cutting metal, files are also used to advantage with other materials, such as hard rubber, plastics, fiber, and wood. Files can make every type of cut, from one fine enough to be compared with polishing to very rough cuts where the main objective is quick removal of excess material.

To meet the demands of industry for a large variety of applications, files are made in many shapes and many cuts. In 1923, the United States Bureau of Standards set forth and named 23 basic designs for files, in place of the endless unstandardized varieties that had existed up to that time. However, there are over a thousand different cuts and sizes of files, and it would be an almost impossible task to explain the uses and characteristics of all of them. The main thing is to select a file that fits the job. Some typical files are shown in Fig. 1.

**SHAPES.** The common shapes of files in cross section are: flat, round, square, three-square (or triangular), half-round, and knife. The outline of the file is either blunt (with parallel sides), taper, or curved, selected to suit the job.

**CUTS.** The cut of files is divided, with reference to the coarseness of the teeth, into rough, coarse, bastard, second cut, smooth, and dead smooth; with reference to the character of the teeth, into single cut, double cut, and rasp cut.

The coarse and bastard cuts are used upon coarser, heavier classes of work, while second cut and smooth are used for finer grades, and for finishing the work started by the coarser files. The four cuts just mentioned are illustrated, in various sizes, in Fig. 2. Rough and dead smooth cuts are seldom used in aircraft work.

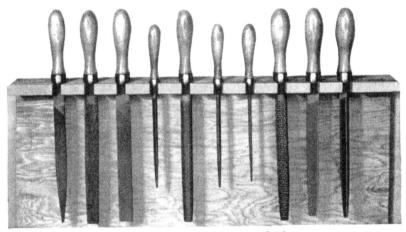

Fig. 1. Some Standard Types of Files
*Courtesy of The Nicholson File Co., Providence, R. I.*

THE FOLLOWING ILLUSTRATIONS SHOW THE ACTUAL COARSENESS OF FLAT AND HAND FILES

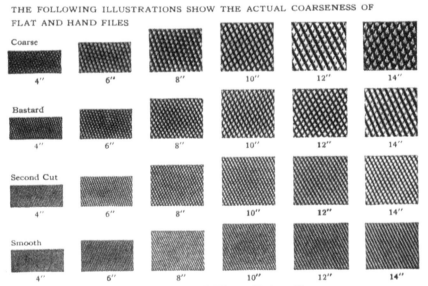

Fig. 2. Four Cuts of Files in Various Sizes
*Courtesy of The Nicholson File Co., Providence, R. I.*

Single-cut files are those in which a single, unbroken course of chisel cuts is made across the surface, the cuts parallel to each

# FILES AND THEIR USES

other, but with an angular rake to the center line of the file. The double-cut file has two courses of chisel cuts crossing each other; the second course usually is finer than the first. The rasp-cut file has separate teeth made by a single pointed tool called a punch. This file has little use on aircraft, except in woodwork. Single- and double-cut files are illustrated in Fig. 3.

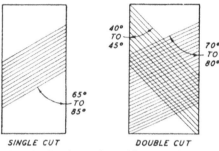

Fig. 3. Single- and Double-Cut Files

Beaver, superior milled tooth, or vixon files have a wide usage in aircraft work, because of their ability to cut fast on soft metals. Due to their curved teeth, they free themselves readily of chips.

**TECHNIQUE OF FILING.** A file that has been used for a time cuts faster and does a smoother job than a brand new file, since any slight irregularity in length of teeth prevents effective cutting until all the teeth are worn to the same length. The best practice is to use new files for dressing down cast iron or larger pieces, or broader areas of steel, before applying them to the narrower surfaces.

**General-Purpose Filing.** The first consideration, after choosing a file of the proper size and shape for the particular job, is how to hold the file correctly. For a large file the handle should be grasped in such a manner that its end will fit into and bring up against the fleshy part of the palm, below the joint of the little finger, with the thumb lying along the top of the handle in the direction of its length; then the fingers will curve about the handle and the ends of the fingers will point upward. The point of the file should be grasped by the thumb and first two fingers of the free hand, and held so as to bring the thumb, as its fatty part presses upon the top of the file, in line of motion, when heavy strokes are required. In filing light work with one hand, the handle should be grasped as just described, except that the hand should be rotated a quarter-turn, bringing the forefinger on top and lying along the handle in the direction of filing. This position allows free action of the hand and wrist, in light work.

You will find that by following these directions the use of the file will be simpler and somewhat easier than if you grasp the tool at random.

Guard against a rocking motion when filing a flat surface. Stand with one foot ahead of the other and make part of the forward stroke by the swing of the body, keeping the motion of the arm to a minimum. Change the course of each stroke slightly to prevent "grooving"; a more even surface results and the work is completed sooner. In filing ovals or round or irregular forms, a rocking motion, resulting from lowering the handle throughout the forward stroke, is best; it distributes the cutting along the full length of file blade.

**Draw Filing.** When a smooth flat surface is desired on work such as steel jigs, dies, flanging blocks, or any piece of steel where file marks need to be removed, the work is accomplished by what is known as *draw filing*. This procedure consists of grasping the file at each end and moving it in a direction at 90 degrees to that followed in general-purpose filing, somewhat after the manner of using a draw knife or spoke shave. All files are not adaptable for use in draw filing; in choosing a file for this work, one with a long angle cut, in mill or second cut, is best. During the forward movement, hold the file at a sufficient angle to cut, rather than scratch, the work. The pressure should be relieved during the back stroke, as in ordinary filing. Work finished by this method is somewhat finer, and the scratches are more closely spaced than by the ordinary use of the same file, since in draw filing the teeth produce a shearing or shaving cut.

### TEN RULES FOR FILING

1. Always use a handle on a file; otherwise the tang is likely to cut the hand or wrist, if the file jams. Make sure the handle is tight.
2. File in cutting direction only.
3. File with a slow, even stroke, using the entire length of the blade.
4. Hold enough pressure on file to keep it cutting. Pressure must be increased as the file dulls.
5. Keep files clean. Use a sharp-edge end grain of a block of pine or

# FILES AND THEIR USES

fir to comb out dirt and chips; in extreme cases, use a file card. A dirty file is a dull file.

6. Keep your files where hard tools, such as chisels and hammers, do not come in contact with them, as such contact would tend to dull or break off the cutting edge. Files are cutting tools, and the blades have been hardened to such an extent that they may easily break if tapped on vise or bench. Never attempt to clean your file by tapping it against anything.

7. The material to be filed should be held or clamped firmly, to prevent the file chattering on the work.

8. Use chalk as a lubricant when filing heavy sections of aluminum or aluminum alloys.

9. Clean the surface of the work before applying the file.

10. When using the round or rat-tail, or the half-round file, use a slicing motion on the forward stroke, to avoid cutting hollows, or a scalloped edge.

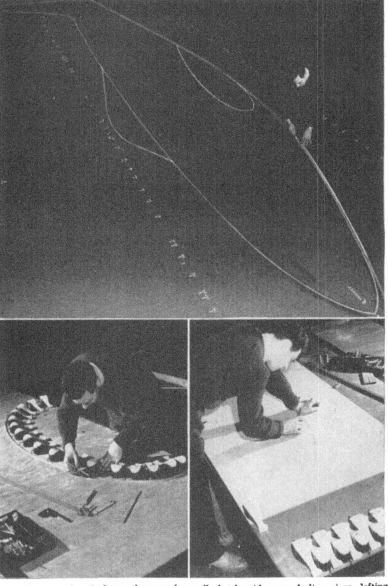

Above. After draftsmen have made small sketch with general dimensions, lofting engineers draw full-sized outline of airplane on loft floor. Note how curved lines are drawn with flexible wood strips held in position by lead weights (or nails) insuring smooth contours. Permanent reference points on loft floor, marked in inches, are used to establish cross section of fuselage.

Lower Left. Cross sections of fuselage at points along its length are next computed. Each cross section is a detailed drawing showing bulkheads, beams, etc. A composite view is drawn full size on plywood as shown here. Note wood strips used to form curves. All lines are scribed to make permanent accurate references.

Lower Right. All detail drawings and templates are made directly from the body plan. In this illustration, loftsman has laid a piece of vellum over the body plan and is tracing scribed lines with a sharp pencil. Detailed structural drawings are then made by the engineering department from this vellum tracing.

*Courtesy of Bell Aircraft Corp., Buffalo, N. Y.*

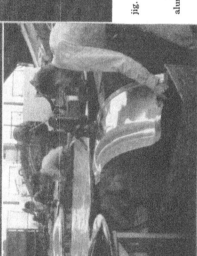
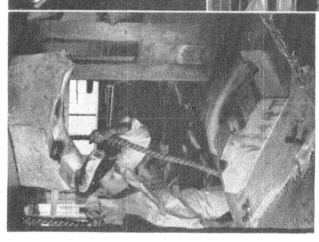

Above. Workman working on part held in a jig. This is in the sheet metal subassembly shop. *Courtesy of Boeing Aircraft Co., Seattle, Wash.*

Left. Rotary shears in operation, trimming aluminum parts fabricated on the hydro-press. *Courtesy of Douglas Aircraft Co., Inc., Santa Monica, Cal.*

Above. Drop hammer operator inspecting sheet metal which has just been formed. *Courtesy of United Air Lines, Cheyenne, Wyoming*

Upper Center. This unique metal stretcher is forming an intricately molded cowling. *Courtesy of The Glenn L. Martin Co., Baltimore, Md.*

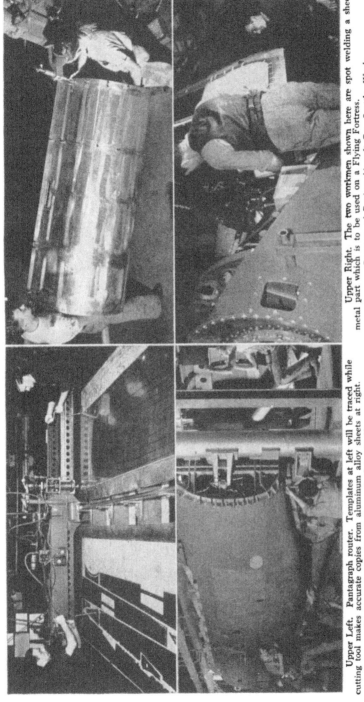

Upper Left. Pantagraph router. Templates at left will be traced while cutting tool makes accurate copies from aluminum alloy sheets at right.
*Courtesy of Bell Aircraft Corp., Buffalo, N. Y.*

Lower Left. Production line for fuselages of attack bombers. The fuselages run along these mechanical lines until ready to be assembled. Such production allows ample working space.
*Courtesy of Douglas Aircraft Co., Inc., Santa Monica, Cal.*

Upper Right. The two workmen shown here are spot welding a sheet metal part which is to be used on a Flying Fortress.
*Courtesy of Boeing Aircraft Co., Seattle, Wash.*

Lower Right. The cabin assembly of the Army Airacobra shown here is swung into place when the fuselage has reached the proper stage of production. This is straight line production.
*Courtesy of Bell Aircraft Corp., Buffalo, N. Y.*

*Chapter III*

# Blueprint Reading

This chapter is not intended to be a course in blueprint reading, but rather an essential, brief review of certain drafting practices, as an aid to reading airplane blueprints.

First of all, blueprints are not necessarily blue. The term was given to the early reproductions (in blue and white) of drawings, and is still used. Now, however, blueprints may be black and white, red and white, or brown and white, as well as blue and white.

After blueprints of drawings have been made, they are sent to the production office. Here they are routed to the various departments that will work upon them. Attached to these blueprints go the shop orders, containing such data as part number, work order number, quantity required, model, completion date required, inspection record, and procedure for routing to other departments for fabrication or assembly.

An engineer goes to one school and learns the standard procedure. Afterward he may work for several different concerns and learn and practice their standard procedures. The mechanic, however, must learn to read all the variations found in blueprints produced by many individuals and firms.

**ABBREVIATIONS.** Before beginning this short study of airplane blueprint reading, a review of the common abbreviations will be helpful. Those most often used in airplane drawings are given here. Abbreviations are not entirely standardized; thus in some cases more than one abbreviation is given for the same term.

| | |
|---|---|
| ABB. .................Abbreviation | APPROX ............Approximate |
| A.B.C. ........After Bottom Center | ASSEM., ASS'Y, OR ASSY..Assemble |
| AC.....................Air Corps | or Assembly |
| AIL. ....................Aileron | A.T.C. ..........After Top Center |
| AL.......Aluminum or Alclad as in | ATMO. ..............Atmosphere |
| 24S–T AL. | AV. ....................Average |
| AL.AL. ...........Aluminum Alloy | |
| ALCOA..Aluminum Co. of America | B.B.C. .......Before Bottom Center |
| ALUM. ...............Aluminum | B.D.C. ........Bottom Dead Center |
| AN......Army, and Navy Standards | B.H.P. ..........Brake Horsepower |

| | |
|---|---|
| BR. HD. | Brazier Head |
| B & S | Brown and Sharpe Gage |
| B.T.C. | Before Top Center |
| | |
| C. | Centigrade |
| CAD. | Cadmium |
| C'BORE | Counterbore |
| CIR. | Circumference |
| ₵. | Center Line |
| CM. | Centimeter |
| COMP. | Compression |
| CP. | Chemically Pure |
| CP. | Circular Pitch |
| CR | Carburetor |
| CS or CSK. | Countersink |
| C'SUNK or CTSK. | Countersunk |
| CU. | Cubic |
| CYL. | Cylinder |
| | |
| D or DIA. | Diameter |
| DEG. or (°) | Degree(s) |
| DEV. | Deviation |
| DF or D. FORG. | Drop Forging |
| DIM. | Dimension |
| DRG. or DWG. | Drawing |
| DURAL | Duralumin |
| | |
| ELEV. | Elevator |
| | |
| F. | Fahrenheit |
| FD | Feed |
| FF | File Finish |
| FIL. | Fillister |
| F.R. | For Reference Only |
| FT. | Foot or Feet |
| | |
| GA. | Gage |
| GAL. | Gallon |
| G.I. | Galvanized Iron |
| GR. | Grind |
| | |
| HD. | Head |
| HEX. | Hexagonal |
| H.P. | Horsepower |
| HR. | Hour |
| HT | Heat Treat |
| H.T.A. | Heavier Than Air |
| | |
| I.D. | Inside Diameter |
| I.E. | That Is |
| IN. or (″) | Inch(es) |
| INST. | Installation |
| INST. | Instrument |
| | |
| LB. | Pound |
| L.H. | Left Hand |
| LONG. | Longeron |
| L.T.A. | Lighter Than Air |
| | |
| M. | Meter |
| MAT., MAT'L, or MATL. | Material |
| MAX. | Maximum |
| MIN. | Minimum |
| MM. | Millimeter |
| | |
| N.C. | National Course |
| N.F. | National Fine |
| NO. or # | Number |
| N.S. | National Special |
| | |
| O.C. | On Center |
| O.D. | Outside Diameter |
| OZ. | Ounce |
| | |
| PAT. | Pattern |
| PCS. | Pieces |
| | |
| R. | Radius or Radii or Round |
| REF. | Reference |
| REQ. | Required |
| REV. | Revision, Revised or Revolution |
| R.H. or R.HD. | Right Hand |
| R.H. | Round Head |
| R.P.M. | Revolutions Per Minute |
| RT. | Rivet |
| RT. BR. HD. | Rivet, Brazier Head |
| RT. CTSK. HD. | Rivet, Countersunk Head |
| RT. FLT. HD. | Rivet, Flat Head |
| RT. R.H. | Rivet, Round Head |
| RUD. | Rudder |
| | |
| SAE. | Society of Automotive Engineers |
| SC. | Screw |
| SEC. or (″) | Seconds |
| SECT. | Section |
| S. FORG. | Steel Forging |
| SPEC. | Specifications |

# BLUEPRINT READING

SPEC. GRAV. ......Specific Gravity
SPHER. ..........Sphere, Spherical
SQ. ......................Square
SQ. IN. or SQ. ".... Square Inch(es)
SQ. FT. or SQ. '..Square Foot or Feet
STA. ......................Station
STAB. ..................Stabilizer
STA'S ....................Stations
STD. .....................Standard
ST'L or S. ..................Steel
S. TUBE.............Steel Tubing
SYM. ................Symmetrical

T.D.C. ...........Top Dead Center
THD. .....................Thread
THDS. ...................Threads
V........Volts, Velocity, or Volume
VAR. ...................Variation
WT. ......................Weight
YDS. ......................Yards
# .......................Number
% .......................Per Cent

**COMPARISON OF TERMS.** It will be interesting, also, to compare a few terms relative to aircraft as used in England, in the United States, and in airplane plants in the United States.

| England | United States | U. S. Industry |
|---|---|---|
| Aerodrome | Airport | Field |
| Aeroframe Fuselage | Fuselage | Hull |
| Aeroplane | Airplane | Plane |
| Air Ministry | Civil Aeronautics Authority | C.A.A. |
| Airscrew | Propeller | Prop |
| Aluminium | Aluminum | 2S |
| Bench Fitter | Bench Mechanic | Tin Knocker |
| Clean | Streamline | Slick |
| Draughtsman | Draftsman | Pencil Pusher |
| Engine Casings or Cowling | Cowling | Cowl |
| Elevators | Elevators | Flippers |
| Flotation Gear | Flotation Gear | Floats |
| N Grider Spars | Warren Truss Spars | Spars |
| Petrol | Gasoline | Gas |
| Royal Air Force | Army, Air Corps, Navy | AC, AN |
| Undercarriage, Alighting Gear | Landing Gear | Wheels |
| Undercarriage Leg | Landing Gear Strut | Strut |
| Vertical Stabilizer | Vertical Stabilizer | Fin |
| Wood Screw | Wooden Propeller | Wood Prop |
| Works | Factory | Plant |

**DESIGNATION OF ALUMINUM ALLOY.** Alcoa wrought alloys are designated by a number followed by the letter S, as 2S.

A letter preceding the alloy symbol means a minor change in composition from that of the basic alloy, as *A17S*.

The letter *H* designates degree of hardness of temper of a not heat-treatable aluminum alloy, based on the maximum amount of cold working. Varying the amount of cold working results in tempers designated as ¼H, ½H, ¾H, or H (full hard).

The letter *O* designates that the alloy is annealed, as *17S-O*.

The letter *T* designates a fully heat-treated condition in a heat-treatable aluminum alloy, as *17S-T*. For example, make from 17S-O (because it can be formed easier when soft) and after it is made, heat-treat (HT) to 17S-T. (Sometimes on blueprints, it is shown *17SO HT ST* without the dash between S and O and between S and T.)

To repeat, remember that the hardness of not heat-treatable aluminum alloy is designated by the letter *H*; the hardness of heat-treatable aluminum alloy is designated by the letter *T*.

The letter *R* in 17S-RT stands for *rolled,* or cold working after heat treatment. Thus the temper which results from strain-hardening an alloy after it has been heat-treated is designated by the symbol *RT*.

**ARMY AND NAVY STANDARD PARTS.** In addition to common abbreviations and terms and alloy symbols, it is necessary to be able to recognize the designations for Army and Navy standard parts, as listed here:

### A (Army), N (Navy) STANDARD PARTS

| | |
|---|---|
| AN3 Bolt–Hexagon Head No. 10 | AN30 Bolt–Clevis 5/8" |
| AN4 Bolt–Hexagon Head 1/4" | AN32 Bolt–Clevis 3/4" |
| AN5 Bolt–Hexagon Head 5/16" | AN34 Bolt–Clevis 7/8" |
| AN6 Bolt–Hexagon Head 3/8" | AN36 Bolt–Clevis 1" |
| AN7 Bolt–Hexagon Head 7/16" | AN73 Drilled Head No. 10 |
| AN8 Bolt–Hexagon Head 1/2" | AN74 Drilled Head 1/4" |
| AN9 Bolt–Hexagon Head 9/16" | AN75 Drilled Head 5/16" |
| AN10 Bolt–Hexagon Head 5/8" | AN76 Drilled Head 3/8" |
| AN12 Bolt–Hexagon Head 3/4" | AN77 Drilled Head 7/16" |
| AN14 Bolt–Hexagon Head 7/8" | AN78 Drilled Head 1/2" |
| AN16 Bolt–Hexagon Head 1" | AN79 Drilled Head 9/16" |
| AN23 Bolt–Clevis No. 10 | AN80 Drilled Head 5/8" |
| AN24 Bolt–Clevis 1/4" | AN81 Drilled Head 3/4" |
| AN25 Bolt–Clevis 5/16" | AN310 Nut–Aircraft Castle |
| AN26 Bolt–Clevis 3/8" | AN315 Nut–Aircraft Plain |
| AN27 Bolt–Clevis 7/16" | AN316 Nut–Aircraft Check |
| AN28 Bolt–Clevis 1/2" | AN320 Nut–Aircraft Shear Castle |
| AN29 Bolt–Clevis 9/16" | AN335 Nut–Plain Hexagonal |

# BLUEPRINT READING

**A (Army), N (Navy) STANDARD PARTS**—*Continued*

AN340 Nut—Machine Screw (NC thread)
AN345 Nut—Machine Screw (NF thread)
AN350 Nut—Wing
°AC365 Nut—Self-Locking
°AC366 Nut—Plate
AN367 Nut—Plate Countersunk
AN380 Pin—Cotter
AN392 Pin—Flat Head 1/8"
AN393 Pin—Flat Head 3/16"
AN394 Pin—Flat Head 1/4"
AN395 Pin—Flat Head 5/16"
AN396 Pin—Flat Head 3/8"
AN397 Pin—Flat Head 7/16"
AN398 Pin—Flat Head 1/2"
AN399 Pin—Flat Head 9/16"
AN400 Pin—Flat Head 5/8"
AN402 Pin—Flat Head 3/4"
AN404 Pin—Flat Head 7/8"
AN406 Pin—Flat Head 1"
AN420 Rivet—90° Countersunk Head—Iron or Copper
AN425 Rivet—78° Countersunk Head—Alum. Alloy
AN426 Rivet—100° Countersunk Head—Alum. Alloy
AN430 Rivet—Round Head—Alum. Alloy
AN435 Rivet—Round Head—Iron (Mild Steel)
AN442 Rivet—Flat Head—Alum. Alloy
AN456 Rivet—Brazier Head
AN502 Screw—Fillister Head (NF thread)
°AC503 Screw—Fillister Head (NC thread)
AN505 Screw—Flat Head (NC thread)
AN510 Screw—Flat Head (NF thread)
AN515 Screw—Round Head (NC thread)
AN520 Screw—Round Head (NF thread)
°AC525 Screw—Washer Head (NF or NC thread)
°AC526 Screw—Brazier Head (NF or NC thread)
AN935 Lock Washer
AN960 Washer—Plain

**AN STANDARD RIVET FORMS.** AN standard aluminum and aluminum alloy rivets are manufactured in five forms, identified by these numbers:

AN425.. 78° Countersunk head   AN430 ..........Round head
AN426..100° Countersunk head   AN442 ..........Flat head
            AN456 ............Brazier head

Drawings of these heads are shown in the chapter on Rivets.

**Rivet Materials.** There are four types of aluminum alloys used in the five AN rivet forms, identified as:

*Type A,* aluminum (2S½H) rivet, not heat-treatable, used only in oil and fuel tanks and for nonstressed joints.

*Type AD,* aluminum alloy (A17S–T), purchased in heat-treated condition and used as such.

---
°The letters AC (Air Corps) indicate that the part is accepted by the Army as standard, but is not a Navy standard part.

*Type D*, aluminum alloy (17S–O), heat-treated to 17S–T before using; used in stressed joints.

*Type DD*, aluminum alloy (24S–O), heat-treated to 24S–T before using.

**Sizes.** All five forms of these rivets are available in $\frac{1}{16}''$, $\frac{3}{32}''$, $\frac{1}{8}''$, $\frac{5}{32}''$, $\frac{3}{16}''$, and $\frac{1}{4}''$ diameters. The lengths are $\frac{1}{8}''$ and longer by 16ths of an inch, depending upon the diameter.

**"AN" RIVET CODE.** The basic AN number is followed by a letter indicating the diameter in 32nds of an inch, followed by a dash and a number indicating the effective length in 16ths of an inch. Only the shank is measured for all rivets except in the case of AN420 (steel or copper), AN425 (aluminum alloy), and AN426 (aluminum alloy), where over-all length including the shank and head is measured, since these rivets have countersunk heads.

*Examples:*

AN420–2–2 equals an iron (mild steel) 90° countersunk head rivet which has a $\frac{2}{32}''$ Dia. and is $\frac{2}{16}''$ long.

AN420C2–2 equals a copper 90° countersunk head rivet which has a $\frac{2}{32}''$ Dia. and is $\frac{2}{16}''$ long.

AN425A4–6 equals an aluminum (2S½H) 78° countersunk head rivet which has a $\frac{4}{32}''$ Dia. and is $\frac{6}{16}''$ long.

AN430D2–4 equals an aluminum alloy (17S–T) round head rivet which has a $\frac{2}{32}''$ Dia. and is $\frac{4}{16}''$ long.

AN442DD3–4 equals an aluminum alloy (24S–T) flat head rivet which has a $\frac{3}{32}''$ Dia. and is $\frac{4}{16}''$ long.

AN456AD4–6 equals an aluminum alloy (A17S–T) brazier head rivet which has a $\frac{4}{32}''$ Dia. and is $\frac{6}{16}''$ long.

**EXPLANATION OF OTHER "AN" (ARMY AND NAVY) AND "AC" (AIR CORPS) STANDARD PARTS. AN Bolts.** The first number after the AN indicates the diameter of the bolt in 16ths of an inch. The number after the dash indicates the length of the bolt in 8ths of an inch for the small size, and in inches and 8ths of an inch for the larger sizes.

*Examples:*

AN3–4 bolt is $\frac{3}{16}''$ in Dia. and $\frac{4}{8}''$ or $\frac{1}{2}''$ long.
AN3–6 bolt is $\frac{3}{16}''$ in Dia. and $\frac{6}{8}''$ or $\frac{3}{4}''$ long.
AN4–7 bolt is $\frac{4}{16}''$ in Dia. and $\frac{7}{8}''$ long.

BLUEPRINT READING 43

AN5–5 bolt is $5/16''$ in Dia. and $5/8''$ long.

AN7–14 bolt is $7/16''$ in Dia. and $1''$ and $4/8''$ or $1\frac{1}{2}''$ long.

Note that the example *AN7–14* does not hold true to the previous examples, since $14/8$ is not $1\frac{4}{8}''$, or $1\frac{1}{2}''$ long. Instead, if two characters are shown after the dash, the first is read as inches, and the second is read as 8ths of an inch. For example, AN7–23 would mean a bolt $7/16''$ in diameter and $2\frac{3}{8}''$ long; AN6–32 would mean a bolt $6/16''$ or $3/8''$ in diameter, and $3\frac{2}{8}''$ or $3\frac{1}{4}''$ long.

**Head Markings.** The markings on bolt heads denote the composition or material used in the bolt. Bolts that are marked on the head with a star, or an x, or an asterisk (°) are made of SAE 2330 (nickel) steel, heat-treated to an ultimate tensile strength of 125,000 lbs. per in.; and they are coded with a dash (AN4–3, for example).

Bolts made from 24S–T dural (aluminum alloy) have indication marks in the form of two small dashes on the head in a manner similar to the 24S–T rivet head marking, except that the rivet markings are raised where the bolt markings are indented. 24S–T bolts are specified in the code by *DD* in place of the dash (AN4DD7).

A plain head on an aluminum alloy bolt indicates that the material is 17S–T dural; however, where bolts have been stocked for a long period of time or removed from an old model plane, the heads may still be marked with the letter *D* which formerly denoted the material as 17S–T dural. In other words, the letter *D* marking on the bolt head is now considered obsolete for 17S–T bolts and they are issued with a plain head. These bolts are coded with the letter *D* however, as in AN4D7.

Bolts which are to be used in elastic stop nuts, AC365°, are supplied without cotter pin holes. The lack of cotter pin holes is designated by an *A* following the dash number. For example, AN4–6A means a steel bolt $1/4'' \times 3/4''$, with no cotter pin hole.

**AN Clevis Bolts.** A clevis bolt has the same system of size designation, except that the code numbers following the AN run from 23 to 30 inclusive and then are AN32, AN34, AN36; these numbers indicate a clevis bolt with a slotted (or Phillips) brazier head. The second number after the AN indicates the diameter of the bolt in 16ths of an inch; AN30 is $10/16''$ or $5/8''$, and AN32 is

---

°Elastic stop nuts are not accepted as standard by the Navy, but are accepted by the Air Corps, hence the AC number.

$12/16''$ or $3/4''$, etc. The numeral after the dash indicates the length of the bolt in 8ths of an inch. For example, AN24-4 is a clevis bolt, $1/4''$ in diameter and $1/2''$ long. Here, also, if two characters are shown after the dash, the first is read as inches and the second is read as 8ths of an inch. AN24-46 is a clevis bolt, $1/4''$ in diameter, and $4 3/4''$ long. For clevis bolts, too, the letter A after the dash number indicates a bolt for use with elastic stop nuts, not drilled for cotter pin.

**AN310 Castle Aircraft Nut.** The numeral which follows the dash indicates the diameter of the nut in 16ths of an inch. For example, AN310-5 is a steel nut, castellated, $5/16''$ in diameter; AN310D4 is an aluminum alloy castellated nut, $1/4''$ in diameter. (Again, note in the first code that the dash as used here indicates steel; in the second code, the D indicates aluminum alloy.)

**AN315 Plain Aircraft Nut.** The numeral after the dash indicates the diameter in 16ths of an inch. The letter L or R after the dash number indicates left- or right-hand threads. The material is steel, unless the letter D appears in place of the dash.

For example, AN315-4L is a steel nut, left-hand thread, $1/4''$ diameter; AN315D6R is an aluminum alloy nut, right-hand thread, $3/8''$ diameter. **Note.** The following Army and Navy items have the same coding as AN315:

AN316 Check Nut     AN335 Plain Hexagonal Nut
AN320 Shear Castle Nut    AC365 Self-Locking (Elastic Stop Nut)
     AC366 Nut Plate (with or without Fiber Insert)

**AN380 Cotter Pin.** Cotter pin designations are similar to those for bolts, except that the number after the first dash indicates the diameter in 32nds of an inch and the number after the second dash, the length in $1/4$ths of an inch. The pin is made of low-carbon steel, cadmium plated, unless the letter C appears in place of the first dash, to indicate that the pin is made of stainless steel. For example, AN380-4-4 is cadmium-plated steel, $4/32''$ or $1/8''$ in diameter and $1''$ long. AN380C-2 is stainless steel, $1/16''$ in diameter, and $1/2''$ long.

**AN515 Round-Head Screw.** AN515-8-8 is a round-head steel screw, cadmium-plated, coarse thread, No. 8-32, $8/16''$ long. AD515D6-7 is an aluminum alloy screw No. 6-32, $7/16''$ long.

**AN520 Round-Head Screw.** AN520 has the same dash number coding as AN515, but has a fine thread.

# BLUEPRINT READING

**DIMENSIONS.** Dimensions on drawings consist of those sizes and distances which are worked to in actual shop operations. They are sufficiently complete so that computations are unnecessary.

General arrangement drawings are dimensioned either in feet or inches, but the indication marks (') or (") are not used. The mechanic simply uses his judgment in determining by comparison whether feet or inches are intended.

Over-all dimensions are kept outside of the intermediate, or detail, dimensions.

Dimensions are not duplicated. No more are given than are required for producing or installing the part, unless an added dimension would be helpful in checking. Such added dimension is followed by the abbreviation *REF.*, indicating it is for reference only.

**TOLERANCES.** When a dimension shows allowable variation, the plus allowance indicates the maximum and the minus indicates the minimum variation allowable. The sum of the plus and minus allowances is called *tolerance*. For example, using $.250 \begin{subarray}{l} +.0025 \\ -.0005 \end{subarray}$, the plus and minus allowances indicate that the part will be acceptable if it is not more than .0025 larger than the .250 dimension specified, or if it is not more than .0005 smaller than .250. Tolerance in this case is .0030.

If the plus and minus allowances are the same, you may see them presented on the blueprint as .250±.0025; this means there is a plus allowance of .0025 and also a minus allowance of .0025. Tolerance is .0050.

Sometimes the maximum and minimum limiting dimensions are shown as $\begin{subarray}{l} .2525 \\ .2495 \end{subarray}$, which means that the dimension of this particular piece or part may not be more than .2525 nor less than .2495. In this case, tolerance is the difference between the two dimensions, or .0030.

Allowances are in either decimal or fractional form. Decimal allowances, such as given in the preceding examples, are used when accurate dimensions are necessary; fractional allowances are sufficient when close dimensions are not demanded. In the title block of many blueprints you will find standard tolerances of ±.010 and ±$\frac{1}{32}$ which apply except where otherwise stated.

**ANGULAR MEASUREMENTS.** The use of angular measurements in degrees (as in Fig. 1.) is avoided, in general, because of

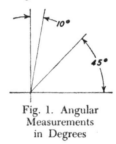

Fig. 1. Angular Measurements in Degrees

the difficulty of obtaining accuracy through such measurement. It is usually preferable to give linear measurements from horizontal and vertical reference lines, as shown in Fig. 2. Here measurements

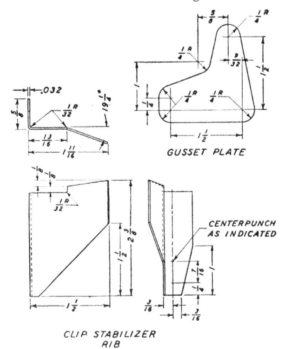

Fig. 2. Linear Measurements from Horizontal and Vertical Reference Lines

are taken from the dimension lines, and points of intersection established for the angular lines. Notice, however, that angle for bend

# BLUEPRINT READING

is given in degrees, which is desirable because the mechanic can check the work with a protractor. In the case of the gusset plate shown in Fig. 2, the whole layout is made by dimensioning from the axes of radii, and the outlines are scribed in tangent to radii.

**DIMENSION LINES AND NUMBERS.** Dimension lines and extension lines are so placed that they will not be confused with the outline of the part. Extension lines start approximately $\frac{1}{16}''$ from the outline and extend about $\frac{1}{8}''$ beyond the dimension line. Dimension lines are broken, so as not to pass through the dimen-

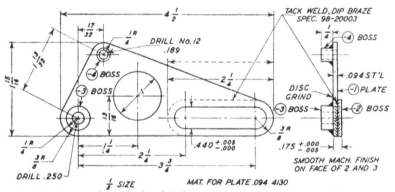

Fig. 3. Approved Methods of Indicating Dimensions Plus an Illustration of the Wrong Method (See 2¼" Elongated Boss Dimension)

sion numbers. The approved method of indicating dimensions is shown in Fig. 3. In this illustration, note that the 2¼" dimension of the elongated boss extends *through* the outline of the drawing, which is considered improper, since it could have been drawn in *above*, as indicated by the dash line. An arrow at each end of the line broken for the dimension numbers serves as further indication that it is a dimension line.

**DIMENSIONING CIRCLES.** The dimension indicating the diameter of a circle is followed by a superior *D*; if the dimension line is inside the circle, the abbreviation is not necessary. A radius dimension is followed by the superior letter *R*. These dimensions for a sphere or a portion of a sphere are followed by the abbreviation *SPHER. D*, or *SPHER. R*.

**LINES.** The heavy, light, and medium line weights are used on finished ink drawings, but few drawings are inked these days.

Pencil drawings ordinarily show only two weights. Fig. 4 gives line symbols approved by the American Standards Association; Fig. 4A shows other symbols in common use.

**Outline of Parts.** The outline of parts is drawn in a heavy,

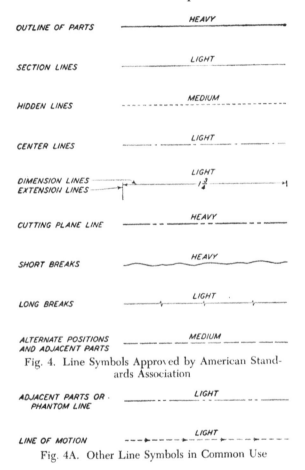

Fig. 4. Line Symbols Approved by American Standards Association

Fig. 4A. Other Line Symbols in Common Use

solid line, and this is used as the trim line, except when accompanied by a *hidden* line as, for example, the dash line showing metal thickness of the clip in Fig. 2.

**Section Lines.** Such lines are used in sectional views to indicate exposed cut surfaces, or to show the kind of metal or material used. Section lines are not used on cross sections of sheet metal unless it is over $1/8''$ thick, and not then unless several parts laid to-

# BLUEPRINT READING

gether might cause confusion. Section lines are so named because they are used to symbolize solid sections of material (as shown in Fig. 9). Section B-B, as shown at the upper right in Fig. 5, is a typical example, designating the material as steel.

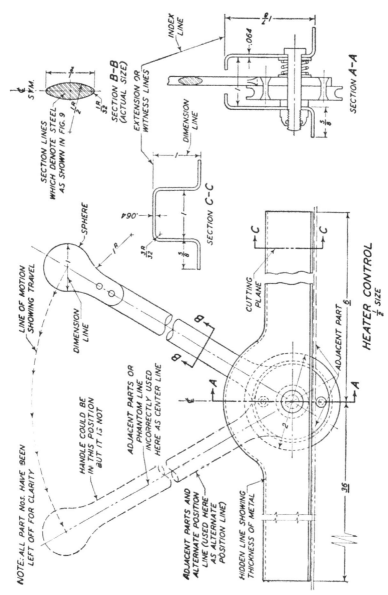

Fig. 5. Typical Aircraft Drawing Showing Various Types of Correct and Incorrect Lines

**Hidden Lines.** These lines, consisting of medium weight dashes, are used to indicate hidden planes or invisible outlines.

**Center Line.** The center line is a line through the center of a part; this line may or may not be the axis of symmetry. If it is an axis of symmetry, it means that both sides are alike and extend an equal distance from the center line. The center line (lines broken by single dashes) is identified by the symbol ℄. Center lines are used for locating dimensions, lightening holes, fittings, rivet and bolt holes, etc. Note in Fig. 5 that the symbol SYM., meaning symmetrical, is not shown on the center line which coincides with A-A, because of the difference in channel length on each side of the center line. In this case the center line is the axis of symmetry only of the significant area of an unsymmetrical part or section.

**Dimension and Extension Lines.** See Fig. 4 for these line symbols and Fig. 5 for their application. A dimension line is a straight line broken to include dimension figures and terminating in arrow heads. It measures the distance between two points, which points are designated by extension or witness lines. The extension lines extend out from these points to set the limits for the dimension line, and they may extend to a counterpart of the dimension line in another view. Extension lines are often broken or interrupted to avoid crossing other lines or parts of drawings.

**Cutting Plane Line.** The cutting plane line (heavy line broken by double dashes) is used to indicate the location of a cutting plane for a sectioned part. In Fig. 5, the correct symbol is used in the cutting plane line C-C, while B-B is considered incorrect, though it is sometimes employed. The center line symbol is shown above the A-A cutting plane line because the draftsman wanted to show a center view.

**Lines Indicating Short and Long Breaks.** The *short breaks* line is used where a part of the drawing has been cut off, or where a short section has been cut out, for lack of space. If the *long breaks* line is used (Fig. 4) it means that a relatively long section has been cut off, or lifted out of a drawing. In Fig. 5, notice the difference in the symbols of the breaks in the channel on the long or 36" side and on the short or 6" side. In dimensioning the channel, the underlined 6 denotes that the dimension line is not actually 6" long. The underscoring of the 36" dimension is incorrect, because a zigzag is used in the dimension line; either the zigzag or underlining would be correct if used singly.

**Alternate Position Lines.** The symbol is a broken line made up of long dashes and is used to show another location of a part which has been moved from the position in which it is drawn.

**Adjacent Parts Lines.** It is often desirable to indicate members that are adjacent but are not actually part of the drawing. The standard symbol for adjacent parts is shown in Fig. 4, and the one shown in Fig. 4A is often used. (Note that the symbol in Fig. 4A is like the cutting plane line except that it is lighter in weight. This lighter weight line is also known as the *phantom line.* An example of this symbol used incorrectly is shown on the handle at the left in Fig. 5.) Notice that at the bottom of Fig. 5 the adjacent parts line is shown on the right by the standard symbol and on the left by the adjacent parts or phantom symbol shown in Fig. 4A, the adjacent part being the skin of a bulkhead to which the combination stiffener channel and control support is fastened. It is easy to read that the part (the bulkhead skin) is not intended to be made from this print, but is shown for reference only.

**Line of Motion.** The direction or progression of the movement of a part is indicated by the symbol shown in Fig. 4A. An illustration of its use is given in Fig. 5.

**ENLARGED DETAILS.** When it is necessary to show a part of a drawing on a larger scale or in more detail, the draftsman may circle that portion of the drawing and give it a letter, as at the lower left in Fig. 6. Such a view should not be rotated from the original position on the drawing, except in specific cases where it is necessary to do so for clarification. If it is rotated, this fact should be made clear. Immediately below the enlarged drawing, the word *View* and the corresponding letter appear, as *View A* in Fig. 6.

Notice that in Fig. 6 the line *B-B* is not a cutting line, and *View B-B* does not show a sectional part, but is merely a special and enlarged view showing the location of the loop on the fuselage.

**SECTIONAL VIEWS.** A sectional view is used when the interior construction cannot be made clear by outside views. It is designated by the word *Section* and a corresponding letter, as *Section B-B*. Letters are used in alphabetical order, and the same letter is not used to designate both a section and a view. For example,

Section A-A and View A would not appear on the same drawing. Should the alphabet become exhausted, the next designation is *A prime* (A') and so on through the alphabet again. A *double prime* (A") would follow Z'.

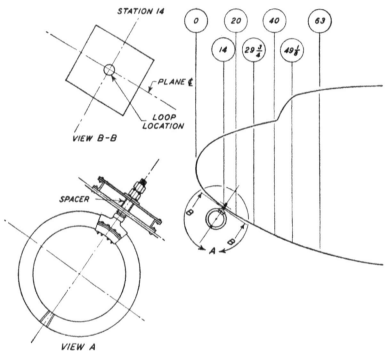

Fig. 6. How Part of Drawing Is Shown at Larger Scale

As illustrated in Fig. 7, where it is desired to show a section of a part, a line is drawn indicating the cutting plane through that portion which is to be shown, and arrows are placed on each end, pointing in the direction in which it is to be viewed. In this case, the letter A is placed at the point of the arrow above and below the view, to indicate that a view of this section, designated as *Section A-A*, is shown at another location on the blueprint.

Sections of structural parts are sometimes shown by another method, as in Fig. 7A, provided this does not interfere with the remainder of the drawing.

Where it is possible, sections and views are placed at the left of the major part of the drawing. Sectional views must be large

# BLUEPRINT READING

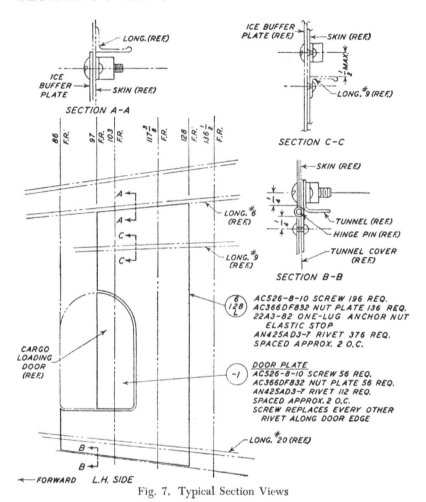

Fig. 7. Typical Section Views

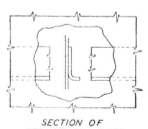

SECTION OF STRUCTURAL PART

Fig. 7A. Another Method of Showing Section of Structural Part

enough, of course, to show all parts clearly. Round sections are indicated broken as in Fig. 8, with the breaks drawn freehand.

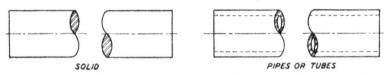

Fig. 8. Sections for Round Items Such as Solid Rods and Pipes or Tubes

**SECTION LINING.** Section lining is generally made with light parallel lines spaced from $\frac{1}{32}''$ to $\frac{1}{8}''$, depending on the size of the drawing and of the part. The lines are drawn at an angle of 45° with the border line of the drawing, unless the shape and position of the part would bring 45° sectioning parallel or nearly so to one of the sides, in which case another angle is chosen. Two

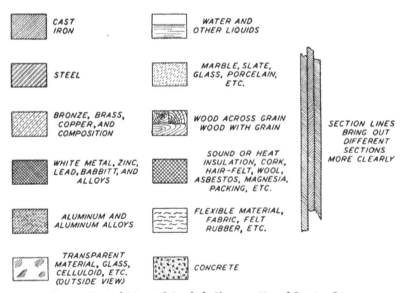

Fig. 9. Typical Material Symbols Showing Use of Section Lines

adjacent parts are sectioned in opposite directions. A third, adjacent to both, is sectioned at a 30° or 60° angle. Where broken to more than one plane, the section lining of a part is generally the same in direction and spacing. On a detail drawing, where only one material is involved, section lining is made with light parallel lines at 45°, as indicated for cast iron in Fig. 9. Assembly drawings generally show section lines in accordance with the material symbols shown in Fig. 9.

Section lining is not used in sections of shafts, bolts, nuts, rods, rivets, keys, pins, and like parts, the axes of which lie in the cutting plane; neither is it used in thin sections such as structural shapes, sheet metal, packing, gaskets, etc.

**DRAWING TO SCALE.** Whenever possible, drawings are made full size, especially in the case of detail parts and small assemblies. Large assemblies or installations are reduced, while very small parts are enlarged. The following scales are more or less standard and are noted on the drawings: TWICE SIZE, FULL SIZE, 1/2 SIZE, 1/8 SIZE, and 1/16 SIZE. The scale of a three-view final assembly drawing is from $\frac{1}{10}$th to $\frac{1}{40}$th actual size.

When all the parts, views, and sections on a drawing are drawn to the same scale, that scale is indicated in the space provided in the title block. Where more than one scale is used on the drawing, the word *NOTED* is placed in the title block, and the scale is then placed under the drawing of each part, view, section, etc., as shown in the following:

| SECTION A-A | STABILIZER ASSY. | VIEW E |
|---|---|---|
| FULL SIZE | 1/4 SIZE | 1/2 SIZE |

The word *scale* seldom appears on drawings.

**TYPES OF AIRCRAFT DRAWINGS.** There are three general types of aircraft drawings, namely: detail drawings, assembly drawings, and installation drawings. (The drawings indicate actual procedure which begins with construction of parts; these parts are assembled into an assembly; the assembly is then installed in the plane.) The title of the drawing definitely classifies it as one of these three types. The title is arranged in two lines, the upper line giving the name of the part, assembly, or installation, and the lower line designating its application. Titles for standard parts, or parts that may have additional applications, are an exception to this procedure, since the title for such parts does not confine their use to a particular assembly or installation.

**Detail Drawings.** If the drawing is of a single part, contains no part circles (parts called for), and does not locate the part in the airplane, it is a detail drawing. Only one word is used in the upper line of the title of such a drawing. The following are illustrative titles of this type. (The upper line names the part, and the lower line states the equipment with which it is to be used.)

| SHAFT | BUSHING | BRACKET |
|---|---|---|
| REEL BOX | ENGINE MOUNT | RADIO SHELF |

**Assembly Drawings.** If the drawing contains part circles (parts called for), indicating either standard or dash-numbered parts, but does not locate the unit in the airplane, it is an assembly drawing. In the upper line of the title of such a drawing, the name of the part is followed by the abbreviation *ASSY.*, as:

| MOUNT ASSY. | MAST ASSY. | LADDER ASSY. |
|---|---|---|
| ENGINE | *PITOT TUBE | CARGO DOOR |

**Installation Drawings.** A drawing showing the location in the airplane of a single part or an assembly is an installation drawing. Parts may be detailed on the drawing, and the drawing may include both an assembly and an installation, but if it locates the part relative to the airplane, only the abbreviation *INST.* follows the name of the part, as in the following:

| PULLEY BRACKET INST. | CONTROL INST. |
|---|---|
| PROPELLER GOVERNOR | HYDROMATIC PROPELLER |

**REVISIONS TO DRAWINGS.** In addition to the design of new parts, it is often necessary to revise existing equipment. The title for such a drawing is as outlined in the preceding, with the abbreviation *REV.* added to the upper line. Illustrations of this type of title follow:

| GUIDE REV. | ROD ASSY. REV. | TANK INST. REV. |
|---|---|---|
| RUDDER CABLE | ENGINE CONTROL | ENGINE OIL |

**THE TITLE BLOCK.** Title blocks vary in shape and size. Each aircraft factory, the Army, the Navy, and every transportation company has its own design of title block. For the most part such blocks are self-explanatory. The title block is the first thing a mechanic must observe as he unfolds a blueprint. It tells him what the part is, the material of which it is to be made, what the scale is, the treatment needed, changes made, etc. Fig. 10 illustrates a typical title block.

**THE CHANGE BLOCK.** The change block is a space gener-

---

* Pronounced Pē'-toe, a part of the air-speed mechanism.

BLUEPRINT READING 57

ally connected to the title block, as in Fig. 10, for listing all the changes made after the print has been released to production, or the shops. For any one of a number of reasons, a blueprint may be returned to engineering to have changes made on the face of the

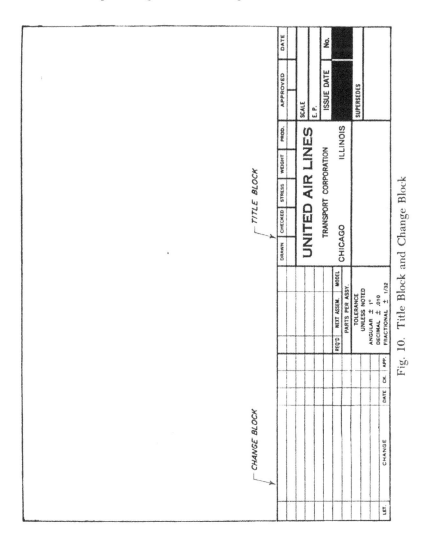

Fig. 10. Title Block and Change Block

original drawing and new prints issued. A brief description of each change is entered in the change block and given a letter, as A, B, C, etc., and the revision is made on the drawing, as near as possible to the point where the change was made. It is usually circled, with an arrow indicating the location.

**IDENTIFICATION NUMBERS.** An identification number is assigned to every detail part, every subassembly, every assembly, and every installation. Parts with these numbers are of two types:

(1) those assigned a number by a company's engineering department, including (a) parts manufactured in its shops, (b) parts manufactured to its specifications by outside concerns, (c) commercial parts modified to suit its requirements, and (d) commercial parts for which there is no manufacturer's part number; and

(2) parts that are identified by manufacturer's part number. In this latter classification are included all purchased parts that are used without modifications.

**THE USE OF DASH NUMBERS.** Dash numbers are used to designate parts to save time and work in the engineering department, as well as to assist the mechanics in assembly. In other words, dash numbers are part numbers. Each manufacturer has his own rules and systems, hence only a general discussion is possible.

(Do not confuse AN codes with dash numbers. As previously explained, AN numbers following dashes indicate size. AN425-2-2 means a steel rivet with $\frac{2}{32}''$ diameter and $\frac{2}{16}''$ length.)

**Parts Exactly Opposite.** One drawing will suffice for two parts, assemblies, or installations that are exactly opposite (mirror images of each other); such parts, assemblies, or installations are referred to as left-hand and right-hand respectively. The left-hand unit is drawn, and the following notation is placed on the face of the drawing, usually above the title block:

LEFT-HAND AS SHOWN, PART NO. 1A–301
RIGHT-HAND OPPOSITE, PART NO. 1A–301–1

Sometimes this notation is simplified in this fashion:
L.H. AS SHOWN
R.H. OPPOSITE –1

The dash number (–1, or –2, or –3, etc.) means that each unit is to be a part of the finished assembly when fastened to it by means of rivets, screws, welding, brazing, etc.

**Parts with Slight Differences.** The following procedure applies to two assemblies which are identical except for a bellcrank,

# BLUEPRINT READING

lever, or any reversed operation which changes one from left- to right-hand. The right-hand part is identified by the assembly part number followed by –1, as 1A–301–1. In this example, the drawing number is 1A–301. If left- and right-hand parts have only such slight differences, the notation on the face of the print would read as follows:

LEFT-HAND AS SHOWN, PART NO. 1A–301
RIGHT-HAND OPPOSITE, EXCEPT AS SHOWN,
PART NO. 1A–301–1

**Detailed Parts.** The drawing number is also the part number, assembly number, or installation number, as the case may be. If parts are detailed on an assembly or installation drawing, they are given a number consisting of the drawing number of the assembly or installation, with a dash number suffixed, as 1A–301–2, meaning part No. 2. If there were another part detailed on the drawing, it would be No. 3. However, on an assembly or installation drawing on which dash-numbered parts have been laid out and dimensioned, only the dash number will be found in part circles on the drawing.

The description accompanying the part circle depends on how the part is detailed. If all information necessary to make the part is given in the body of the drawing without the aid of views and sections, the material, the gage (if applicable), and treatment is given at the part circle, as shown in Fig. 11. Here .040 designates the metal thickness, and 24 is the aluminum alloy composition number. The letter S tells us the metal is in the wrought state, and

Fig. 11. Manner of Presenting Parts and Quantity on Drawing

O means soft temper. *HT* means to heat-treat, and *S-T* means that the metal has been heat-treated from soft to full hard temper. However, if the part is detailed separately on the assembly or installation drawing, this information, plus the scale (full, ½, ¼) is given immediately below the detail, instead. In such case, the name of the part and the wording *SEE DETAIL* are placed at the part circle, as shown in Fig. 19A, part No. 4; and the following is shown under the detail in Fig. 19B:

HEADER PART NO. 4
.032 D. 17SO H.T. 17ST.

Since in this case the detail is the same size, third size, as the rest of the drawing, there is no scale given. (As a rule, blueprints are not shown third size; Figs. 19A and 19B were so reduced in order to fit the allotted space in this book.) Also note that the draftsman omitted the dashes in the aluminum designations on Fig. 19. This is done in some shops.

**Detail for Another Drawing.** When a detail part or unit, assembly, or installation is called for on another drawing, the number may be placed in a circle accompanied by one word (the name of the unit) as shown at (A) in Fig. 12. If the unit or part is used

Fig. 12. Identification by Dash Numbers Showing Quantity

more than once, a number is placed near the top of the circle, indicating the quantity required. Or if no circle is used, the number, name of part, and quantity required may be shown as at (B) in Fig. 12. The arrow points to the location for the unit, as with the circled –1, –2, –3 in Fig. 3. (Note that the circles here are small. The larger circles are used on the circle diagram type blueprint, such as Figs. 11, 19A, and 19B. Diagrammatic type of part numbering is used by Boeing; smaller circles are used by Douglas, Lockheed, etc.)

**Left- and Right-Hand Parts Detailed.** When left- and right-hand parts are detailed separately on an assembly or installation drawing, the notation below the detail is as shown here:

ANGLE
L.H. AS SHOWN, PART NO. 2
R.H. OPPOSITE, PART NO. 3
.032 24S–0 HT S–T
FULL SIZE

**Details with Slight Differences.** If left- and right-hand parts are exactly opposite except for slight differences and where the differences can be clearly shown and noted on one drawing, the notation below the detail would read as follows:

ANGLE
L.H. AS SHOWN, PART NO. 2
R.H. OPPOSITE, EXCEPT AS SHOWN, PART NO. 3
.032 24S–0 HT S–T
FULL SIZE

Parts called for by dash numbers on an assembly or installation drawing are usually not called for by part number on any other drawing.

**Left- and Right-Hand Parts on Same Assembly.** If left- and right-hand parts are used on the same assembly or installation, they are called for independently, or the circles are grouped as shown in Fig. 13. Notice that here the part circles contain not only dash

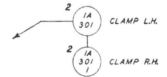

Fig. 13. Way of Identifying
Left- and Right-Hand Parts
Showing Quantity

numbers as they would if the parts were detailed on the print, but complete part numbers for the left-hand clamp, and the same number with a dash 1 (1A–301–1) for the right-hand clamp. This indicates that these clamps have been made from another drawing, and they are called for at the parts stockroom by the mechanic for use on the portion of the assembly to which the arrow points. (Note that the dash is omitted in this diagram-type circle.)

**Left- and Right-Hand Parts on Opposite Assembly.** If the left hand and the right hand parts are used on opposite assemblies or installations—that is, if these left- and right-hand clamps are to go on left- and right-hand assemblies respectively—they are called for as shown in Fig. 14, rather than as in Fig. 13. The method of Fig. 13 might be confusing to the mechanic, especially when a left-hand clamp may be used on a right-hand part, or a right-hand clamp may be used on a left-hand part, or both may be used on each part.

This method of calling for left- and right-hand parts also applies to dash-numbered parts, meaning those parts that are detailed on a drawing and designated by a dash number affixed to the drawing number. See parts 5 and 6 on Fig. 19A.

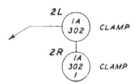

Fig. 14. Identifying Left- and Right-Hand Parts When Assembled on Opposite or Right- and Left-Hand Assemblies

**Subassemblies.** A subassembly is a unit (such as a rudder, bulkhead, stabilizer, wing, center section, food box, etc.), which consists of a group of parts that are to be assembled from parts made from other drawings than the one used in assembling. Sometimes a simple subassembly, such as the parts list illustrated in Fig. 15, is detailed on the assembly drawing, except, for example, the angle shown at (B) in Fig. 15, which carries a part number

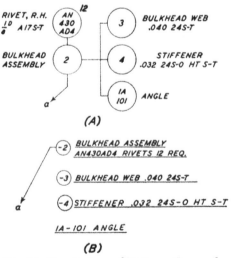

Fig. 15. Two Systems of Listing a Group of Parts and Material for a Subassembly

BLUEPRINT READING 63

and would be detailed on another drawing. Here No. 3 and No. 4 are dash numbers which go to make up No. 2, also a dash number. Thus, the bulkhead assembly is the subassembly of a larger assembly.

A subassembly is given an identification number. When called for on an assembly or installation drawing with its component parts, it is arranged as shown at (A) in Fig. 15. Notice the position of the parts that are used for attaching the subassembly, in this example the rivets. Since circle 2 is the subassembly, the arrangement here indicates that the rivets are to be used to fasten the parts 3, 4, and *1A–101* together after they have been bolted in place and holes have been drilled. Thus a story is told to the mechanic by the diagramming of the attaching lines between the circles. This usually applies also to the more ordinary method, shown at (B) in Fig. 15, which is followed where the use of diagrammatic part circles is not standard drafting procedure.

**Hidden or Obscure Parts.** Fig. 16 shows the method of indicating location of hidden parts, or parts that are not easily found.

Fig. 16. Typical Method of Showing the Location of Hidden Parts or Parts in Locations Hard to Find

**ARMY AND NAVY, AND COMMERCIAL, IDENTIFICATION NUMBERS.** As previously stated, Air Corps standard part numbers are preceded by the letters *AC*. If the part is an Army and also a Navy standard, the letters *AN* are used. As a rule, parts that are listed in the Army Air Corps and the Navy standard books are known by these numbers, not by the manufacturer's numbers. In calling for Army and Navy parts, the complete number is shown except for rivets; for these the length or number indicating length is not always given. Accompanying the name beside the part circle of parts called for, appears the size, material, etc., as in Fig. 17; although as a matter of fact the circle itself contains nearly the same information as given outside it. Decode the numbers in the first circle and you have:

AN4–10

AN = Army and Navy Spec

4 = 1/4" hexagonal bolt (Aircraft)
10 = 1" long

(The 28 at the right means 28 threads to the inch.)

Commercial parts that are used without modification (alteration) and that are not Army and Navy standard are known by the manufacturer's numbers. The manufacturer's name always accom-

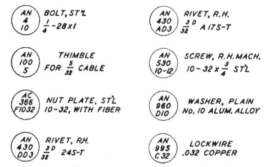

Fig. 17. Showing Dual Method of Identifying Parts When Using Army and Navy and Air Corps Standards

panies such a part number; for example, Parker tube fittings not in the AN book are called for by Parker numbers.

**BOLTS.** In specifying bolt holes, there are three general classes of fits:

1. Standard fit, shown in the tabulation which follows, is the recommended drill size for all general work.

2. Clearance fit, also shown in the tabulation, is used only for such items as clamps, in which the hole may be punched or drilled.

3. Bearing, or light-drive fit, is used only in isolated cases where a very snug fit is necessary. Usually this requires a reamed hole, and the size may vary depending upon the thickness of the material. For instance, in sheet metal the hole diameter may be from .001" to .002" under the bolt's indicated diameter, while for material of greater thickness than .125", the hole may be the same as the bolt diameter. In producing this type of fit it is often necessary to use a special reamer, or one that has been stoned to the correct diameter for the bolt; in other words, to drill the hole undersize and ream to fit. On a drawing, the diameter of the reamer is not specified, but is left to the judgment of the mechanic.

**CHOICE OF DRILL.** To help the mechanic choose the proper drill for making holes, when he is working without blueprints or where drill sizes are not specified, the following tabulations are given:

**Drill Sizes for Bolts—Standard and Clearance Fits**

| Bolt Size | Bolt Diameter, In.* | Drill Size for Standard Fit | Drill Size for Clearance Fit |
|---|---|---|---|
| 10 | .189 {+.0000 / −.0025} | 11 (.191) | 8 (.199) |
| 1/4 | .249 {+.000 / −.003} | 1/4 (.250) | G (.261) |
| 5/16 | .3115 {+.000 / −.003} | 5/16 (.3125) | 21/64 (.328) |
| 7/16 | .4365 {+.0000 / −.0035} | 7/16 (.4375) | 29/64 (.4531) |
| 1/2 | .499 {+.0000 / −.0035} | 1/2 (.500) | 33/64 (.5156) |

\* AN bolts are manufactured undersize approximately to these dimensions which allow for plating plus enough clearance to be inserted in a drilled hole of the indicated size. It is impossible to place a true ¼″ bolt in a true ¼″ hole without heavy driving when the material is at room temperature.

**Drill Sizes for Machine Screws**

| Machine Screw Size No. | Nominal Diameter, In. | Drill Size Standard, No. |
|---|---|---|
| 0 | .060 | 52 (.0635) |
| 1 | .073 | 48 (.076) |
| 2 | .086 | 43 (.089) |
| 3 | .099 | 37 (.104) |
| 4 | .112 | 32 (.116) |
| 5 | .125 | 30 (.1285) |
| 6 | .138 | 28 (.1405) |
| 8 | .164 | 19 (.166) |
| 10 | .190 | 10 (.1935) |
| 1/4 | .250 | F (.257) |

Usually, clevis bolts and flat-head pins are used in reamed holes. In calling for a reamed hole on a drawing, the diameter is given in decimals with the required tolerance.

**Drilled Holes.** Where the hole for a bolt or a machine screw is given, it is usually specified as follows:

DRILL ⅝   DRILL NO. 8 (.199)   DRILL F (.257)
2 HOLES                         3 HOLES

If a number or letter drill is called for, as Drill No. 8 (.199), the decimal in parentheses indicates the size of the drill in thousandths of an inch. Letter drills are made in sizes from No. 1, the largest, to No. 80, the smallest for practical work. Sizes can be procured up to No. 125, but these are too small for use outside the instrument department. Letter drills continue from the No. 1 drill, A, through Z. The A drill is slightly smaller than $15/64''$, and a Z drill is $7/1000$ths larger than a $13/32''$ drill. Far larger holes, we have to depend on fractional and millimeter (MM.) drills.

**Reamed Holes.** When a hole size is followed by a plus or minus as,

$$\text{REAM} \quad .125 \quad \begin{cases} +.0005 \\ -.0000 \end{cases}$$

it means that the hole can be larger by .0005 (a half-thousandth) than .125'' (⅛''), but that it cannot be any smaller (minus .0000) than .125, or ⅛''. Open a micrometer to a half-thousandth of an inch and see how much this is.

**Punched Holes.** It should be noted that punched holes are given to the nearest $1/16''$.

PUNCH ⅛

**Countersunk Holes.** Here a No. 10 drill is called for, which is $1935/10000''$ in diameter.

DRILL NO. 10 (.1935)
CSK 82° × ⅜D.
4 HOLES

An 82° countersink is called for, to countersink the drilled hole until the top is ⅜'' in diameter. Four holes are to be drilled. The accompanying tabulation shows correct diameters of countersunk holes for different sizes of machine screws:

| Size of Machine Screw | Countersink |
|---|---|
| No. 4 | 7/32'' Dia. |
| 6 | 17/64'' Dia. |
| 8 | 21/64'' Dia. |
| 10 | 3/8'' Dia. |
| 1/4 | 31/64'' Dia. |
| 5/16 | 21/32'' Dia. |

## BLUEPRINT READING

**DESCRIPTIONS OF MATERIAL.** All blueprints do not necessarily carry, for example, the word *steel*. The mechanic or student must choose the material by Society of Automotive Engineers' numbers, as 1025 or 4130. With the exception of skin-flush rivets, or assemblies in which it is obvious that flush rivets are desired, there usually is a cross section of the countersink rivet showing where the head is to be. On large assemblies the locations of a few rivets are shown with the proper spacing; the marking, drilling, and centering of the holes for the remainder are left to the mechanic. The same applies to aluminum, and aluminum alloys. The Alcoa designations, such as 52S, 17S–T, 24S–O, and 3S, also must be understood before the mechanic can select the proper materials.

**VIEWS AND THEIR RELATION TO EACH OTHER.** Now that you have become familiar with sections and symbols and have learned to read the symbols of lines, the common abbreviations, and part numbers, there is one more fundamental principle in connection with airplane blueprints which you must learn. You must acquire a thorough understanding of the various views of an object, and their relation to each other on the drawing. For most airplane parts, two or three views are sufficient to allow the mechanic to visualize well enough to lay out the work on metal.

**Orthographic Projection.** Working drawings of objects are made by the aid of projection principles known as *Orthographic Projection*, sometimes called *Orthogonal Projection*. Views of the object are projected onto two planes bisecting at right angles. This makes four angles, numbered and named *first angle, second angle, third angle,* and *fourth angle*, Fig. 18 at (A). This diagram illustrates and explains both the first angle and third angle systems of orthographic projection.

Fig. 18 at (B) shows the arrangement in Third Angle Orthographic Projection. The front view is always *below* the top view. In other words, the top view has been swung up into the vertical plane with the front view, as shown in the diagram in (B) at left. This is the system of drawing used in the United States.

Fig. 18 at (C) shows the arrangement in First Angle Orthographic Projection. In this system, the front view is always *above* the top view. In other words, the top view has been swung *down* into the vertical plane with the front view, as shown in the diagram

# AIRCRAFT SHEET METAL WORK

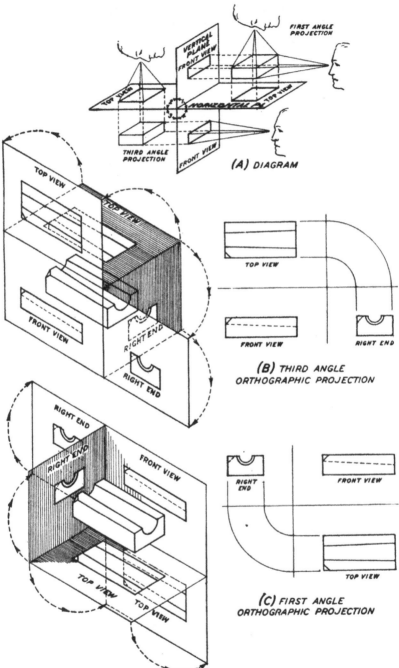

Fig. 18. Explanation of Orthographic Projection

# BLUEPRINT READING

in (C) at left. This system of projection is used in England and in other foreign countries.

Note that in both Third and First Angle Orthographic Projection, all views are swung into the vertical plane in which the front view lies. Note further that irrespective of the angle used, the facts about the object (such as shape, dimensions, etc.) do not change. However, the system used determines the *position* of the views.

As you read these descriptions, the subject may not seem entirely clear, but you will be agreeably surprised at the ease with which, after a little study and practice, you can interpret the views.

Before attempting to make a pattern from it, study your blueprint carefully, looking for any discrepancies that may have been missed when the drawing was checked. A mechanic must follow the drawing and use the exact dimensions thereon. If he notices any mistake in the blueprint, he should consult his superior so that the print may be corrected before the work proceeds.

**MISCELLANEOUS NOTES.** Usually, new drawings are scaled within plus or minus $1/32''$ of construction dimensions. Where changes in drawings have made them out of scale, however, dimension figures are usually *heavily underlined.*

Installation and assembly drawings are made as if viewed from the left-hand side of the airplane with the airplane headed toward the left.

In the development of the larger airplanes in the factories, there is less use of detailed drawings for laying out hull or fuselage, the bulkheads, and other portions of the airplane which are much too large for full-scale drawings on paper. Instead, the method called *lofting*, an old shipyard technique, has been adopted by airplane manufacturing engineers. Here the mechanic simply works from dimensions scaled full size on layout boards in the lofting department; these boards furnish curves and contours for layout of hull, fuselage, wings, empennage, etc. For this reason, approved repair bases sometimes must obtain replacement parts direct from the factory, because no blueprints exist. In some instances, the corresponding part from a new machine, or the removed damaged part, can be scaled and copied satisfactorily.

**BEND RADII.** When a formed part has several bends, only

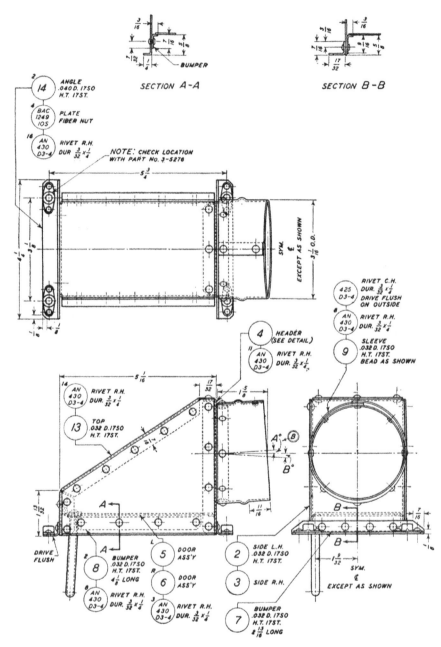

Fig. 19A. Airplane Drawing Showing Heater Box Assembly

# BLUEPRINT READING

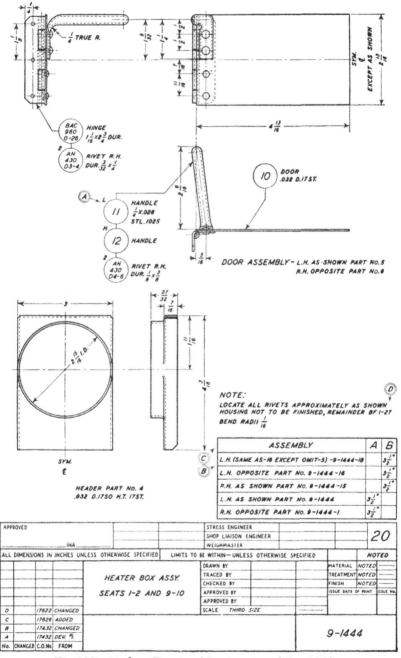

Fig. 19B. Airplane Drawing (Continuation of Fig. 19A)

one of which has the radius dimensioned, we assume all the radii are equal. However, usually ALL BENDS EQUAL or ALL BEND RADII NOT DIMENSIONED ARE (followed by proper value), appears on the drawing.

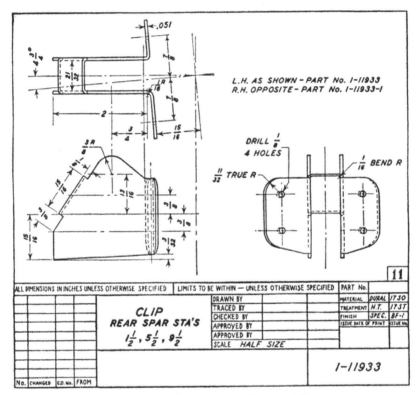

Fig. 20. Airplane Drawing Showing Clip

The phrase DRILL 1/8 (.125) 3 HOLES THROUGH on a blueprint usually refers to a tube or channel, and means to drill through more than one leg of a channel, both walls of a tube, etc.

The size of rivet holes is not always specified. The mechanic must drill hole sizes for corresponding rivets, as follows:

|  |  |
|---|---|
| 1/16" Rivet | No. 50 Drill |
| 3/32" Rivet | No. 40 Drill |
| 1/8" Rivet | No. 30 Drill |
| 5/32" Rivet | No. 20 Drill |
| 3/16" Rivet | No. 10 Drill |
| 1/4" Rivet | G Drill |

# BLUEPRINT READING

When notes are used on a drawing, they are most often placed just above the title block; usually they are numbered consecutively from the bottom up. The reason for this order of numbering is because the draftsman will make the first note just over the title block, and give it the number *1*. If a second note is added, it has to be placed above the first note, and it is given the number *2*. However, all draftsmen do not follow this procedure, and notes must be expected all over the face of a blueprint.

Figs. 19A, 19B, and 20 show typical airplane drawings. Some of the following Questions and Answers are based upon these drawings.

### QUESTIONS AND ANSWERS

Note: The following questions refer to Figs. 19A and 19B

**1. What is the part number of the finished heater box?**

*Answer.* The part number of the finished heater box is 9–1444.

**2. What is the part number of the left-hand side of the heater box? The right-hand side?**

*Answer.* The part number of the left-hand side is 9–1444–2. Right-hand, 9–1444–3. See print.

**3. What is the name of the part 9–1444–5? Of 9–1444–6?**

*Answer.* Left- and right-hand door assemblies, respectively.

**4. Are all the single figures in circles, dash numbers?**

*Answer.* Yes, they are dash numbers, and the part numbers to be stamped on each part would be 9–1444–?, with the number in the circle (dash number) in place of the question mark. If a part of an assembly is so complicated that it necessitates another drawing (subassembly), that drawing number will be 9–1444–13, or whatever the dash or individual part number is. A dash number represents an individual part number.

**5. Is the heater box a subassembly?**

*Answer.* Yes. It is a subassembly because it is only a single part of the complete heating system assembled on the airplane, which consists of many ducts, control valves, branch **Y**s, etc.

**6. What does the Section A-A show?**

*Answer.* It shows the location of and dimensions for making the bumper or seal for the door of flapper valve.

**7. How many bumper angles are there in the heater box assembly?**

*Answer.* Three. There is one bumper angle for each side and one for the end. None, of course, is needed on the end sealed by the hinge.

**8. Is the bumper angle made of soft material (17S–O)?**

*Answer.* Yes, at the circle enclosing dash number 8, the material called for is: .032 D. 17SO H.T. 17ST. 4⅛ long; the bumper at dash number 7 is also made of soft material. Throughout, this book uses the standard method of the Aluminum Company of America for designating aluminum alloys, as 17S–O or 17S–T. Notice that on this print the draftsman did not use the dash between the S and O. The D. stands for Dural and is not necessary though it is sometimes used. 17S–O *is* Dural. Notice, too, that the length for this angle is shown here, while the dimensions for the cross section are shown at Section A-A.

These instances are pointed out as illustrations of the need of reading blueprints with an open mind. You are likely to find such irregularities on the face of any blueprint, and you will need to use your own judgment as to whether such irregularities are actual mistakes, or simply varying ways of presenting data.

**9. Is the cross section of the bumper angle the same for the end as for the sides?**

*Answer.* No. As shown at Section B-B, the riveted leg of the angle shows 9/16" over-all; the side angle, at Section A-A, shows 7/16".

**10. What would tell you how to space the nut plates in the base of the heater box?**

*Answer.* The centers of the nut plates are located by the dimensions 3⅛"x 5¾" in the top, or plan, view.

**11. At what angle is the sleeve set on the heater box?**

*Answer.* This depends on what dash number you have on your work order when starting to lay out the pattern. Notice the chart above the title block which tells us what angle to set the sleeve under A and B. For part number 9–1444, as shown, the sleeve is set at an angle A, or 3½° above the center line, while for part number 9–1444–15 it is set at an angle B, or 3½° below the center line.

**12. In the plan view, the center line is said to be symmetrical except as shown. Where is this variation shown?**

*Answer.* This variation is shown in part No. 5 and No. 6, the door assembly handle location, which is the only unsymmetrical feature of the box. This handle is also the deciding factor in making the heater box left- or right-hand.

**13. What is the difference between the assembly as shown and assembly number 9–1444–18, in addition to the difference in the angle of the sleeve, part No. 9?**

*Answer.* As shown in the chart above the title block, the heater box does not have a door assembly.

**14. In assembling the heater box, where is the flange on part No. 13 in relation to left-hand part No. 2—on the inside or outside?**

*Answer.* On the inside, as indicated on the side view by the invisble outline (dashed line) of the flange.

**Note: The following questions refer to Fig. 20**

**15. Why are the radii called for as *11/32 TRUE* R?**

*Answer.* Because you cannot see the true radius of any corner of either flange or ear. The base of this clip lies at 4¾° from the face of the blueprint. Notice that you can see the thickness of one edge of the ear at the left, but that the right edge is designated by a dotted line.

**16. How many reference lines are there in the upper view of the drawing?**

*Answer.* Four reference lines intersect at one point. Each set crosses at

# BLUEPRINT READING

90°, but they are rotated 4¾° from each other. A fifth line, which extends through the axis of the bend radius of the tab in the upper view and down through the lower view, is used as a reference line.

**17. How is the outline of the clip dimensioned?**

*Answer.* Partly by drawing lines through points dimensioned on the drawing, and partly by drawing a line through a point and tangent to a radius that is drawn from a dimensional point as an axis.

**18. What is the part number of the right-hand part? Is it shown?**

*Answer.* The part number of the right-hand part is 1-11933-1. It is not shown and is opposite to the left-hand part shown.

**19. At what radii are all bends made in the clip?**

*Answer.* All bend radii are 1/16".

## GENERAL QUESTIONS

1. What is meant by a *line of motion?*
2. How is a hidden line made?
3. Define a Third Angle Orthogonal Projection.
4. Define a First Angle Orthogonal Projection.
5. Can a blueprint be scaled for dimensions in layout work?
6. What designates a drawing out of scale?
7. For what reason are dash numbers used?
8. When left- and right-hand parts are exactly opposite, which part is shown on the drawing?
9. What is the difference between reamed and clearance holes?
10. How would you know whether to use steel or aluminum alloy in making a part from a blueprint?
11. What is meant by *24S–O?*
12. What is the difference between *24S–O* and *24S–T?*
13. What does the letter *R* designate in *17S–RT?*
14. Does an *A17S–T* rivet have to be heat-treated before using?
15. Does a *17S–T* rivet have to be heat-treated before using?
16. What does *AN4–4* mean?
17. What alloy is used in a rivet called for thus: *AN430D4–4?*
18. (a) How long is a bolt which is called for by the number *AN4–7?* (b) What is the diameter of the same bolt? (c) Of what kind of steel is this bolt made? (d) What is the approximate composition of this steel? (e) How is this bolt identified? (f) How do you know it is not an aluminum bolt? (g) What is the tensile strength of this bolt after heat-treatment, or is it heat-treated?
19. (a) What is an *AN24–4?* (b) What is an *AN24–4A*, where used?
20. What is an *AC365* and why the *AC?*
21. What is a witness line and how is it used? Explain your answer using drawing shown in Fig. 2.
22. What does an underlined dimension figure mean?
23. Is the axis of symmetry always the center line?
24. Is the center line always the axis of symmetry?
25. Define the following: 2S; 2S½H; 17S–O HT to 17S–T.

Above. Sheet metal is put under white painted sheets of the photographic reproductions of original contours; holes are drilled through white sheet and sheet metal, outlining template.
*Courtesy of The Glenn L. Martin Co., Baltimore, Md.*

Left. Engineer is drawing a small sketch of a plane with dimensions for over-all length, diameters, and locations of principal parts.
*Courtesy of Bell Aircraft Corp., Buffalo, N. Y.*

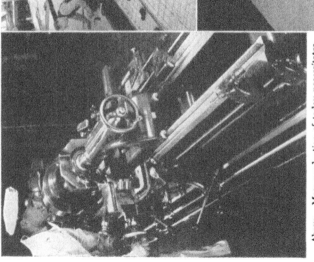

Above. Mass production of today necessitates accuracy such as can be achieved by this expert mechanic and a precision lathe. His lathe has 27 speeds and feeds of from .002 to .118.
*Courtesy of United Air Lines, Cheyenne, Wyo.*

Upper Center. These workmen are measuring and laying out the full scale lines of an airplane on white painted sheets.
*Courtesy of The Glenn L. Martin Co., Baltimore, Md.*

*Chapter IV*

# Measuring and Tools for Measuring

Measuring is the process of determining length, area, or volume. The unit of length is an arbitrary distance, the shortest distance between two points determined by steel scale or rule. The United States Government preserves a platinum bar on which the standard unit of length, the meter, is indicated and from which all other units are derived by comparison. There are many kinds of instruments designed for measuring; only those most important to the aircraft sheet metal mechanic are discussed here.

**WHY A MECHANIC MUST KNOW HOW TO MEASURE.** In aircraft work, the parts that go into an airplane are made in different departments, and in many cases in different plants or even different parts of the country. As one example, a company in Chicago makes wings for planes which are manufactured in Los Angeles. All of the parts, therefore, must be made to accurate measurements, or they will not fit when assembled on an airplane. Furthermore, replacement parts or assemblies are often needed for damaged or wornout parts, and whether or not the new part or assembly will fit depends upon accurate measurements. Thus, measurement is one of the most vital factors in aircraft construction, entering into every process, and influencing the quality and rate of output as well as production costs.

**TOLERANCE.** It is important, also, to know just how accurate a measurement should be. It would not be economical, for example, to build a part to dimensions plus or minus 0.005" if that part would serve the purpose just as well with a $\frac{1}{32}''$ tolerance. The blueprint describes the part or assembly completely, giving all the necessary dimensions; either in the title block or on the blueprint near the dimension figures you will find the permissible error, or the limit of inaccuracy for each part. Every mechanic should work as closely as practicable to dimension, however, because even with a given tolerance of $\frac{1}{32}''$, that limit might be exceeded when several parts are assembled together. The limits of error are thus stated because absolute accuracy is impossible, and the greater the accuracy demanded, the greater the costs of production. Thus, in laying out dimensions on a template, your scriber should be sharp and your work within the limits of accuracy on the blueprint.

**KINDS OF MEASUREMENT.** The first thing to decide in making any measurement is the unit in which the result shall be expressed. Of the many linear units—including the inch, foot, yard, and rod—the inch, with the fractions or decimals thereof, is the most convenient for use in mechanical work. The following are the kinds of measurement that will be used by a sheet metal worker.

**a. Linear.** Measurement of lines or the distance between two points. Use a *tape line* for relatively long distances, the *steel scale* for shorter distances, and the *micrometer* for very short and accurate dimensions. Since no one can make an absolutely correct measurement, we speak of measurements in close work as within plus or minus so many thousandths of an inch, or some fraction of an inch.

**b. Square.** The surface of four unit sides; used, for instance, in measuring the necessary skin (sheets of material) for the surface of a given job.

**c. Cubic.** Cubic measurement is that of volume or mass. Used in calculating displacements, weights, and capacities of different parts of a plane.

**d. Weight.** This is one of the most important measurements in aircraft building. The calculated weights of all parts or components are added together to determine the weight of a plane while the airplane itself is still in the blueprint stage. Also, before the airplane is built, it is known approximately how this weight will be distributed over the entire airplane.

**e. Circular.** Since a circle is a curved line passing through all points in a plane at a given fixed distance from a fixed point of the plane, all measurements of a circle are figured from the dimension of the radius; that is, all the measurements of a circle—area, diameter, or circumference—can be figured from the length of the radius. The angle of degrees is found by an instrument called a *protractor* which has equal spaces, 360 to a circumference, marked and numbered for easy reading.

MEASURING AND MEASURING TOOLS 79

**f. Temperature.** In sheet metal work, temperatures are read from the Fahrenheit scale for the most part. Temperatures for heat treatment, melting points, etc., are given in so many degrees Fahrenheit (F.).

**MEASURING TOOLS. Steel Rules.** It is important to own steel rules (referred to as a scale by the mechanic) that conform to the accurate standards determined by the United States Government. The rule shown in Fig. 1 is correctly marked for quick

Fig. 1. Three-Inch, Easy-to-Read, Spring-Tempered Steel Rule, Graduated in 8ths, 16ths, 32nds, and 64ths of an Inch
*Courtesy of The L. S. Starrett Co., Athol, Mass.*

reading; such a rule can be purchased in nearly any length or cross-sectional size you might desire. Those made of stainless steel are most popular, especially in seacoast areas.

**Steel Measuring Tapes.** Where correct measurement of long distances is required, nothing gives such close results as a steel tape. The temperature standard for steel tape is 68° F., and the coefficient of expansion, as determined by the United States Bureau of Stand-

Fig. 2. Spring Dividers with Automatic Closing Spring Nut for Quick Adjustment
*Courtesy of The L. S. Starrett Co., Athol, Mass.*

ards, amounts to not more than $0.007/_{74}''$ per degree on a 100' tape. Steel tapes are easy to read and are made of stainless steel, furnished in leather cases with push-button handle openers. Do not apply more than ten lbs. tension in using tape; keep it dry and clean.

**Dividers.** Dividers like those shown in Fig. 2 are equipped with an automatic closing nut which makes quick adjustment possible. The dividers are used as a measuring tool or go-between from steel rule to work, transferring and marking off dimensions; they are also used for checking dimensions, or scribing arcs and circles.

**Trammel Points.** This instrument is used simply as an extended pair of dividers, for checking and transferring measurements.

**Combination Set.** This instrument, Fig. 3, is indispensable to

Fig. 3. Combination Set with Center Square, Left; Protractor, Center; and Combination Square, Right
*Courtesy of The L. S. Starrett Co., Athol, Mass.*

a sheet metal worker. The units may be adjusted at any point along the blade by a locking screw. The set includes a spirit level and an emergency scriber, which last should be used only when no other scriber is at hand. The combination square (on the right in Fig. 3) has two surfaces, one at an angle of 90° and one at an angle of 45° to the blade. The protractor (in the center in Fig. 3) has the head extended on both sides of the blade, greatly increasing its usefulness over those protractors in which the head extends only on one side, because, with the former, angles may be transferred from either side of the frame without resetting. The centering attachment (on the left in Fig. 3) is not much used by the aircraft sheet metal worker, but as it is inexpensive it should be included, for the instances in which it is needed.

**Steel Protractor.** A steel protractor must be added to an

# MEASURING AND MEASURING TOOLS

apprentice's tool kit before he can accurately lay out angles of different degrees on a template. The protractor with a rectangular head, Fig. 4, is preferred because it has four working faces; this

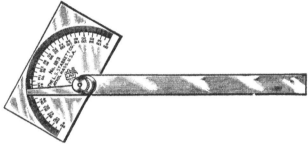

Fig. 4. Steel Protractor, Essential Tool for Sheet Metal Craftsmen in Laying Out Angles to Proper Degrees
*Courtesy of The L. S. Starrett Co., Athol, Mass.*

makes it very handy in checking degrees from a surface plate, or in brake work. A 6" blade is adjusted to the desired number of degrees, and set firmly by a slight turn of the nut. The sooner a mechanic can acquire a steel protractor, the better.

**Drill Gage.** It is often hard to read the numbers on twist drills, by reason of damaged shanks, poor marking, or no marking on the smaller sizes. If a drill of the wrong size is used, it may be broken, and twist drills are very expensive tools to replace. In addition, a drill of the wrong size may spoil the work. Therefore, a drill gage like that shown in Fig. 5 should be in every mechanic's collection

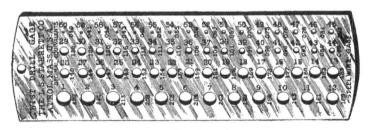

Fig. 5. Twist Drill and Steel Wire Gage
*Courtesy of The L. S. Starrett Co., Athol, Mass.*

of tools to insure using a drill of the correct size. This gage gives the number of the drill to fit each hole, and the dimensions of the hole in thousandths of an inch.

**Micrometer Caliper.** The micrometer is used to measure the gage (thickness) of sheet metal, the wall thickness of tubing, bolts, pins, etc. For accurate determination of any thickness dimension, a micrometer is indispensable. By studying the following instructions and referring to the micrometer illustrated in Fig. 6, the use of the micrometer caliper may be easily learned.

*How to Read a Micrometer Caliper.* The spindle $C$ is attached to the thimble $E$ on the inside, at the point $H$. The part of the spindle which is concealed within both the sleeve and the thimble is threaded to fit a nut located in the part of the frame $A$ which is attached to the sleeve. With the frame held stationary the thimble is revolved by the thumb and finger; the spindle $C$ which is attached to the thimble revolves with it, and moves through the nut in the frame, toward or away from the anvil $B$. The article to be measured is placed between the anvil and the spindle. The measurement of the opening between the anvil and the spindle is shown by the lines and figures on the sleeve and the thimble.

The pitch of the screw threads is 40 to an inch on the concealed part of the spindle. Therefore one complete revolution of the spindle moves it longitudinally one-fortieth (or 0.025) inch. The sleeve is marked 40 lines to the inch, corresponding to the number of threads on the spindle. When the caliper is closed, the beveled edge of the thimble coincides with the line marked $0$ on the sleeve, the $0$ line on the thimble agreeing with the horizontal line on the sleeve. Open the caliper by revolving the thimble one full revolution, or until the $0$ line on the thimble again coincides with the horizontal line on the sleeve: the distance between the anvil and the spindle is then $1/40$ or $0.025''$, and the beveled edge of the thimble will coincide with the second vertical line on the sleeve. Each vertical line on the sleeve indicates a distance of $1/40''$. Every fourth line is made longer, and is numbered $0, 1, 2,$ or $3$, etc. Each numbered line indicates a distance of four times $1/40''$, or $1/10''$.

The beveled edge of the thimble is marked in 25 divisions, and every fifth line is numbered, $0, 5, 10, 15, 20,$ and $25$. Rotating the thimble from one mark to the next moves the spindle longitudinally $1/25$th of one complete revolution, or $1/25$th of $.025$, which equals $.001''$. Rotating it two divisions indicates $.002''$, etc. Twenty-five divisions will indicate a complete revolution, $1/40''$ or $0.025''$.

# MEASURING AND MEASURING TOOLS

To read the caliper, therefore, multiply the number of divisions visible on the sleeve by 25, and add the number of divisions on the bevel of the thimble from 0 to the line coinciding with the horizontal line on the sleeve. For example, on the tool illustrated in Fig. 6, there are 7 divisions visible on the sleeve. Multiply

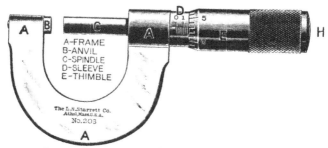

Fig. 6. Micrometer Caliper with Parts Numbered
*Courtesy of The L. S. Starrett Co., Athol, Mass.*

this by 25, and add the number of divisions shown on the bevel of the thimble, 3. Thus, the micrometer is open $7 \times 25 = 175 + 3 = 178$ one-thousandths inch or .178″.

Fig. 7. Toolmakers' Surface Gage Used by Aircraft Sheet Metal Craftsmen for Marking
*Courtesy of The L. S. Starrett Co., Athol, Mass.*

**Surface Gages.** The laying out of work, or the making of scribe marks at a given height from some face of the work, or the continuation of lines around several surfaces may be accomplished by the aid of a surface gage such as shown in Fig. 7. The aircraft sheet metal worker often uses this tool on a cast-iron surface plate for locating or checking rows of rivet holes, bolt locations, trim lines on flanges, etc. To do this work, the surface gage has been equipped to hold an adjustable scriber to a pivoted upright arm on the base.

For making some types of jigs, or for scribing a cutting or other line on a steel channel or fitting, parallel to some part of it, simply adjust the scriber to the desired height from the surface plate, using the scale for measuring. With the fitting, as well as the base of the surface gage, resting firmly on the surface plate, the scriber is now moved in such a way as to scribe the desired lines. The surface gage may be used as a height gage, as well.

**Center Punch.** Although a supply of ordinary small center punches must be included in every mechanic's tool kit, one of the automatic type, such as shown in Fig. 8, is very convenient in close

Fig. 8. Automatic Center Punch with Adjustable Stroke
*Courtesy of The L. S. Starrett Co., Athol, Mass.*

layout work for punch marking points from measurements, as it does not require the use of both hands. This punch is made in three sizes, and extra points may be had at small cost. The punch embodies a mechanism which automatically strikes a blow of any required force when the punch is set in position by the mechanic.

Fig. 9. Transit Level
*Courtesy of The L. S. Starrett Co., Athol, Mass.*

**Transit Level.** This instrument, Fig. 9, is used for lining up jigs of large size from reference lines. For example, a fuselage jig is made in sections for different stations of the airplane. (Stations are reference points, usually in inches, measured from the nose back to the tail or from the root of the wing to the wing tip or from the center of the center section outward. For example, station 72 in the fuselage would be 72" from the nose of the plane. However, sometimes the stations are numbered by the circumferential stiffeners and wing ribs which are so many inches apart as stations.) Reference lines are scribed on the sections and as they are assembled they are lined up by the transit level, an instrument having a telescope and cross hairs with which the reference lines on the jig, or an extended reading on a scale, may be noted. This instrument is also used for checking the rigging of an airplane. The transit level is a shop tool and therefore not needed by the average mechanic. It is included here because it is an instrument used in measuring.

**Plumb Bob.** This piece of equipment is another aid to measurement, one employing the force of gravitation. It is not confined to use in jig work and rigging, as is the transit level. The plumb bob is often used for locating fittings, bulkheads, etc., in the building or repairing of airplanes.

**Straight Edge.** There are two kinds of straight edges used in aircraft plants. One of these is of steel, with or without a beveled edge, made in lengths from $\frac{1}{2}$" to 10', in thicknesses varying from $\frac{3}{32}$" to $\frac{1}{4}$", and in widths from $1\frac{9}{32}$" to $3\frac{1}{2}$". The other is the wood straight edge, made to any desired dimensions. The steel straight edge is a precision instrument and is used where lines are to be scribed straight or where surfaces must be tested or checked for precision; it is also used for making jigs and locating subassemblies in their proper places. Straight edges of wood are used in many operations where the precision demanded is not so great as in repair and assembly work.

To make a straight edge of wood, first decide upon the length needed. The width and thickness should be governed by the length, somewhat in accordance with this tabulation:

| Length, Inches | Thickness, Inches | Width, Inches | Laminations |
|---|---|---|---|
| 48 | $\frac{1}{2}$ | 3 | 3 or 5 |
| 72 | $\frac{5}{8}$ | $3\frac{1}{2}$ | 3 or 5 |
| 96 | $\frac{3}{4}$ | 4 | 5 |
| 120 | 1 | 5 | 5 |

86                AIRCRAFT SHEET METAL WORK

The laminae (layers) should be of quarter sawed maple and mahogany, used alternately, with the maple placed at each edge for service. In cutting each piece, allow for planing the sides smooth. The laminae are glued together with casein glue and clamped to a straight edged object, such as an I-beam or angle iron, to get them approximately straight as the glue sets. The straight edges are then machined by jointer, and given the final straightening with a hand jointer or plane. The edges should be parallel. They may be checked for straightness by laying on a board, marking a line along the length then turning the other side of the straight edge to the line. Deviations will be seen at once and can be corrected by the jointer. Use waterproof varnish on the wood. The alternation of light- and dark-colored wood helps prevent the tool being misplaced or mistaken for a piece of lumber and destroyed.

Chapter V in this book deals with templates laid out from blueprints. These pictures show another procedure using photography. At top, a part is being cut by following lines of a photographic reproduction made on the construction material. At bottom, parts so cut out are checked directly on photographic reproduction of original drawing.
*Courtesy of The Glenn L. Martin Co., Baltimore, Md.*

*Chapter V*

# Layout of Templates and Bench Work

The purpose of this chapter is to teach you how to lay out templates from blueprints, how to make typical airplane sheet metal parts from templates, and how to perform typical bench work.

It would be impossible in a work of this scope to give all the procedures followed in present-day mass production, where many new ideas—such as photography, paper templates, transfers, etc.—are being tried out almost daily. We will therefore consider the typical procedure: you are employed in the aircraft industry as a sheet metal worker and your foreman will hand you a shop order, attached to a blueprint. The shop order gives certain instructions concerning the making of the part, while the blueprint gives the information needed for making the template. (A template is the pattern used for marking the outline, hole locations, bend locations, etc., of the part or parts onto sheet metal—also spelled *templet*.)

There are four steps in the procedure to be followed in laying out templates and making parts, namely:

1. Studying shop orders
2. Studying blueprints which accompany shop orders
3. Making templates from blueprints
4. Making parts from templates

Before beginning your consideration of these steps, however, it might be well to remember that with this study you begin your training toward qualifying yourself for the rating of aircraft master mechanic. The extent to which this study will help you to that end, of course, depends upon the thoroughness and intelligence which you bring to it. These apparently simple principles and problems must be completely mastered before you can proceed to more complicated instruction and work. It is well to keep in mind always that as soon as you enter employment in this industry, whether in manufacturing, or in commercial or Army or Navy maintenance work, you enter a position of grave responsibility, no matter how minor your duties may appear. You and your fellow mechanics are responsible for keeping planes in the air; you are responsible not only for the lives of human beings but also for the property of the industry or of the nation.

Because of these factors, you should approach this study in all seriousness; and three work habits should become second nature to you from the start. These are:

1. Do every job as perfectly as you know how. (There are two classes of mistakes; one caused by carelessness, one by ignorance. You will be excused only because of ignorance.)

2. Never conceal a mistake.

3. Be absolutely sure you understand your instructions, then execute them to the highest degree of accuracy.

**SHOP ORDER.** The shop order, an example of which is shown in Fig. 1, is usually a small printed form issued with a drawing. This form is almost self-explanatory. In the space designated *Description* the name of the part appears; this name will correspond with the name in the title block on the blueprint. The name of the shop also is shown—in this case *Sheet Metal* because it is an order to make a sheet metal part. There is also an additional direction given for the paint department. If there were work to be done by other departments, such as plating or welding, each department would be added under the *Shop* column, and there would be as many shop orders as shops involved. The order shown in Fig. 1 is made in duplicate only, and the duplicate copy is given to the paint shop with a check mark designating that shop in the same manner as this shop order is checked for *Sheet Metal*. Each time the work assigned to any shop is finished, the shop orders are routed to the inspection department with the part. Here the material, dimension, temper (hardness), and workmanship are all carefully checked before the part is sent to the next shop designated.

On the face of the shop order opposite the shop name and work directions, a space is provided for the date of completion, an important feature that aids considerably in smoothness of production. Of course the hours estimated in the budget space, or the length of time allotted for the job, may be more or less than the actual hours the work takes. After the work is finished, the actual

# TEMPLATE LAYOUT AND BENCH WORK

## SHOP ORDER

**DESCRIPTION:** CABLE CONNECTOR

**PART NUMBER:** 1A-100254

**WORK ORDER NUMBER:** 1234

**USED ON:** DUAL Carburetor feed valves

**QUANTITY REQUIRED:** One

**MODEL:** DC-3

**ISSUED BY:** (signature)

**APPROVED BY:** (signature)

**DATE ISSUED:**

**DATE RECEIVED:** 3/4

**BUDGET HOURS / ACTUAL**

| SHOP | WORK TO BE DONE | TO BE COMPLETED BY |
|---|---|---|
| Sheet Metal | Make | 12-30- |
| Paint Dept. | Paint P27 Alum. Primer | |
| | | |
| | | |
| | | |
| FINAL DELIVERY | | |

| NO. | DATE | INSP. | REC'D BY | POSTED |
|---|---|---|---|---|

**SUBMITTED FOR INSPECTION**

Fig. 1. Typical Shop Order Form Filled Out for Making Cable Connector Shown in Fig. 7.

work time is inserted in the space provided, by the accounting department.

**BLUEPRINTS.** After the mechanic has studied the shop order, he proceeds to the blueprint which comes with it and which contains all the information necessary for making the part or parts called for on the shop order.

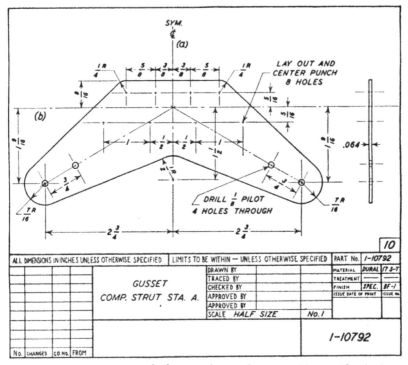

Fig. 2. Representing Typical Blueprint for Airplane Part (Gusset Plate); Outline Is Not Dimensioned But Centers for Radii Are

Be sure to study the blueprint until you are certain you understand all the dimensions and data given. Give careful attention to the title block, which will indicate what kind of material is to be used, and at what scale the drawing was made.

**Example.** Refer to Fig. 2 and follow the reasoning which will give the information you need. Ask yourself:

1. How thick is the material to be used in making the gusset? *Answer.* It is indicated at the right on the edge view as .064.

# TEMPLATE LAYOUT AND BENCH WORK 91

2. What kind of material?

*Answer.* It is called for in the title block as Dural 17S-T.

3. Are any holes to be drilled?

*Answer.* Yes, they are called for on the face of the print.

4. Are any bends to be allowed for, in the template?

*Answer.* No, the template will be flat and will look like the outline on the drawing.

You will learn more about blueprint reading in a later section of this chapter, *Laying Out Templates on Sheet Metal.*

**TEMPLATES.** As you have learned, a template is a flat pattern of a part. For economy and efficiency, it is necessary to make a template of each sheet metal part that goes into an airplane. Templates must be made within a maximum tolerance of $10/1000$ths (.010) of an inch, or plus or minus $5/1000$ths (.005) which is less than a half of $1/64''$.

Each time a new model of airplane is built, a vast amount of template work must be done. When the templates for that model are completed, they are filed away in case there should be orders to build another series of that particular model, or in case it should be necessary to furnish exact duplicates of any parts of the plane that may be needed for repair or replacement.

When template sheets are painted for the purpose of making a sharp contrast to the scribing marks, the coat on the template should be very thin and have a good bond to prevent chipping along the scribed lines. In factories, duplicate templates are often made, for reference or for use in lofting. These duplicates are usually marked *master template* or are painted a different color; they are seldom used by the mechanic.

The template or pattern usually is made of materials cheaper than the aluminum alloys. Some materials used are tin plate, terneplate, galvanealed iron (trade name for dull zinc-coated iron), galvanized iron or black iron, especially for templates of larger size. In laying out the templates in this text, however, the cost of material will be negligible, because the parts are so small. It is recommended that the students cut out a supply of pieces, of assorted sizes and various thicknesses, somewhat as shown in Table 1. These may be cut from scrap salvaged from airplane factories, identified with stencil or stamp, and placed in stock. This practice will eliminate confusion in classes, since students may obtain from stock the size and gage of the proper alloy. The projects will thus be expedited.

Table 1—Recommended Sizes to be Stocked, Cut from 2S, 17S–O, 17S–T, 24S–O and 24S–T

| Gage | Dimension of Material, Inches | | | | |
|---|---|---|---|---|---|
| .025 | 3x4 | 4x4 | 4x6 | 6x6 | 8x8 |
| .032 | 3x4 | 4x4 | 4x6 | 6x6 | 8x8 |
| .040 | 3x4 | 4x4 | 4x6 | 6x6 | 8x8 |
| .050 | 3x4 | 4x4 | 4x6 | 6x6 | 8x8 |
| .064 | 3x4 | 4x4 | 4x6 | 6x6 | 8x8 |

**LAYOUT FLUIDS.** It is hard to identify the scribing marks on bright metals such as the aluminum alloys, especially when the templates are more complicated. For this reason some sort of layout fluid is often used. Sheets of metal which have been given a coat of blue linoil or of P27 (zinc chromate primer) for protection need no layout fluid. Neither layout fluid nor primer is required for terneplate or galvanealed iron. Black iron or galvanized iron should be primed with a suitable primer; P27, which is yellow in color, is recommended.

In laying out jigs or tools on polished steel, a solution of blue vitriol is used. This displaces a thin layer of iron with copper, giving an effect of copper plating which makes a good background for layout work. Brush the solution on and let it stand a few minutes until the copper color appears, then rinse off with water. Blue vitriol may be purchased in hardware and drug stores.

**THE BENCH.** The craftsman who works at the bench will never find his job uninteresting, since usually it is here that parts are projected from the blueprint to the template, and from the template to the actual part. There is great satisfaction in watching the transformation of a part.

Bench mechanics prefer a polished table with a maple top which is hard enough to prevent metal chips from becoming imbedded in it. Heavy paper is sometimes used when large sheets are being worked with, to prevent scratching of the metal.

**A Few Practical Rules for the Bench Mechanic.** If the work

# TEMPLATE LAYOUT AND BENCH WORK

you are doing can be done efficiently in a sitting position, sit down; if you can accomplish more standing, stand up.

Do not use your bench top as a drill block. Place a piece of board under the work to be drilled.

Do not use your bench drawers for a stockroom. Turn your surplus stock of screws, bolts, material, and equipment into stock or other disposition.

Keep your vise clean and oiled. Keep it free of burrs, scars, and defects by the use of a file and an emery cloth.

Keep your tools in order, sharpened, and in a workmanlike state of repair.

Keep the top of your bench clear of unnecessary objects which will hinder your work.

Be sure to keep the aisles clear. A cluttered aisle is always a safety hazard.

**ACCURACY.** Bear in mind always that accuracy is a measure of a mechanic's ability and that it keeps to a minimum the waste of time and material. Half an hour spent studying a job through, or mentally measuring its problems, or considering the possibility of easier methods of procedure usually is better than half a day working at it on the rush. Rushing things through is the surest way to commit many sad errors.

In working the exercises on the following pages, you are expected to adhere to dimensions as closely as is customary in manufacturing parts for the Army and Navy. All parts which go into the making of airplanes must be accurate to within less than $1/64''$. Practically the only exception to this is where closer allowances[*] are required, as in the direction "ream hole $7/8 {+.002 \atop -.002}$." Since there is a tolerance here of only $4/1000$ths, we must use a drill $1/64''$ under $7/8$, or $55/64$, afterward reaming the hole to $7/8$.

**TEMPLATE DRAFTING AND CUTTING.** Before starting the projects outlined in this chapter, we will consider some of the more important aids to the simplification of our work, for we do not want to do anything the hard way if there is an easier or shorter method.

---
[*] *Allowances* are expressed in plus and minus (+ and −), the plus meaning larger and the minus smaller than the given dimension. *Tolerance* is the sum of the plus and minus allowances.

In transferring dimensions from blueprint to template, marks are made with the scribe or scratch awl. The scribe is used also in marking metal that is to be fabricated by use of the template. Where bend lines or reference lines are to be made, a soft pencil is used. A prick punch (a tool about 4" long and $5/16$" in diameter with a tapering sharp point) is used for transferring hole locations, if the holes to be drilled are about size number 60 (No. 60 drill). This prevents the distortion and enlarging of the pilot hole in the pattern, which might be caused by use of the center punch. The marks made with the prick punch may afterward be center punched.

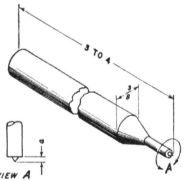

Fig. 3. Duplicating Punch; Depth of Point, *a*, Must Be Less than Thickness of Template

Care must be taken not to strike the punch too heavily with the hammer, thus driving the point through the metal. This bad practice can be avoided by using a duplicating punch (Fig. 3) if a larger hole is drilled in the template, say $3/32$" or $1/8$".

**Development of Template.** The proper place to begin the development of the template is at a reference line. Reference lines may be vertical or horizontal or at angles, as base lines, center lines, or axes of symmetry. If the drawing indicates the part has an axis of symmetry, this is the best line from which to begin projection, as only one setting of the dividers is necessary for marking dimensions on both sides of the reference line.

Many blueprints of small parts have neither center line nor reference line (other than one edge or side of the part itself) so you must decide where to begin to develop that part according to the relation of the dimensions within the drawing. To scribe a reference line on the template material, set the head on your combination square to allow for clearance when the work is projected toward the edge of the sheet. Using the square head for a guide on the edge of the sheet, mark two points which include the length of the line needed, and mark the reference line by using a straightedge, as indicated at (A) in Fig. 4.

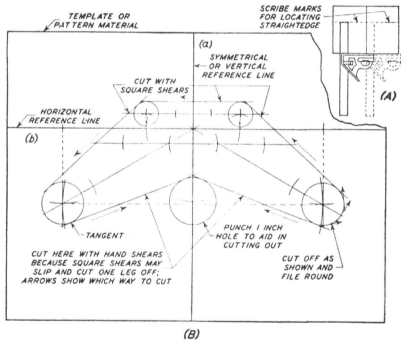

Fig. 4. Layout Which Has Been Transferred from Blueprint to Pattern Material; When Cut Out This Will Be a Pattern or Template

If there is a vertical reference line at 90° to the horizontal reference line on the face of the blueprint, this must also be transferred to the pattern. On large layouts the square must not be used for scribing lines at right angles to each other by using the edge of the sheet as a guide. Instead you must use the dividers (or trammel points) to bisect a line at 90°; see Fig. 5. From a predetermined point on the horizontal line, A, scribe an arc at B and C with dividers or trammel points intersecting the reference line on each side of the axis A. Open the dividers to greater radius, and using B as the center, strike an arc at D and at E. With the same setting use C as the center, and strike an arc intersecting the arcs just scribed at D and E. With a straight edge, scribe a line through the intersection points, and you have a vertical reference line. On smaller layouts it is not necessary to scribe arcs at E; merely mark the vertical line through DA.

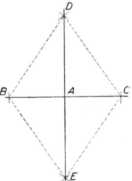

TO BISECT A GIVEN LINE
Fig. 5. Typical Way to Lay Out Reference Lines at a 90° Angle or How to Bisect a Given Line by Another Line Making 90° Angles

**Transfer of Dimensions.** Transfer all dimensions from the scale (rule) to the template with a pair of rigid dividers. Place one leg of the dividers at one inch, and allow for that inch in the scale reading. Where dimensions are too long for the dividers, use trammel points or a beveled-edge scale. If you use the scale from your combination set for transferring dimensions, hold the edge to the work, to prevent deviation in measurement due to sighting through the scale thickness.

Simple templates may be scaled from the edge of the template sheet which was previously squared in the squaring shears. Check the template sheet for squareness, and place it so that it overhangs the edge or corner of the bench, and is weighted, so as to leave both your hands free. When the combination square is used for marking dimensions, set the head the number of inches required and scribe at the end of the scale.

# TEMPLATE LAYOUT AND BENCH WORK

**Scribing Radii.** In using the dividers for scribing radii, the beginner often makes the mistake of pressing too heavily upon the points, thus causing the legs of the dividers to spread. If the divider points are kept sharp, you will not need to bear down heavily in order to obtain a clear cutting line.

**Cutting Out Templates.** In cutting out either template or work, care must be exercised to follow the line closely, thus avoiding the necessity for excessive filing. Straight lines may be cut on the squaring shears, while curved lines, holes, etc., are cut with left- and right-hand aircraft snips.

**CUTTING OUT PARTS USING TEMPLATES.** When small parts are being made in great quantity, either in a factory or at a repair base, the metal may be blanked or roughed out in the squaring shears and pressed together in stacks of about two inches thick, with corners tack welded. Then from the template the top piece of each stack is scribed, and all the pieces in the stack are sawed out and filed on a Doall saw in two operations. The choice of a production method of this kind over handwork depends largely on the time element; that is, on whether or not the job can be done by hand in less time than it would take to set it up for machine work. In any case, the production method will be used when the material for the parts is too heavy to be cut by hand.

The punch press is often used to cut and form small fittings in large quantities. In some factories the cutting of larger sections, prior to forming or fabrication, is done by means of a router. The router is used for relatively large sections whose size makes the cost of punch dies prohibitive. In this process, a stack of sheet metal of the required gage is clamped or bolted on a large table. If holes in the pieces are called for, drilling templates are used for the drilling, then they are removed and replaced with the router jigs. The sheet metal stack (sheet pack) is cut by holding the router guide against the jigs; a roughing collar placed around the cutter holds the cutter $1/32''$ away from the template. This allows $1/32''$ around the profile (edges of the template and stack). For later finishing, the roughing collar is removed and a $1/32''$ cut is taken around each profile. The router head is mounted on a rotating arm so that it can be used over any part of the table.

**FORMING PARTS.** Pieces made of aluminum alloy S–O

materials must be formed, after the cutting operation. This may be done by using the bending brake, power brake, drop hammer, hydraulic press, punch press, trip hammer, draw bench, corrugating rolls, or any of the various methods now employed. The selection of the machine and the method of forming depend upon the shape and size of the pieces, and the quantity needed. Where only one part is needed, as for replacement repair, or for experimental purposes, the finishing will be done by hand by using flanging blocks clamped in the vise, or by pounding the desired shape out in a sand bag and surface plate. Hand-finishing methods are likely to be used by the student, since he is required to make only one each of the several parts or assemblies.

**LAYING OUT TEMPLATES ON SHEET METAL. Example 1.** Let us suppose we have just received a blueprint, shown in Fig. 2, from which we are to make a pattern on a piece of 24-gage galvanized iron. The part is a gusset used in fabricating a compression strut in an airplane wing. We estimate that a piece about 6"x8" will be large enough for the pattern. Since there are quite a few lines in this drawing, our layout marks on the bare metal might be confusing, so we decide to use a layout fluid. P27 dries fast and makes a good background for scribing, so we brush or spray on a smooth thin layer of this paint. While the paint is drying we study the blueprint further.

We see that it has two reference lines—(a) which is vertical and is a symmetrical line and (b) which is horizontal. We also see that most of the dimensions are to be taken from these reference lines, and that the outline of the fitting is to be scribed tangent to five circles whose radii are shown. The title block tells us the material to be used for the part is Dural 17S–T, and the edge view indicates the gage as .064.

By this time the paint must be dry on the piece of metal we prepared for our template, so we are ready to begin our layout. We have decided to lay out the reference lines first, and since the piece of galvanized iron may not have absolutely square corners and parallel sides, we must do all the dimensioning from one side. The 6"x8" sheet is large enough so that the horizontal reference line may be scribed in the center, or above the center of the sheet; see (A) at the upper right in Fig. 4. The combination square is set to the desired length and locked with the set screw, and two scribe marks are made (see (A) in Fig. 4) using the end of the scale as a guide for scribing. The scribe should have a long tapered sharp point.

The next step is to loosen the set screw and pull the scale through the square head far enough to reach across the width of the pattern and scribe the vertical line, which is done by holding the square head firmly against the edge of the sheet and scribing the line along the scale approximately in the center of the sheet. When this is finished, pull the scale free of the head, and use it as a straightedge to scribe the horizontal line through the scribe marks previously made, as indicated by the dotted line in (A), Fig. 4. This completes our reference lines; we are now ready to start laying out dimensions.

The $2\frac{3}{4}''$ dimensions, one on each side of the symmetrical line $(a)$, are good places to start, since they aid in locating the center for the $\frac{7}{16}''$ radii. Set your dividers at exactly $2\frac{3}{4}''$ by placing one leg at the $1''$ mark on the scale, and the other leg on the $3\frac{3}{4}''$ mark. Now, place one leg on the vertical reference line $(a)$, approximately $1\frac{9}{16}''$ down from the intersection. ($1\frac{9}{16}''$ is the distance from the horizontal reference line down to the center for the $\frac{7}{16}''$ radii, from which centers the ends of the legs of the gusset will be determined.) Scribe an arc on each side of the symmetrical line $(a)$. Next, use the square and scribe a reference line tangent to each arc, which will be $2\frac{3}{4}''$ from, and parallel with, the symmetrical line $(a)$. The square should extend a short distance across the horizontal reference line $(b)$, as indicated by dotted lines at $(B)$ in Fig. 4. Place one leg of the dividers at this intersection of the extended tangent line and the horizontal reference line and, using a $1\frac{9}{16}''$ setting, intersect the arcs previously made. This establishes the location of the centers for the $\frac{7}{16}''$ radii.

Repeat this process to locate the centers for scribing the other radii, and then set your dividers for scribing the various sized circles.

*Note.* It is not necessary to scribe complete circles, of course. However, in pattern drafting of this kind it is better to make all scribed lines extend through to the outside of the pattern and use extended arcs. This saves time where an extended line is used for reference elsewhere on the layout. Also, if you do not follow this practice, you may not guess close enough; then you will have to waste time by extending your scribe lines farther.

The outline of the template is now scribed tangent to the various circles, using a straightedge or your scale. (When cutting out the pattern, it is easy to tell which lines the cutting is to follow.)

Lay out the center-punch holes (see Figs. 2 and 4) with the dividers, setting the divider legs at proper distance by use of a scale. Do not lay out center-punch holes by using the first mark to locate the next; reset the dividers and use the reference line to dimension each center-punch location. This dimension is found by adding the dimensions to give you the proper distance from the reference line. This is recommended because if your dividers are off .004″ in laying out the first hole, by the time you have stepped off 25 holes, the last hole would be off .100″, or over $3/32$″. Rivet spacing, however, may be stepped off from the preceding mark.

The blueprint in Fig. 2 reads *LAY OUT AND CENTER PUNCH 8 HOLES*, but actually you must lay out 12, because the 4 pilot holes must be laid out and center punched before they can be drilled.

The inside radius of $1/2$″ in the crotch (lower boundary at center line) must be center punched at the axis of the circle (see Fig. 2), because it is easier to punch a 1″ diameter hole in this location on the punch press than to try cutting around the radius with the hand shears. After this hole is punched, cut out the template on the straight lines and snip off around the radii as indicated on the end of the right-hand leg of the gusset, Fig. 4. Never attempt to cut exactly around the radius of a heavy gage part of this type. Snip off as close as practical and finish with a smoothing file. This filing operation will be easy if the gusset is placed in a smooth-jaw vise so that the radius just protrudes. File cross-wise with a cylindrical motion throughout each stroke.

As you have observed, the layout of this gusset was a matter of locating the centers of five circles which you afterward connected with straight lines (tangents) to form the boundary.

**Example 2.** Perhaps we should consider, also, how to lay out a simple part where the boundary is given and you have to find the axes or centers for the radii. Shown in Fig. 6 is just such a part which may be laid out at the corner of a sheet 3″ x 4″ in size.

# TEMPLATE LAYOUT AND BENCH WORK 101

The part is so simple that we will not take time to apply any layout fluid. Simply make sure, by use of your combination square, that the lower right-hand corner of your stock is square. Set the square at 1 11/16", lock, and from the end of the scale scribe the outside boundaries for both ½" dimensions; see Fig. 6. Be sure to make

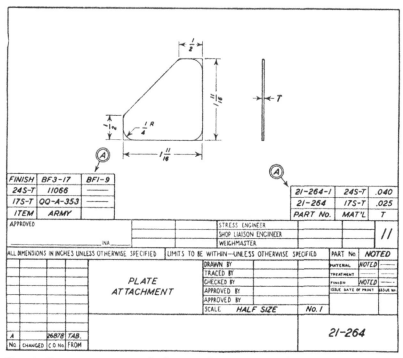

Fig. 6. Representing Typical Blueprint for Airplane Part (Plate Attachment); Outline Is Dimensioned

the scribe mark longer than just the ½", because there may be a later use for the extension of this line. Reset the combination square at exactly ½" and intersect the lines just scribed, as in (A), Fig. 7. By intersecting the crosses with a straightedge, you can scribe the diagonal line, which leaves only the arcs at the corners to be scribed.

Small templates of this kind may be cut out on the squaring shears before the arcs are scribed; thus, with edges to measure from, the combination square may be used as an aid in determining centers for the radii.

To find the centers for radii in the corners which are square (90°), Fig. 7 at (A), follow the procedure outlined in the previous paragraphs for finding the centers for the radii from the reference lines; however, in this instance use the boundary lines instead of reference lines. Set your combination square for ¼" and mark a point on each boundary line ¼" from the square corner with which you have chosen to work. Then, set your dividers for ¼"

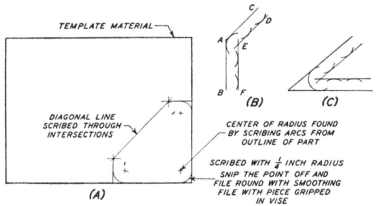

Fig. 7. (A) Example of Template Laid Out on Piece of Pattern Material by Using Dimensions from Blueprint; (B) Method of Finding Center for Radius When Angle Is More Than 90°; (C) When Angle Is Less Than 90°

and scribe an arc from each of the points established. The intersection of these arcs is the center for the ¼" radius of the arc in the square corner. Follow the same procedure for each of the remaining square corners.

Refer to (B) of Fig. 7; to find the centers for radii in the corners which are more than 90°, set your dividers for the required radius, in this case ¼". Place one leg of the divider at any point along A-B and scribe an arc. Repeat this procedure from another point along A-B. Draw line E-F tangent to these arcs and parallel to A-B. Keeping your dividers at ¼", establish two like arcs from two points on A-C. Next, draw line D-E tangent to these arcs and parallel to C-A. Point E at the intersection of D-E and E-F is the center for the ¼" radius of the arc in the corner greater than 90°. This same procedure is followed when the angle is less than 90°; see (C) of Fig. 7. (This procedure would also be accurate when there is a

# TEMPLATE LAYOUT AND BENCH WORK

90° angle; however, it is a more lengthy process than the other method described.)

To complete the cutting operation on this pattern, snip off the corners, and file smooth to the arc lines.

These directions should assist you in working the projects in the following pages. Practice with your combination square, adjusting it to various 64ths of an inch, in order to become thoroughly acquainted with the scale and to acquire skill with your fingers.

## QUESTIONS

1. How would you drill a hole through flat stock on the bench without drilling into the bench top?
2. What size drill would you use for a hole $.375 \begin{array}{l} +.000 \\ -.002 \end{array}$?
3. In the layout of a bulkhead, how would you construct horizontal and vertical center lines?
4. How would you lay out on a template a $2\frac{23}{64}''$ dimension from a blueprint?
5. What tool or tools would you use in cutting out an irregular hole in a pattern?
6. What is the purpose of a shop order?
7. Where are parts routed after completion, and what accompanies them?
8. What should you do with a blueprint, before starting to lay out the template?
9. What kind of metals are used in making templates?
10. Which template materials do not need to be treated with layout fluid?
11. What does tangent mean?
12. How do you lay out reference lines on a drawing?
13. What tool would you use in marking a hole location through a pattern?
14. What is the difference between a template and a pattern?
15. How would you cut around a sharp radius on heavy material?
16. How would you start layout of a pattern; that is, from what would you start your dimensioning?
17. What will happen if you bear down too heavily on the dividers when striking an arc?
18. What would determine whether you should cut out a number of parts by hand, or use the Doall?
19. How would you prepare a stack of parts to be cut on the Doall?
20. Should soft or unheat-treated materials be formed before or after cutting?
21. What is P27?
22. Is center line, as shown in Fig. 2, a symmetrical line? Why?
23. What kind of material is called for on the print in Fig. 2? Is it hard or soft?

24. How many center punch locations are to be laid out from this print?

25. How long a time has been allotted by production department for making the part shown in Fig. 8, indicated on the shop order in Fig. 1?

**NOTE.** In working the projects of this chapter, we will lay out the part on the material of which the part itself is to be made. In this way, our finished product actually is either a template or a finished part. This will allow us to acquire more practice in layout, since the template is used as a pattern for making duplicate parts. We will practice production work later, when making left- and right-hand parts.

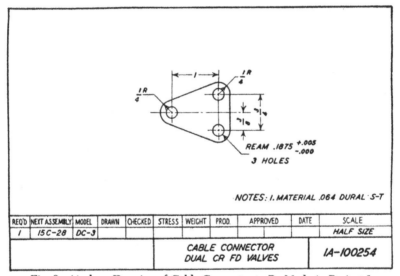

Fig. 8. Airplane Drawing of Cable Connector to Be Made in Project 1

### PROJECT 1. MAKING A CABLE CONNECTOR
(See Fig. 8)

**Objectives**
1. To teach proper use of layout tools
2. To teach the importance of working to close dimensions
3. To give practice in accurate hole drilling

**Tools and Equipment**
1. The following layout tools:
   6" steel scale
   Scriber
   Dividers
   Prick punch
   Ball peen hammer
   Combination square
2. Vise
3. 10" smooth flat file
4. 11/64" drill
5. 3/16" reamer and handle
6. Drill press
7. Combination snips

# TEMPLATE LAYOUT AND BENCH WORK

**Material**

1. Dural 17S–T  2"x 2"x .064
2. Layout fluid

*Note:* A good, fast-drying primer such as P27 is satisfactory for use as layout fluid on aluminum alloys to make the scribed marks clearer. However, it is not always necessary to use a layout fluid on such small patterns; it is called for here merely to acquaint the student with the proper procedure for larger scale drawings.

**Procedure**

1. Coat the metal on one side with the layout fluid, taking care to cover the surface thoroughly. Use a quick-drying primer; P27 is recommended. Brush or spray on a thin coat, and let it dry for about 15 minutes.

2. Lay out the fitting in accordance with drawing, Fig. 8. Center punch hole locations. The part is laid out as explained for the gusset, in Fig. 2. The outline is scribed tangent to the arcs of three circles (not shown on Fig. 8) which have the 1/4" radii. These arcs which make a part of the outline of the cable connector are scribed from centers laid out from the blueprint.

3. With combination snips, cut along straight sections and snip off around arcs. When cutting along outline, do so to the right, as indicated by the arrows in Fig. 4.

4. File the snipped portions to the scribed lines, making sure to hold the flat file surface at a 90° angle from the surface of the metal. In other words, hold the file straight up and down.

*Note:* A long-angle lathe file is excellent for finish-filing aluminum. Files of this type may have a pitch of 14 to 20 teeth per inch, and the teeth should be cut to a side-rake angle of about 45° to 55°. The side-rake angle provides a more efficient cutting action by producing a slicing motion.

5. Smooth the edges by draw filing; that is, stand in line with the edge and, holding the file by the handle or shank in the left hand and the tip in the right hand, use a draw motion lengthwise on the edge, in the same way a draw knife is used.

6. Drill holes with 11/64" drill and ream to 3/16".

*Note:* The reason for not using a 3/16" drill here, although drawing calls for a hole reamed to .1875", which is 3/16", is because we are allowed only +.005; that is, we must not make the holes more than .005 larger than dimension. If we used a 3/16" drill, we might drill a hole .010 oversize, depending on how the drill was sharpened. Therefore it is best to use an 11/64" drill and ream to 3/16". Since an 11/64" hole is relatively large, it might be better to drill a pilot hole with a 1/16" drill which is easier to keep in a center-punch mark than the larger drill. If the drills used by all the students are sharpened correctly, the holes as well as the outlines should line up when the parts are stacked together.

Chuck the 3/16" spiral or straight-flute reamer in a ratchet-handle tap wrench and ream the 11/64" holes to 3/16". Hold the reamer at an angle of 90° to the work, and do not force it too hard or chattering may occur, making a bad hole.

*Note:* No cutting compound is necessary for this job, because of the softness of the metal and the shallow holes.

7. Have fitting checked by your instructor.

## AIRCRAFT SHEET METAL WORK

**PROJECT 2. MAKING A REINFORCING WASHER**
(See Fig. 9)

**Objectives**

1. To teach the proper methods of laying out holes with the protractor
2. To teach the process of countersinking for flush rivets

**Tools and Equipment**

The same as in Project 1, except the hand drill with a 1/8" drill and a 3/8"x82° countersink; plus a protractor

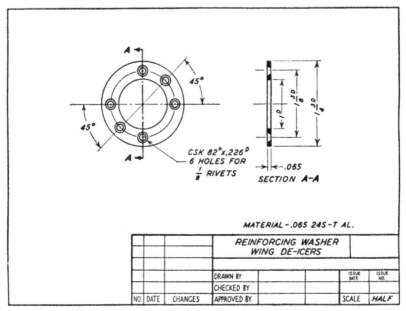

Fig. 9. Drawing of Reinforcing Washer to Be Made in Project 2

**Material**

Dural, 24S-T Al. 3"x 3"x .065

**Procedure**

1. In this project you are to lay out and center punch the locations for rivet holes, in accordance with drawing, Fig. 9. Scribe a vertical and a horizontal reference line and use the intersection as a center from which to lay out the inside and outside boundaries of the washer. Divide the diameters, shown at right in Fig. 9, by 2 to get the radii; set your dividers on the scale; and scribe the three large circles as shown, Fig. 9. Four of the center-punch marks for holes are found at the intersections of the two reference lines with the circle on which the centers of the rivet holes are to be located.

To find the location of the two holes at 45° from the horizontal reference line, place the scale of your combination set in the protractor head, and set

# TEMPLATE LAYOUT AND BENCH WORK 107

the protractor at 45°. Using the same edge of the pattern from which you laid out the reference lines, scribe the 45° line intersecting the center axis. Now place the scale blade back in your square head and check your work. (You will see that one face of your square head is fixed at 45°, and therefore it is not necessary to use the protractor at all in this particular exercise. In addition, you could use the dividers to step off these hole locations, because the points you are locating are equidistant from the reference lines. If the holes were located more or less than 45° from any reference line, however, the use of the protractor would be the proper procedure for laying out the holes.)

2. Punch a 1" hole in the center. This is a simple operation in which you center punch the intersection of the reference lines and place the center point of the punch in center-punch mark and pull down the lever. There are many different kinds of hand-operated punches; have your instructor show you how to operate the one you are to use.

3. Cut out washer with combination snip. This is done as previously explained. Place in smooth-jaw vise and file outline round, smooth, and even.

4. Drill a 1/8" hole through the metal and then countersink to .226" diameter. In drill press, using a flat block of wood to back up the washer on the drill table, use a 78° 1/8" countersunk rivet to gage when you have countersunk the hole to just the right depth. The head diameter of the rivet mentioned is exactly .226".

5. Remove any burrs caused from drilling or countersinking, by rubbing against fine emery cloth on small block of wood.

6. Have your instructor check the work.

### PROJECT 3. MAKING A FRICTION WASHER
(See Fig. 10)

**Objectives**
1. To teach the layout of fittings to close dimensions
2. To teach the use of a drill for cutting smooth holes

**Tools and Equipment**
1. Layout Tools
   Steel scale
   Scriber
   Prick punch
   Ball peen hammer
   Square
   Center punch
2. Vise
3. Flat file, smooth
4. Combination snips
5. Twist drills, 21/64", 45/64"
6. Drill press
7. Heavy C clamp

**Material**
Dural 17S–T, 4"x 4"x .064

**Procedure**
1. Lay out and center punch hole locations in accordance with drawing, Fig. 10.

2. Clamp piece between two small boards with heavy C clamp which can be used for a handle, thus preventing injury while drilling.

3. If your stock does not carry a 45/64" drill sharpened for thin sheet metal, you must sharpen one, as shown at C, Fig. 4, Chapter 1. Be sure the point is longer than the cutting lips or the drill will not hold itself in one place on the work. After chucking the drill in the press, place a block of wood to back the washer on the drill-press table. The 21/64" hole may be made with a drill sharpened for general use, but the washer should be clamped on a hardwood block for this operation. This helps hold the point of the drill after it penetrates the metal, thus preventing chatter.
4. File edges and remove burrs.
5. Check the work with your instructor.

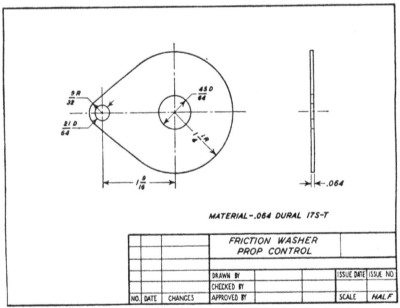

Fig. 10. Drawing of Friction Washer to Be Made in Project 3

## PROJECT 4. MAKING A BELL CRANK
(See Fig. 11)

**Objectives**
1. To teach proper method of layout and reading blueprints
2. To teach proper method of band sawing and filing
3. To teach process of drilling and reaming to size

**Tools and Equipment**
1. Layout tools
2. Vise
3. Hack-saw frame and flexible back blade, with only the teeth hardened
4. Bandsaw

*Note:* If a bandsaw is not available, a hack saw will have to be used; since an ordinary hack-saw blade is too wide for turning curves as sharp as

# TEMPLATE LAYOUT AND BENCH WORK

are necessary in sawing out this section, we will use a flexible back blade, with only the teeth hardened. It is easy to see how far the hardened area extends. Take an old pair of circle snips and cut the flexible area away, leaving plenty of stock at the holes, to prevent breakage. The result will be a blade about 3/16" wide with 18 teeth to the inch; this will allow us to cut around curves.

    5. Files,   8" half round, second cut
                 10" bastard
                 12" fine

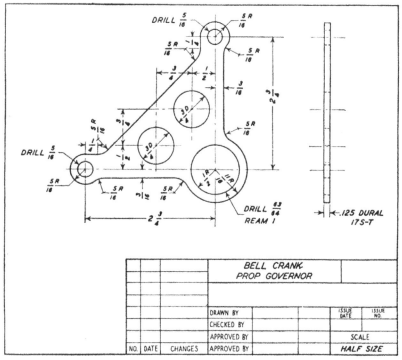

Fig. 11. Drawing of Bell Crank to Be Made in Project 4

    6. Drills,  5/16" average ground
              3/4" sheet metal
              63/64" sheet metal
    7. Reamer, 1"
    8. Drill press and drill vise
    9. Layout fluid

**Material**
    Dural 17S–T. 4"x 4"x .125

**Procedure**
    1. Paint one side of material with smooth coat of zinc chromate (P27) or some such layout fluid.

# AIRCRAFT SHEET METAL WORK

2. Lay out and center punch all hole locations, as shown in Fig. 10.

3. Drill the holes before cutting out, to aid in clamping in the drill vise. Use a 5/16" drill sharpened in the usual manner; for the large holes use a sheet metal drill sharpened as at C, Fig. 4, Chapter I.

4. In sawing out the bell crank, keep outside of the edge line 1/64" to 1/32". In sawing out a part made of aluminum alloy from 1/16" to 1/4" thick, the saw speed should be 1,200 feet per minute.

5. File at 90° to the flat surface until you have reached the outline, using the 10" bastard, and the 8" half round for the radii. Draw file edges to remove the file marks.

6. Check work with your instructor.

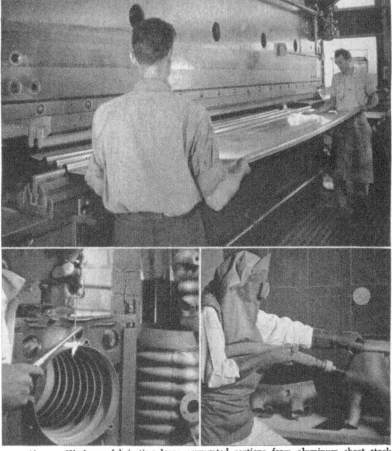

Above. Workmen fabricating large corrugated sections from aluminum sheet stock on a power brake.
*Courtesy of Douglas Aircraft Co., Inc., Santa Monica, Cal.*

Lower Left. Pieces of sheet metal were corrugated and bent before this workman could weld them into boilers for the steam heating system on Mainliners. Lower Right. Other bent parts are being sandblasted in the sandblasting room of a sheet metal department.
*Courtesy of United Air Lines, Cheyenne, Wyoming*

*Chapter VI*

# Pattern Development for Bends

In connection with previous chapters, none of the projects have called for any part which included a bend, such as an angle, channel,

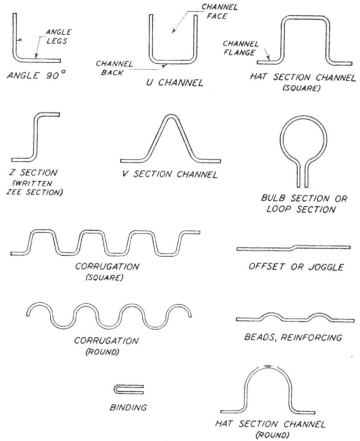

Fig. 1. Some Bends Made on Bending Brake

offset, etc., as shown in Fig. 1. Before you can begin any project involving bends, you must first learn the method used for making bend allowances. For instance, if you wanted to make an angle 1″ by 1″, it would be natural for you to suppose that this would require a piece of metal 2″ long, bent in the middle. If this procedure were followed, however, you would find upon checking the legs of your angle with a scale that they would be too large and would measure considerably over 1″ on each side. The excess metal would vary on metals of different thicknesses and with different radii.

**BASIC PRINCIPLES OF ALLOWANCE FOR BENDS.** Where a template for a part containing several angles of bends is prepared for layout, we must use the bend allowance to determine the proper area for the pattern. Mechanics who have been "guesstimating," or using the trial-and-error method, should modernize themselves by studying this chapter carefully and by putting its teachings into practice. The method is taught in vocational schools today, therefore most of the student apprentices are familiar with its use.

Manufactured angles may be obtained sharp or square, but angles of sheet metal must be formed on a radius. The minimum radii to which aluminum alloy may safely be formed are:

$$S\text{--}O \text{ material, } 1 \times T$$
$$S\text{--}T \text{ material, } 2.5 \times T$$

Usually a radius greater than the minimum is used. A radius of twice minimum should be used where strength in the radius is a factor to be considered.

**Terms Used.** In order to understand this important phase of pattern development, you must become acquainted with the principles of bend allowance. Fig. 2 is a diagram illustrating these principles, and the following paragraphs explain the terms used.

*Bend Allowance (BA)* is the length of metal required to form a bend. This is determined by the length of the *neutral axis*, the neutral axis being the imaginary line in the metal where compression on the inside of the bend changes to tension on the outside of the bend; see (C) of Fig. 3.

*Bend Line (BL)* is the point where the curve of the bend and the unbent metal meet, i.e., the exact place where the metal begins to bend.

*Mold Line (ML)* is the outside dimension line (witness line) where the bend starts.

# PATTERNS FOR BENDS

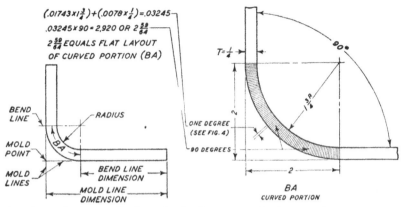

Fig. 2. Left, Terms Used in Bend Allowance and Their Application; Right, Illustration of Working Formula for Finding Stretch-Out

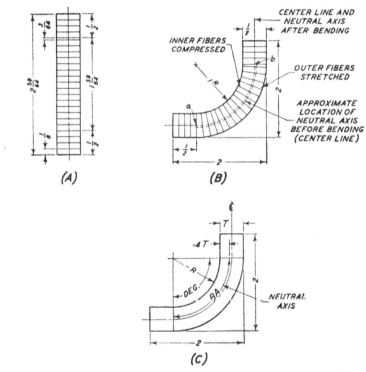

Fig. 3. (A) Edge View Stretch-Out of Angle; (B) Movement, and Location $a$-$b$, of Neutral Axis When Material Has Not Reached Its Elastic Limit But Is Being Held in This Position, Assuming This Could Be Done; (C) Movement of Neutral Axis When Material Has Exceeded Elastic Limit and Stays in Bent Position

*Mold Point (MP)* is the intersection of the two mold lines or witness lines.

*Radius (R)* is a straight line from the center of a circle or sphere to its periphery. A bend is formed using a given radius. The given radius is always to the inside of the metal thickness.

*Thickness (T)* means thickness of metal, or gage.

*Thickness Plus Radius (TR)* is used in working BA formula.

**Bend Line Dimensions.** To find the dimensions of an angle, the angle must be divided into three sections (the two bend line dimensions and the BA) and the dimension found for each section. It can be seen from Fig. 2 that, when the mold line dimension is given, the bend line dimension is found by subtracting the sum of the radius and the metal thickness from the mold line dimension. In this manner, for any given angle we can get the bend line dimension on each side, leaving the third dimension (determined by finding the bend allowance) to be added to the bend line dimensions.

**Bend Allowance Dimension.** As previously stated, it is necessary that very close dimensions be followed in working sheet metal in the aircraft industry. The third dimension left for us to find, in order to complete the layout of an angle, can be figured accurately within one or two thousandths of an inch by using the following *empirical formula.* (An empirical formula is one that is arrived at by experiment, rather than by theory.) In figuring bend allowance, take $(.01743 \times R) + (.0078 \times T)$, multiplied by the number of degrees in the bend. The number of degrees means the angle of the bend, such as the 90° in Fig. 2.

Thus, $.01743R + .0078T = BA$ for 1°, and $(.01743R + .0078T) \times 90 = BA$ for a right angle or 90°.

(Remember that $R$ equals radius of bend, and $T$ equals thickness of sheet in inches.)

The bend allowance chart given in Table 1 includes factors for various radii and the commonly used metal thicknesses, worked out so that any bend allowance may be determined simply by finding the factor under the gage (thickness or T) and opposite the radius or R, and multiplying this factor by the given degrees. This table is a great timesaver for both student and mechanic. Nevertheless, you should be familiar with the bend allowance formula,

PATTERNS FOR BENDS

Table 1—Bend Allowances for 1°

| R\T | .022 | .028 | .032 | .035 | .040 | .049 | .051 | .064 | .065 | .091 | .095 | .120 | .128 | .187 |
|---|---|---|---|---|---|---|---|---|---|---|---|---|---|---|
| 1/32 | .00072 | .00076 | **.0079** | .00082 | .00140 | .00093 | | | | | | | | |
| 1/16 | .00126 | .00131 | .00135 | .00136 | .00195 | .00147 | **.00149** | **.00159** | .00160 | | | | | |
| 3/32 | .00180 | .00185 | .00188 | .00191 | .00249 | .00202 | **.00203** | **.00213** | .00214 | **.00234** | | | | |
| 1/8 | .00235 | .00240 | .00243 | .00245 | .00249 | .00256 | .00258 | **.00268** | .00269 | **.00289** | .00292 | | | |
| 5/32 | .00290 | .00294 | .00297 | .00300 | .00304 | .00311 | .00312 | .00322 | .00323 | **.00343** | .00346 | | | |
| 3/16 | .00344 | .00349 | .00352 | .00354 | .00358 | .00365 | .00367 | .00377 | .00378 | **.00398** | .00401 | .00420 | | |
| 7/32 | .00398 | .00403 | .00406 | .00409 | .00412 | .00419 | .00421 | .00431 | .00432 | **.00452** | .00455 | .00475 | | |
| 1/4 | .00454 | .00458 | .00461 | .00463 | .00467 | .00474 | .00476 | .00486 | .00487 | .00507 | .00510 | .00529 | **.00317** | |
| 9/32 | .00507 | .00512 | .00515 | .00516 | .00521 | .00528 | .00530 | .00540 | .00541 | .00561 | .00544 | .00584 | **.00372** | |
| 5/16 | .00562 | .00567 | .00570 | .00572 | .00576 | .00583 | .00584 | .00595 | .00596 | .00616 | .00619 | .00638 | **.00426** | |
| 11/32 | .00616 | .00620 | .00624 | .00626 | .00630 | .00637 | .00639 | .00649 | .00650 | .00670 | .00673 | .00693 | **.00481** | .00691 |
| 3/8 | .00671 | .00675 | .00679 | .00681 | .00685 | .00692 | .00693 | .00704 | .00705 | .00725 | .00728 | .00747 | **.00535** | .00745 |
| 13/32 | .00725 | .00730 | .00733 | .00735 | .00739 | .00746 | .00748 | .00758 | .00759 | .00779 | .00782 | .00802 | **.00590** | .00800 |
| 7/16 | .00780 | .00784 | .00787 | .00790 | .00794 | .00801 | .00802 | .00812 | .00813 | .00834 | .00837 | .00857 | .00644 | .00854 |
| 15/32 | .00834 | .00839 | .00842 | .00844 | .00848 | .00855 | .00857 | .00867 | .00868 | .00888 | .00891 | .00911 | .00699 | .00908 |
| 1/2 | .00889 | .00893 | .00896 | .00899 | .00903 | .00910 | .00911 | .00921 | .00922 | .00943 | .00946 | .00965 | .00753 | .00963 |
| 17/32 | .00943 | .00948 | .00951 | .00953 | .00957 | .00964 | .00966 | .00976 | .00977 | .00997 | .01000 | .01019 | .00808 | .01017 |
| 9/16 | .00998 | .01002 | .01005 | .01009 | .01012 | .01019 | .01020 | .01030 | .01031 | .01051 | .01055 | .01074 | .00862 | .01072 |
| 19/32 | .01051 | .01055 | .01058 | .01061 | .01065 | .01072 | .01073 | .01083 | .01084 | .01105 | .01108 | .01127 | .00917 | .01126 |
| 5/8 | .01107 | .01111 | .01114 | .01117 | .01121 | .01128 | .01129 | .01139 | .01140 | .01160 | .01161 | .01183 | .00971 | .01179 |
| 21/32 | .01161 | .01166 | .01170 | .01171 | .01175 | .01182 | .01183 | .01193 | .01194 | .01214 | .01218 | .01237 | .01025 | .01235 |
| 11/16 | .01216 | .01220 | .01223 | .01226 | .01230 | .01237 | .01238 | .01248 | .01249 | .01269 | .01272 | .01292 | .01080 | .01289 |
| 23/32 | .01269 | .01273 | .01276 | .01279 | .01283 | .01290 | .01291 | .01301 | .01302 | .01322 | .01326 | .01345 | .01133 | .01344 |
| 3/4 | .01324 | .01329 | .01332 | .01335 | .01338 | .01345 | .01347 | .01357 | .01358 | .01378 | .01381 | .01401 | .01189 | .01397 |
| 25/32 | .01378 | .01383 | .01386 | .01389 | .01392 | .01399 | .01401 | .01411 | .01412 | .01432 | .01435 | .01456 | .01245 | .01453 |
| 13/16 | .01433 | .01438 | .01441 | .01443 | .01447 | .01454 | .01456 | .01466 | .01467 | .01487 | .01490 | .01510 | .01298 | .01507 |
| 27/32 | .01487 | .01491 | .01494 | .01497 | .01501 | .01508 | .01509 | .01519 | .01520 | .01540 | .01543 | .01563 | .01351 | .01562 |
| 7/8 | .01542 | .01548 | .01550 | .01552 | .01556 | .01563 | .01565 | .01575 | .01576 | .01596 | .01599 | .01619 | .01407 | .01615 |
| 29/32 | .01596 | .01601 | .01604 | .01606 | .01619 | .01617 | .01619 | .01629 | .01630 | .01650 | .01665 | .01673 | .01461 | .01671 |
| 15/16 | .01651 | .01655 | .01659 | .01661 | .01665 | .01672 | .01674 | .01684 | .01685 | .01705 | .01708 | .01728 | .01516 | .01727 |
| 31/32 | .01705 | .01709 | .01712 | .01715 | .01718 | .01725 | .01727 | .01737 | .01738 | .01758 | .01761 | .01781 | .01569 | .01780 |
| 1 | .01760 | .01765 | .01768 | .01770 | .01774 | .01781 | .01783 | .01793 | .01794 | .01814 | .01817 | .01837 | .01625 | .01833 |
|  |  |  |  |  |  |  |  |  |  |  |  |  | .01679 | .01889 |
|  |  |  |  |  |  |  |  |  |  |  |  |  | .01734 |  |
|  |  |  |  |  |  |  |  |  |  |  |  |  | .01787 |  |
|  |  |  |  |  |  |  |  |  |  |  |  |  | .01843 |  |

Note: The values given are based on the Empirical Formula and are the BA for 1°. BA = (value from table) × (number of degrees of bend). Values omitted from the table are not to be used, as the bends are too sharp for satisfactory production. Values in bold face type, as .00079 are not to be used for hard Dural but are all right for soft Dural.

and will probably want to know how it works; now is a good time to take it apart and see how it is made, so that you will have a clear idea as to its workability.

**BEND ALLOWANCE FORMULA.** The values given (.01743R+.0078T) are empirical figures, meaning they were found experimentally rather than mathematically or scientifically. Mathematically we find that the *arc length of 1°* in a circumference of a circle having a *1" radius* is 3.1416($\pi$)×2 (twice the 1" R, or Dia.) divided by 360 (the number of degrees in a circle) or .01745". By multiplying .01743 (trial-and-error experiments having proved that .01743 is a better figure than .01745) by the radius, times the given number of degrees in the arc, we get the stretch-out of a line with no thickness. However, sheet metal has thickness that compresses and stretches when bent. The bend allowance formula, therefore, must function for various metal thicknesses in a way that will give the length of the neutral axis after the desired bend is made. As stated before, the neutral axis is the imaginary line in the metal where compression on the inside of the bend changes to tension on the outside of the bend. The flat pattern must be so designed that its length is the same as the length of the neutral axis after the bending operation. We shall consider the neutral axis before learning how to arrive at the other empirical value, .0078. This value, when multiplied by the metal thickness and added to the product of .01743×R, gives just the additional length to make the proper bend allowance for one degree, because the .0078 takes into consideration the greater radius required to reach the neutral axis.

**Neutral Axis.** It is not necessary that a mechanic know the location of the neutral axis. However, since the ambitious student usually wants to know the answers to such questions, and since constructive discussion will help toward an understanding of what happens to metal under bending, the following explanation is given.

**Question. Why is the neutral axis not in the center of the metal thickness in the radius of a bend, as it is in the metal before being upset (bent)?**

*Answer.* In a straight or curved sheet, rod, or beam, the neutral axis is in the exact center so long as the metal is not bent beyond its elastic limit. The *elastic limit* is that point beyond which bent metal will not return to its original form, but exhibits permanent changes in structure, size, etc. Compression and tensile strength in wrought alloys are equal, so when a bar, sheet, or beam is bent within its elastic limits, the metal on the inside of the bend is displaced, which causes an increase in volume through compression; at the same time, there is a proportionate decrease on the outside of the bend through stretch. The rule is that the neutral axis, at any section in a beam subject to bending only, passes through the center of gravity of that section. In this case the neutral axis moves to the inside of the bend as the mass moves, but is still in the center of the cross section as indicated at *(B)* in Fig. 3. However, in forming angles we must bend the metal *beyond* its elastic limits, so that it will retain its shape when released from the bending machine. When metal is bent to very sharp corners or small radii, the bent section is 10 to 15% less in thickness than before bending. This is because the metal moves more easily in tension than in compression, since after metal moves beyond its elastic limit, the strength of the metal on the inside of the bend (compressive

# PATTERNS FOR BENDS

strength) becomes greater than the strength of the metal on the outside of the bend (tension strength). For the same reason, the neutral axis of the metal moves in toward the radius axis.

Therefore, in figuring the allowance, as shown at (C) in Fig. 3, to be made for the bend up to an inside radius, R, of 2 or 3 times the metal thickness, T, the length may be figured closely as along a neutral line at .4T, (or .4 of the metal thickness) out from the inside radius. Thus, as shown at (C) in Fig. 3, BA is found for any angle by the following equation:

$$BA = (R + .4T \times 2) \times \pi° \times \frac{\text{number of degrees}}{360}.$$

The factor .4T locating the neutral axis is variable, according to radius, condition of metal, and method of applying forces in making the bend. This variation is from .35 to .45T. The figure used in the aircraft industry, however, is .45T. Adding .4T or .45T to R gives the radius to the neutral axis, as shown in Fig. 4. The second empirical value of the BA formula is found by taking the factor .01743R times .45T which equals .0078.

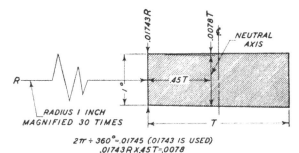

$2\pi \div 360° = .01745$ (01743 IS USED)
$.01743 R \times .45 T = .0078$

Fig. 4. Enlarged View of One Degree of a Bend Showing Radius and Radius of Neutral Axis

Radius to neutral axis × 2 = diameter of circle.
Diameter × 3.1416 = circumference at neutral axis.
Circumference divided by 360 = length of 1° at neutral axis. (Circumference divided by, say, $\frac{90}{360}$ equals length of 90° at neutral axis.)

*Example:*
Find BA in a 90° angle made of .040 17S–O, with a 1/8" R
.040 × .45 = .018" distance from inside radius point to radius of neutral axis.
Then, 1/8" R = .125 + .018 = .143 or R of neutral axis.
.143 × 2 = .286 = twice radius (or diameter).
$\frac{.286 \times 3.1416}{360} = .00249$ bend allowance for 1°.
.00249 × 90 = .224 = BA for 90°.

Check this figure with Table 1. In actual practice, this problem would be worked as follows:
(.01743 × .125) + (.0078 × .040) = .00249, for 1°.

---
° $\pi$ equals 3.1416.

The edge view and stretch-out at *(A)*, and the angle at *(C)*, in Fig. 3, show how a piece is dimensioned for a 2" angle with a 1" radius, made from 1/2" thick material. The BA is found as follows: $(.01743 \times 1) + (.0078 \times 1/2) \times 90 = 1.9197$, or approximately $1^{59}/_{64}$".

At *(A)* and *(B)* in Fig. 3, the bend allowance is divided into sections of exactly 1/8" to show what happens through the process of bending. Notice that the spacing along the neutral axis has remained at 1/8" spacings; also notice how these have spread apart on the outside, and closed in on the inside, of the bend; and how the neutral axis at *(C)* has shifted to the inside of the bend from the center line.

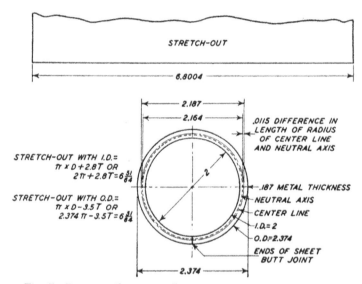

Fig. 5. Diagram Showing Difference Between Center Line and Neutral Axis of a Circle and Giving Formulas for Determining Stretch-Out When Inside and Outside Diameters Are Given

**TUBES, PIPES, AND CYLINDERS.** The diagram in Fig. 5 shows a rolled-up cylinder 2" I.D. (inside diameter), together with stretch-out and dimensions needed for making the cylinder.

In making tubes, pipes, or cylinders from sheet metal where close fits are desired, as in fitting a fabricated tube over a cast boss or junction box, the stretch-out may be determined by the same empirical formula as that used for finding the bend allowance for an angle, as follows:

.01743R + .0078T × 360 (number of degrees in full circle) = stretch-out.

# PATTERNS FOR BENDS

The empirical formula is preferable; however the following formulas may also be used.

Where the dimension given is the *inside* diameter, the formula is $(2.8 \times T) + (3.1416 \times Dia.)$. The metal thickness, $.187 \times .8 = .5236$, amount added for BA. $3.1416 \times 2 = 6.2832 = 360°$ of $1''$ R; $6.2832 + .5236 = 6.80$—stretch-out of cylinder.

Where the dimension given is the *outside* diameter, the formula is $(3.1416 \times Dia.) - (3.5 \times T)$.

These values are arrived at as follows: multiplying $.0078 \times 360$ gives 2.808, the factor used with inside diameter. Subtracting 2.808 from $2 \times 3.1416$ equals 3.4752, or 3.5, the factor used with outside diameter.

Notice that on Fig. 5 the stretch-out is given as 6.8004. This was determined by using Table 1.

**THE STRETCH-OUT.** Fig. 6 illustrates the various requirements necessary for developing a pattern for an angle, including the correct method of using the bending brake to form it. Note where $e$ at $(D)$ designates metal clearance between the clamping bar and the bed of the brake; also where $f$ at $(D)$ designates clearance between the bending leaf and the upper jaw when the angle is bent up.

The following questions and answers may help you understand how a pattern is developed:

**1. Viewing the dimensioned angle at (A) in Fig. 6, what is the first dimension we want?**
*Answer.* $a$ to $b$.

**2. This dimension is not given. How do we get it?**
*Answer.* By subtracting TR (the sum of the metal thickness, .064, and radius, $1/8''$) from $1''$ (dimension to the mold line). The dimension $c$ to $d$ is obtained in the same way.

**3. What other dimension do we need?**
*Answer.* $b$ to $c$, or bend allowance, is the remaining dimension to be found.

**4. How do we find this dimension?**
*Answer.* Refer to Table 1 and find the factor under .064 (metal thickness) and opposite 1/8 radius.

**5. What is this number?**
*Answer.* The number is .00268.

**6. What must be done with this number?**
*Answer.* It must be multiplied by 90, because there are 90° in the angle in the drawing.

**7. What is the result?**

*Answer.* The result is .241, or approximately 15/64.

**8. How do you arrive at the 15/64?**

*Answer.* We looked on a decimal equivalent table, and 15/64 was the nearest fractional equivalent, in 64ths, to .241.

**9. Why do you take the fractional equivalent in 64ths?**

*Answer.* So we can lay the stretch-out on a piece of metal with our scale; our scale is graduated in 64ths of an inch.

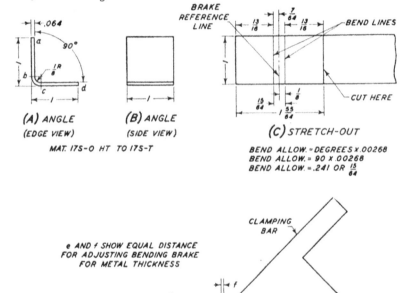

Fig. 6. (A) Edge View of a 90° Angle Showing Metal Thickness, Degree of Bend, Radius; (B) Side View of Same Angle; (C) Stretch-Out of Same Angle; (D) Cross Section of Bending Brake Showing Clearances $e$ and $f$ Which Provide for Metal Thickness

**10. What was your dimension from $a$ to $b$?**

*Answer.* The dimension was 13/16, the remainder after subtracting 3/16 from 1″.

**11. Where did you get the 3/16?**

*Answer.* TR=3/16, because .064 is 1/16 and when added to the 1/8 radius it gives 3/16, the distance from the mold line to the bend line.

*Note:* When drawings are dimensioned to the bend line, as at (A) in Fig. 7, we need only add bend allowance between the two other dimensions.

# PATTERNS FOR BENDS

**Procedure:** Now go ahead and lay out the angle as shown in Fig. 6. When the pattern has been developed, mark the brake reference line with a sharp pencil as shown in the stretch-out at (C)

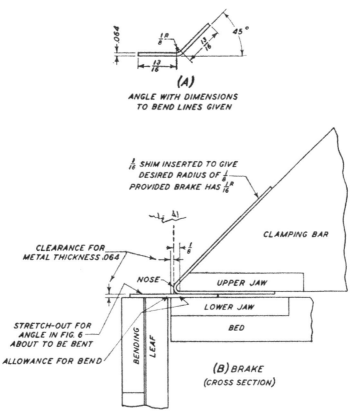

Fig. 7. (A) Drawing of an Angle with Dimensions to Bend Line Given; (B) Diagram Showing Cross Section of Bending Brake with a Piece of Metal Located for Bending

in Fig. 6. This line must be radius distance (in this case, $\frac{1}{8}''$) out from the bend line on the end which is inserted under the jaw of the brake. The reference line is held out flush with the nose of the upper jaw, as the eye sights at 90°; see Fig. 7, at (B). Note the $\frac{1}{16}''$ shim that had to be placed around the nose of the brake to build it up from a $\frac{1}{16}''$ to a $\frac{1}{8}''$ radius.

# AIRCRAFT SHEET METAL WORK

## PROJECT 1. CHANNEL, TABULATED

Make those assigned by instructor. (One of each in Fig. 8 is recommended.)

**Objectives**

1. To learn the proper method of allowance for bends

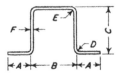

| | | | | | | | |
|---|---|---|---|---|---|---|---|
| 1A-123-4 | $\frac{3}{8}$ | $\frac{1}{2}$ | $\frac{1}{2}$ | $\frac{1}{16}$ | $\frac{1}{16}$ | .025 | 17 S-T |
| 1A-123-5 | $\frac{1}{2}$ | 1 | 1 | $\frac{1}{16}$ | $\frac{1}{16}$ | .051 | 17S-O HT S-T |
| 1A-123-6 | $\frac{3}{8}$ | $\frac{5}{8}$ | $\frac{3}{4}$ | $\frac{1}{16}$ | $\frac{1}{16}$ | .040 | 17S-O HT S-T |
| 1A-123-7 | $\frac{3}{8}$ | 1 | $\frac{5}{8}$ | $\frac{1}{16}$ | $\frac{1}{16}$ | .025 | 24 S-T |
| 1A-123-8 | $\frac{1}{2}$ | $\frac{3}{4}$ | $1\frac{1}{8}$ | $\frac{1}{16}$ | $\frac{1}{16}$ | .032 | 17 S-T |
| 1A-123-9 | $\frac{3}{8}$ | $\frac{3}{4}$ | $\frac{3}{4}$ | $\frac{1}{8}$ | $\frac{1}{8}$ | .064 | 17S-O HT S-T |
| 1A-123-10 | $\frac{1}{2}$ | $\frac{7}{8}$ | $\frac{7}{8}$ | $\frac{1}{8}$ | $\frac{1}{16}$ | .064 | 17S-O HT S-T |
| 1A-123-11 | $\frac{1}{2}$ | $\frac{7}{8}$ | $\frac{7}{8}$ | $\frac{1}{16}$ | $\frac{1}{16}$ | .032 | 17S-O HT S-T |
| 1A-123-12 | $\frac{1}{2}$ | $\frac{1}{2}$ | $\frac{1}{2}$ | $\frac{1}{16}$ | $\frac{1}{16}$ | .032 | 24S-O HT S-T |
| 1A-123-13 | $\frac{3}{8}$ | 1 | $\frac{3}{16}$ | $\frac{1}{16}$ | $\frac{1}{8}$ | .064 | 24S-O HT S-T |
| 1A-123-14 | $\frac{3}{8}$ | 1 | $\frac{5}{16}$ | $\frac{1}{16}$ | $\frac{1}{8}$ | .032 | 17 S-T |
| 1A-123-15 | $\frac{3}{8}$ | $1\frac{1}{4}$ | $1\frac{1}{8}$ | $\frac{1}{16}$ | $\frac{1}{16}$ | .064 | 17S-O HT S-T |
| 1A-123-16 | $\frac{3}{8}$ | $\frac{3}{8}$ | $\frac{3}{8}$ | $\frac{1}{16}$ | $\frac{1}{16}$ | .032 | 17S-O HT S-T |
| SECTION NO. | A | B | C | D | E | F | AL. ALLOY |

NOTE: MAKE ALL MODELS 1" LONG

| | | CHANNEL TABULATED | | | 1A-1426 | |
|---|---|---|---|---|---|---|
| | | DRAWN BY | | | ISSUE DATE | ISSUE NO. |
| | | CHECKED BY | | | | |
| NO. DATE | CHANGES | APPROVED BY | | | SCALE | HALF |

Fig. 8. Channel Tabulation of a Hat Section

2. To learn the proper method of dimensioning between bends for the stretch-out

**Tools and Equipment**

1. Layout tools
2. Bending brake

**Material**

Aluminum alloy—noted in tabulation in Fig. 8.

# PATTERNS FOR BENDS

**Procedure**

1. Lay out the hat section channel called for in the tabulation.
2. Follow the procedure illustrated in Fig. 6 and explained under the section, *The Stretch-Out*, in this chapter.
3. Mark all bends starting from one end, and brake in the same rotation.
4. Note both the metal thickness and bend radii to be subtracted from over-all dimensions to obtain stretch-out of section involved between bends.
5. Present the finished part or channel to instructor for inspection and comments.

Measurements should be within close tolerance. With practice you will find yourself using a micrometer for checking with results of surprising accuracy, say within plus or minus .002" to .004" between angles.

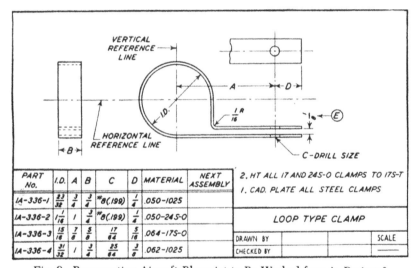

Fig. 9. Representing Aircraft Blueprint to Be Worked from in Project 2

6. Make channels with soft alloys—heat-treat to full hardness.

**Caution**

Do not forget which way to insert the part in the brake, in relation to the way you have laid out sight marks for locating in the brake, as shown at *(B)* in Fig. 7.

## PROJECT 2
### (Preliminary Explanation)

**Loop Type Clamp.** To make the loop type clamp, Fig. 9, first observe the dash number suffixed to the part number given you by the instructor.

For example, we will say your work order reads:

"Make one clamp, Part No. 1A–336–3."

We find in the table beside the title block the dimensions for dash 3 clamp which has an inside diameter (I.D.) of 15/16, width (at *B*) of 5/8, distance (at *D*) of 5/16, (at *A*) 7/8, and material .064 17S–O Dural. The hole size (at *C*), also called edge distance, is 17/64.

*Note:* This instance proves the fallacy of a mechanic scaling a drawing to get dimensions. You will find that your finished product will not look anything like the drawing as to size. *Never scale a blueprint to obtain dimensions.* You are likely to find a dimension line 11" long with the figure 12 as dimension; or an arc with a 12" compass setting calling for a 14" radius. Allowance must be made, too, for paper shrinkage through the processing of the blueprint.

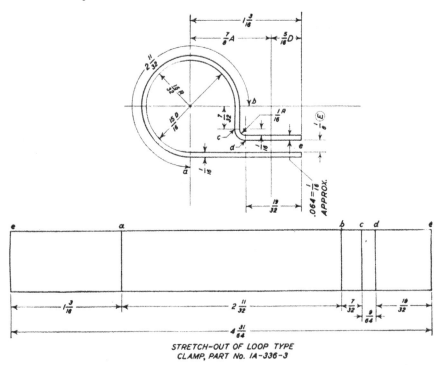

STRETCH-OUT OF LOOP TYPE
CLAMP, PART No. IA-336-3

Fig. 10. Mechanic's Shop Sketch of Fig. 9 for Figuring Various Dimensions (May Be Drawn Free-Hand But Is Not Here Because of Importance of Illustrating Dimensions)

In laying out a pattern for the clamp shown in Fig. 9, where the inside of the loop is only 3/4, or 270°, of a circle, the first thing to do is to make a shop sketch as shown in Fig. 10. Such a sketch may be drawn freehand; this has not been done in this instance because of the importance of illustrating dimensions. Since the diameter is 15/16, the radius is 15/32. The factor for finding the bend allowance, with metal thickness of .064 and radius of 15/32, is shown in Table 1 to be .00867 for one degree.

Then, the distance from *a* to *b*, lower view, Fig. 10, is .00867×270= 2.340", or approximately 2–11/32".

# PATTERNS FOR BENDS

The distance $e$ to $a$ is 7/8 plus 5/16 $(A+D)$ or 1-3/16.

The distance $b$ to $c$ is the most difficult dimension to compute from the drawing, Fig. 9. We know the dimension we want is that from the bend line $b$ (shown in shop sketch, Fig. 10, where the horizontal reference line intersects the clamp at $b$) to the bend line at $c$, which intersects the 1/16 radius horizontally. Then by adding the 1/8" (circle $E$) to the R (1/16) and T (1/16, or .064) we get the sum of 1/4", which when subtracted from the radius (15/32) of the loop, leaves 7/32 for this dimension.

The bend allowance $c$ to $d$ is determined by referring again to Table 1 for the factor under T (.064) and opposite R (1/16); this factor is .00159. The degrees of this bend are not given, but we know it must be 90° or it would be stated otherwise. Therefore $.00159 \times 90 = .14310$. Referring to the table of decimal equivalents, we convert this .14310 into the nearest fraction in 64ths, or 9/64ths.

The last dimension needed is that from $d$ to $e$, which is found by a method similar to that used in finding the dimension $b$ to $c$; in this case subtract the sum of the loop radius (15/32) metal thickness (1/16) and bend radius (1/16) from the sum of 7/8 (at $A$) and 5/16 (at $D$), leaving a remainder of 19/32, the distance from $d$ to $e$.

We now have all the required dimensions so that they can be laid out on a pattern from scale readings. The width $B$, as shown for 1A-336-3 in Fig. 9, is 5/8", so we must cut a piece of metal to this width, and longer than is necessary for the total length. The bend of 1/16 radius is made in a brake as shown in Figs. 6 and 7, and the loop is formed around a 7/8" tube held in the vise after the 1/16 radius bend is made.

### PROJECT 3. MAKING A REINFORCING CLIP

**Objectives**

1. To learn the proper method of laying out a channel from a third angle orthogonal projection
2. To learn the use of the bend allowance formula
3. To learn the technique of using the bending brake

**Tools and Equipment**

1. Layout tools
2. Drill 5/8 sheet metal—Drill #26
3. Bending Brake

**Material**

1. Dural 17S-T, 2"x 3"x .032

**Procedure**

1. Lay out pattern on material as shown in Fig. 11.
2. Drill all holes before bending.
3. Bend in the brake.
4. Check all holes to make sure you have followed the proper procedure in measuring, making bend allowance, and figuring the mark locations which were used in bending. The 5/8 Dia. hole center should measure 3/8" from the inside of the channel, as it is not in the center. Hold the clip with the back on a surface plate to measure the No. 26 holes for center with the

scale end on the surface plate. The holes should be centered at 3/16", plus T, or 7/32".

5. Review text and Fig. 6 of this chapter. Note where dimensions are given, in relation to metal thickness of the clip in Fig. 11. Instead of channel width being dimensioned to mold lines, it is dimensioned to the inside of the metal thickness. Such small items, if overlooked, will lead to rejection of your finished part.

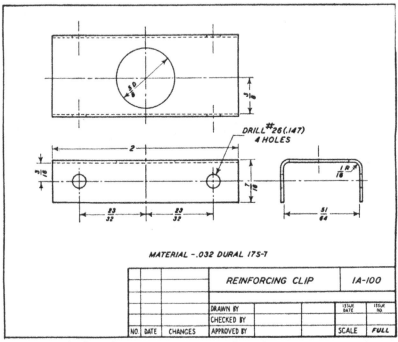

Fig. 11. Third Angle Projection Drawing of Airplane Part to Be Made in Project 3

## PROJECT 4. MAKING A REINFORCED LOOP BRACKET CLAMP

**Objectives**
1. To learn layout practice
2. To learn technique of bend allowance layout
3. To learn blueprint reading and pattern projection

**Tools and Equipment**
1. Layout tools
2. Vise
3. Bending brake
4. Drill #8
5. Aircraft snips (left and right)
6. Assortment of files

**Material**
1. Dural 17S–O, HT to 17S–T when finished. (Gage noted.)
2. 1" round bar, to be used in forming.

# PATTERNS FOR BENDS

**Procedure**

1. Lay out clamp from drawing in Fig. 12.
2. Lay out hole locations and drill before forming.

*Note:* No hole is drilled in the end of the attached bracket; this will be done when used on assembly.

3. Brake up channel part.
4. Place the 1" round bar or shaft in the vise, and form the loop by hand around it.

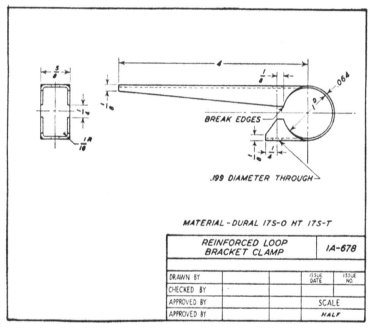

Fig. 12. Drawing of Airplane Part to Be Made in Project 4

5. Have clamp inspected and heat-treated.

### PROJECT 5

Make clamp for square tubing, as in Fig. 13, using the method employed in Project 4, except that where, in Project 4, the 1" bar was used for forming the loop, in this instance a wood block formed to the proper size, or a piece of square tubing, is utilized instead.

### PROJECT 6

Make the loop type clamp, preliminary explanation of which appeared in Project 2.

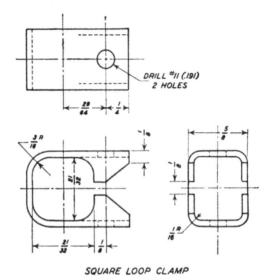

SQUARE LOOP CLAMP
MATERIAL .064 DURAL 17S-O HT 17S-T

Fig. 13. Drawing of Square Loop Clamp to Be Made in Project 5

## PROJECT 7. FABRICATED TUBES, ONE OF WHICH SLIPS INSIDE THE OTHER

**Objectives**

1. To learn the application of the stretch-out in laying out cylinders or tubes, Fig. 14, by method outlined in text
2. To practice good workmanship

**Tools and Equipment**

1. Layout tools
2. Square shears
3. Rolls
4. Hammer (riveting or ball peen)
5. Stake (beakhorn)
6. #30 Drill
7. Drill motor (electric or air driven)
8. 72° countersink
9. Hand shears

**Material**

1. Strip of .064 17S-O dural 2"x 14"
2. Strip of .064 17S-O dural 1 5/8"x 14"
3. 8 A17S–T 78° CTSK 1/8x5/16 rivets.

**Procedure**

1. Find the stretch-out of the inside cylinder, which is 2" wide as shown in Fig. 14, by multiplying the diameter (3") by 3.1416 and adding 2.8 x metal thickness (.064), which will give you 9.4248+.1792, or 9.604, or approximately $9^{19}/_{32}"$, the length of the neutral axis.

# PATTERNS FOR BENDS

The outside cylinder, which is 1⅝" wide, is laid out by multiplying 3⅛ (the diameter, because it is larger by two metal thicknesses of .064 than the inside diameter) times 3.1416, and adding 2.8 x T (metal thickness), making it 9.996 or, say, 10" long.

2. Do not cut off the ends until you have run the strips through the rolls because the ends are to be used for the splice material.

3. After the strips are rolled to approximately 3" Dia., cut off the end and lay out a splice plate with a length four times edge distance. Edge distance is at least two times rivet hole size; since we are to use 1/8" rivets, the splice measuring four times 17/64 in Fig. 14 is above the minimum. Center punch the plate over the beakhorn stake to prevent its losing its curved shape and cut to length on the mark laid out.

4. Use the plates to lay out holes in cylinder after the plates have been laid out from dimensions as shown in Fig. 14, and drilled; then drill holes through cylinder, using C clamp to hold pieces together.

5. As countersunk rivets are used, the outside cylinder must have rivet holes countersunk on the inside, and the inside cylinder must be countersunk on the outside, to allow cylinders to be slipped together after riveting.

6. Rivet the splice plates on the inside of inside cylinder, and on the outside of outside cylinder, as in Fig. 14. *Note:* If we made the cylinders to exact size, they would not fit together. Notice that we gave and took a few thousandths of an inch in changing decimals to fractions through the decimal

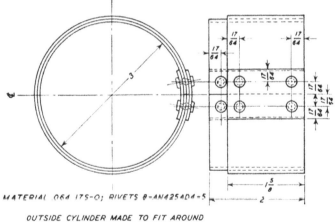

MATERIAL .064 17S-O; RIVETS 8-AN425AD4-5

OUTSIDE CYLINDER MADE TO FIT AROUND
OUTSIDE OF INSIDE CYLINDER WITH
INSIDE DIAMETER OF 3 INCHES

NOTE: MAKE SEPARATELY AND SLIP TOGETHER

Fig. 14. Drawing of Two Cylinders, One to Slip Inside of Other

equivalents table, keeping in mind the necessary adjustment because of this fact. There must always be clearance for fitting objects together, unless the outside sleeve is to be heated, and slipped over in heat-expanded state, shrinking to a tight fit as it cools off.

## PROJECT 8. TEST

**Fabricating a tube that can be slipped into a casting, such as shown in Fig. 15**

### Objective

A test of the student's ability to lay out a tube which is to fit the inside diameter of the casting in Fig. 15.

### Tools and Equipment

The same as those used in Project 7

### Material

1. .040 x 1"x 10" strip of 17S-O Dural
2. 4 rivets—3/32"x 3/16" CTSK A17S-T

### Note 1

The material to be used is 40/1000" (.040") thick. The problem must be worked from the circumference obtained by using the 2¼" diameter dimension.

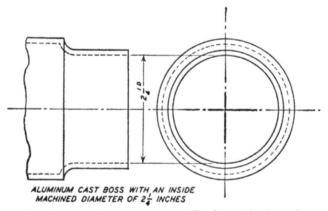

Fig. 15. Cast Aluminum Boss or Collar from Which a Fabricated Tube Is to Be Made to a Sliding Fit

### Note 2

Finding the stretch-out of a cylinder which is to have a specified outside diameter has not occurred in previous projects but is explained in text.

### Procedure

1. Figure dimensions including splicing plate; give data on paper.
2. Explain how stretch-out is figured so as to allow approximately 6/1000" (.006") in length for close fit clearance.
3. When you have answers, check them against answers in following alphabetical-figure code in which letters stand for numbers, A=1, B=2, etc.

*Answer:* The clearance is derived from choosing F and EI/FD for F.IBHE.

4. Form the tube.
5. Rivet splice on inside.

Above. These automatic riveting machines are used in fabrication of certain small subassemblies to save time.
*Courtesy of Douglas Aircraft Co., Inc., Santa Monica, Cal.*

Left. This workman is counterpunching rivet holes preparatory to flush riveting.
*Courtesy of Lockheed Aircraft Corp., Burbank, Cal.*

Above. The workman here is holding two sections of corrugation which he has spliced together using the power riveter.
*Courtesy of Lockheed Aircraft Corp., Burbank, Cal.*

Upper Center. Workmen riveting a portion of the skin of an Airacobra wing.
*Courtesy of Bell Aircraft Corp., Buffalo, N. Y.*

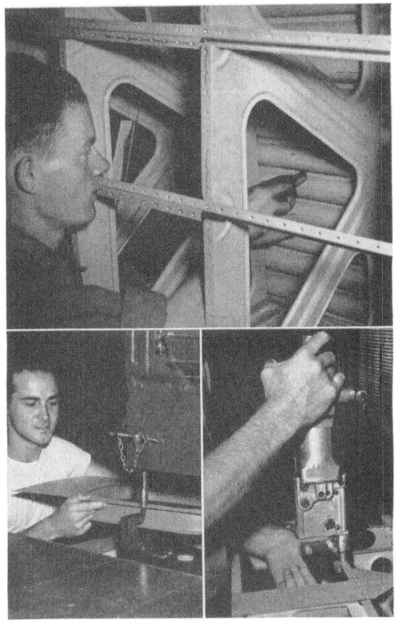

Top. Bucker is working on a center section, bucking rivets between the corrugation with a special attachment for the bucking bar. The corrugation takes the compression loads in wing during flight.

Lower Left. Worker operating a punch press which can be used for counterpunching and riveting sheet metal.

Lower Right. This workman is installing countersunk rivets with a squeeze riveter.
*Courtesy of Lockheed Aircraft Corp., Burbank, Cal.*

*Chapter VII*

# Rivets and Riveting

Riveting is the most common method of fabricating aluminum, particularly the structural alloys which depend upon heat treatment for their high mechanical properties. In smaller parts aluminum alloy rivets are always used, except for the nonstructural parts such as cowling partitions, diaphragms, etc., which are spot welded. The aluminum 3S shells used as oil tanks, gas tanks, or any tank required to hold liquids, are first riveted to the baffles with the riveted end extending on the outside of the tank, which is later fused to the shell. For the thicker sections used in larger structures, either aluminum alloy or steel rivets may be used, the engineers' choice depending upon various requirements and factors such as strength, size, etc. Steel rivets are stronger than aluminum alloy rivets; however, their use is limited to those applications which can be protected adequately against corrosion. In aircraft, the aluminum alloy rivets are preferred because of their weight-saving and corrosion-resisting qualities.

Aluminum alloy rivets are available in any of the wrought aluminum alloys—2S (commercially pure aluminum), 3S, A17S–T, 17S, 24S, and 53S. With the exception of 17S and 24S, all are driven cold in the condition in which they are received from the manufacturer. The four named last are always used in the heat-treated temper, since only in that temper do they have the desired mechanical properties. The heat treatment of 53S is covered by patents owned by the Aluminum Company of America. Aluminum alloy rivets are furnished with various types of heads, shown in Fig. 1. Button, round, mushroom, brazier, and flat-head rivets are furnished with a small fillet at the junction of shank and head for added strength, which prevents fatigue cracks. The radius of this fillet is equal to about one-tenth of the shank diameter, with a minimum of 0.01". The No. 5 or brazier head rivet is not used in Army and Navy planes of new design. This is rivet AN455 and is being replaced by AN456. Sizes for this rivet are given in Fig. 1. Rivet

# AIRCRAFT SHEET METAL WORK

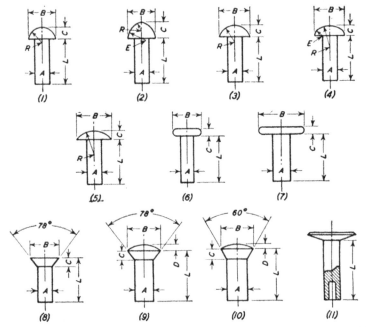

| No. | Kind | Width B | Head Depth C | Head Radius R | Edge Radius E | Oval Depth D |
|-----|------|---------|--------------|---------------|---------------|--------------|
| 1 | Button Head | 1.75A | 0.75A | 0.885A | ...... | ....... |
| 2 | High Button Head°—Amer. Std. | 1.50A+ 0.031 | 0.75A+ 0.125 | 0.75A— 0.281 | 0.75A+ 0.281 | ....... |
| 3 | Round Head | 2A | 0.75A | 1.042A | ....... | ....... |
| 4 | Mushroom Head | 2A | 0.625A | 1.634A | 0.50A | ....... |
| °°5 | Brazier Head | 2.50A | 0.50A | 1.8125A | ....... | ....... |
| 6 | Flat Head | 2A | 0.40A | ....... | ....... | ....... |
| 7 | Tinners' Rivet | 2.25A | 0.30A | ....... | ....... | ....... |
| 8 | Csk Head | 1.81A | 0.50A | ....... | ....... | ....... |
| †9 | Csk Oval Head | 1.81A | 0.50A | 1.7656A | ....... | 0.25A |
| ‡10 | Csk Oval Head | 1.577A | 0.50A | ....... | ....... | 0.187A |
| 11 | Tubular Shank | This rivet made with several sizes of heads. | | | | |

° The high button head supplied in sizes 1/2″ and larger.
°° This AN455 rivet has been replaced by AN456.
† For sizes up to and including 7/16″ diameter.
‡ For sizes 1/2″ and larger.

### Sizes of Rivet AN456, which Replaces Rivet AN455

| Dia. | 3/32 | 1/8 | 5/32 | 3/16 | 1/4 |
|------|------|-----|------|------|-----|
| B | .156 | .235 | .312 | .390 | .468 |
| C | .031 | .047 | .063 | .078 | .094 |

Fig. 1. Common Types of Heads Supplied on Aluminum Alloy Rivets
*Courtesy of Aluminum Company of America, Pittsburgh, Pa.*

# RIVETS AND RIVETING

heads on the new rivets are slightly smaller; for example, the head on the new $\frac{1}{8}''$ size is about the same as the head on the old $\frac{3}{32}''$ size.

Table 1—Bearing Strengths of Aluminum Alloy Plates and Shapes Representative of Material Having Typical Properties

| Alloy | Bearing Yield Strength, Lbs. per Square Inch | Bearing Ultimate° Strength, Lbs. per Square Inch | Alloy | Bearing Yield Strength, Lbs. per Square Inch | Bearing Ultimate° Strength, Lbs. per Square Inch |
|---|---|---|---|---|---|
| 2S–O | 13,000 | 23,000 | 24S–T | 68,000 | 123,000 |
| 2S–¼ H | 15,000 | 27,000 | 27S–T | 65,000 | 117,000 |
| 2S–½ H | 17,000 | 30,000 | | | |
| 2S–¾ H | 20,000 | 36,000 | 52S–O | 29,000 | 52,000 |
| 2S–H | 24,000 | 43,000 | 52S–¼ H | 34,000 | 61,000 |
| 3S–O | 16,000 | 28,000 | 52S–½ H | 37,000 | 67,000 |
| 3S–¼ H | 18,000 | 32,000 | 52S–¾ H | 39,000 | 70,000 |
| 3S–½ H | 21,000 | 38,000 | 52S–H | 41,000 | 74,000 |
| 3S–¾ H | 25,000 | 45,000 | 53S–W | 33,000 | 59,000 |
| 3S–H | 29,000 | 52,000 | 53S–T | 39,000 | 70,000 |
| A17S–T | 43,000 | 77,000 | 61S–W | 35,000 | 63,000 |
| 17S–T | 60,000 | 108,000 | 61S–T | 45,000 | 81,000 |

° These values should be used only when the edge distance measured from the center of rivet hole in the direction of stressing is not less than twice the diameter of the rivet hole. For smaller edge distances values should be reduced proportionately.

*Courtesy of Aluminum Company of America, Pittsburgh, Pa.*

Table 1 gives the bearing strength of typical aluminum alloy plates and shapes; Table 2 gives the ultimate shear and bearing strengths of driven rivets made from certain of these alloys.

The 2S and 3S rivets are soft rivets and are driven cold as received. They may be stored indefinitely without their strength or driving characteristics being affected.

A17S–T rivets are always furnished in the heat-treated condition and are driven cold as received. They are used principally as a substitute for small 17S rivets because they eliminate some of the extra operations required for 17S rivets, such as frequent heat treatment and cold storage. While the strength of driven A17S–T rivets is not equal to that of driven 17S rivets, as seen in Table 2, it is enough so that the A17S–T rivets have been found satisfactory for many uses, particularly in the fabrication of wing and fuselage skin. Rivets of A17S–T may be stored indefinitely with no change in properties.

Rivets made of alloy 17S must be heat treated to obtain maximum strength. The heat treatment consists of subjecting the rivets to a temperature of 930°F. to 950°F. for a period of from 5 to 30 minutes, depending on the size and the quantity being heated, then quenching immediately in cold water. Following this, the

Table 2—Average Ultimate Shear Strength and Bearing Strength for Driven Rivets

These values are for rivets driven with cone-poin heads. Rivets driven with heads requiring more pressure may be expected to develop slightly higher strengths.

| Rivet | Driving Procedure | Shear Strength, Lbs. per Square Inch | Bearing Strength†, Lbs. per Square Inch |
|---|---|---|---|
| 2S | Cold, as received | 11,000 | 33,000 |
| 3S | Cold, as received | 14,000 | 42,000 |
| A17S–T | Cold, as received | 30,000 | 90,000 |
| 17S–T | Cold, immediately after quenching | 34,000° | 102,000 |
| 53S–T61 | Cold, as received | 23,000 | 69,000 |
| 53S–T | Cold, as received | 26,000 | 78,000 |
| 17S–T | Hot, 930° to 950°F. | 33,000° | 99,000 |
| 53S–T61 | Hot, 960° to 980°F. | 18,000° | 54,000 |
| Steel | Hot, 1700° to 1900°F | 45,000 | 135,000 |

°Immediately after driving, the shear strengths of these rivets are about 75% of the values shown. On standing at ordinary temperatures they age-harden to develop their full strengths, this action being completed in about four days.

†These bearing strengths are to be used only if they are less than the corresponding bearing ultimate strength for the plates or shapes in which the rivet is used, as shown in Table 1.

*Courtesy of Aluminum Company of America, Pittsburgh, Pa.*

rivets harden gradually at room temperature, obtaining their full strength in about four days. During the first one or two hours after quenching, the rivets are relatively soft and may be driven cold. If 17S rivets are allowed to age more than two hours at room temperature after quenching, they are too hard to drive and must again be heat treated. Heat treatment may be repeated as often as desired without injury to the rivets.

Aging can be retarded considerably by storing the quenched rivets at low temperatures. Rivets remain soft enough for driving for about 36 hours if stored at 32°F. immediately after quenching. By using solid $CO_2$ (dry ice) or mechanical refrigeration, much lower storage temperatures can be maintained, and the driving period prolonged almost indefinitely. At a temperature of 50° below zero Fahrenheit, for example, rivets remain soft enough for driving for a period of two weeks or more.

When doing repair work away from a factory or repair base, with no heat-treating equipment available, 17S rivets must be heat treated and shipped to the job packed with dry ice in an insulated container.

In aircraft work, riveting is limited to the smaller sizes of rivets and is always done cold. Extensively used in aircraft construction are 17S-T rivets, driven immediately after quenching. Driven with the small flat head ordinarily used in aircraft construction, these rivets develop an ultimate shear strength, about four days after being driven, of 34,000 lbs. per sq. in., based on the area of the hole. For design purposes, however, the shear strength is assumed to be 30,000 lbs. per sq. in.

Where a shear strength higher than that of 17S-T is needed, 24S-T rivets are sometimes used. Driven as soon as quenched, after being subjected to a temperature of 910°F. to 930°F., they develop within about one day an ultimate shear strength of approximately 44,000 lbs. per sq. in., based on the area of the hole. For design purposes, however, the shear strength is assumed to be 35,000 lbs. per sq. in. Since the rate of age-hardening of 24S-T rivets is more rapid than that of 17S-T rivets, it is necessary to drive them more quickly after quenching; the elapsed time should be in the neighborhood of 5 to 10 minutes for best results. Rivets of 24S-T are more difficult to drive than those of 17S-T. They are not regularly carried in stock and sometimes cannot be obtained as promptly as rivets in the other alloys.

Rivets of A17S-T or 53S-T, driven in the condition in which they are received, are being used more and more in aircraft construction to avoid shop heat treating, refrigeration, and similar operations involved in the use of 17S-T and 24S-T rivets. These rivets develop ultimate shear strength somewhat less than that of 17S-T, as shown in Table 2, but are found to be quite satisfactory for many purposes. They drive as readily as 17S-T rivets and much more easily than 24S-T rivets.

When unpainted Alclad 17S-T or Alclad 24S-T sheet is used, 53S-T rivets are the most satisfactory from the standpoint of resistance to corrosion, first because they have the best inherent resistance of any of the rivets mentioned and, in addition, they have substantially the same electrolytic potential as the Alclad coating. Under severely corrosive conditions, the heads of A17S–T, 17S–T, and 24S–T rivets tend to be protected electrolytically by the Alclad coating.

Anodic oxide finishes materially improve the resistance of aluminum alloys to corrosion, and rivets for use in aircraft construction are usually supplied with °Alumilite 205 finish. Where the yellow color of this finish is objectionable, Alumilite 204 is sometimes specified.

While the driving tends to break the oxide film, joints made with 17S–T rivets which were given Alumilite 205 finish before driving have been subjected to the standard salt spray test required for anodic oxide coatings by both the U.S. Army and Navy air services without showing signs of corrosion.

Reheat treating rivets as many as fifteen times has been found to have no appreciable effect on the protection afforded by the Alumilite 205 treatment, provided the rivets do not come in contact with the molten nitrate of the hot solution.

Identification markings have been adopted for use on the heads of aircraft rivets of the following alloys: 17S–T, A17S–T and 24S–T. These markings are shown in Fig. 2. No identification marks have yet been adopted for rivets made of other alloys.

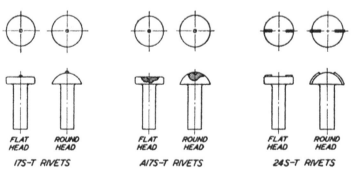

Fig. 2. Typical Identification Marks for Aluminum Alloy Rivets
*Courtesy of Aluminum Company of America, Pittsburgh, Pa.*

Certain sizes of aluminum alloy rivets can now be supplied with the shank ends chamfered as shown in Fig. 3. This cham-

---
° Patented process, Aluminum Company of America.

# RIVETS AND RIVETING

fer simplifies the insertion of the rivet into the hole, particularly with automatic riveting machines.

Rivets of alloys 17S–T, A17S–T and 24S–T are supplied with an anodic coating if such a coating is specified. This coating im-

A = Shank Diameter  B = Diameter of End
C = Length of Radius

Formulas
C = 0.25 × A
Radius = 0.3125 × A

| Diameter of Rivet, In. | B In. | Diameter of Rivet, In. | B In. |
|---|---|---|---|
| 1/16 | 0.047 | 5/32 | 0.118 |
| 3/32 | 0.071 | 3/16 | 0.137 |
| 1/8 | 0.092 | 1/4 | 0.185 |

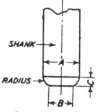

Fig. 3. Rivets with Shank Ends Chamfered for Easy Insertion
*Courtesy of Aluminum Company of America, Pittsburgh, Pa.*

proves the resistance of the rivets to corrosion and also provides a better surface for painting.

Rivets to be used with Alclad alloys are sometimes ordered with heads burnished to match more nearly the appearance of the Alclad coating. Burnished heads are only slightly discolored by subsequent heat treatment of the rivets.

**LENGTHS OF RIVETS.** The length of rivets required for forming a flat or countersunk head depends upon the grip, or total thickness of metal through which the rivet is to be driven,

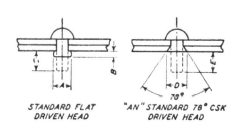

| Dia. of Rivet | A | B | C | D | E |
|---|---|---|---|---|---|
| 1/16 | 3/32 | 1/32 | 3/32 | 1/8 | 5/64 |
| 3/32 | 1/8 | 3/64 | 9/64 | 11/64 | 3/32 |
| 1/8 | 3/16 | 1/16 | 9/64 | 15/64 | 3/32 |
| 5/32 | 7/32 | 5/64 | 3/16 | 9/32 | 1/8 |
| 3/16 | 1/4 | 3/32 | 7/32 | 11/32 | 1/8 |
| 1/4 | 11/32 | 1/8 | 9/32 | 29/64 | 3/16 |
| 5/16 | 27/64 | 5/32 | 5/16 | 9/16 | 3/16 |

Dimensions, Inch

STANDARD FLAT DRIVEN HEAD

"AN" STANDARD 78° CSK DRIVEN HEAD

Fig. 4. Amount of Rivet which Should Protrude before Forming of Head Begins

clearance between rivet and rivet hole, the alloy, and the form of head. Under the letter C in the tabulation in Fig. 4, you will find, for each size of rivet, the approximate length which should protrude through the metal for driving a flat head. In column $E$, the length of the protruding portion needed for driving countersunk heads is shown. In either case, allowance must be made for sheet thickness.

**RIVET HOLES.** Rivet holes in aluminum alloys may be punched, drilled, or subpunched and reamed to size. The last-named method is preferable, especially if the reaming is done in assembly to give the holes exact coincidence. The clearance which is to be allowed in the holes depends largely on the class of work. It is easier to drive rivets cold when the clearance is small. If a loose fit is used, it will be hard to hold the rivets straight, and eccentric heads may result. The best clearance is the smallest one which will allow the rivet to be inserted easily without delay. Drill size No. 50 is usually used for $1/16$ rivets, No. 40 for $3/32$, No. 30 for $1/8$, No. 20 for $5/32$, and No. 10 for $3/16$.

**THE SHEET METAL CRAFTSMAN AND RIVETING.** In the building of present-day airplanes, mass production methods have progressed to a point where a mechanic in a factory may very seldom have to do his own riveting; or again he may do nothing but riveting for long periods of time. At any rate, it is essential that he know how to take care of all the operations leading up to the final act of heading the rivet.

The simplest assembly job that is encountered, probably, is the installation of fittings, such as pulleys, brackets, the various fittings needed for securing cabin furnishings, and the like. Even though they appear to be of minor importance, great care should be given to their installation, for such small items are very likely to become almost inaccessible in the completed airplane, so that additional attention, either in revision or inspection, will add greatly to operation expense; in addition, there is the possibility of creating a hazard.

The design requirements of a riveted joint are taken care of by the engineer; they include the shape and size of the joint, shape and size of rivets, rivet alloy, and spacing; all of which is conveyed to the craftsman through the blueprints. Just as important is the riveting itself, which must produce all the strength characteristics on which the engineer depends, trusting that the craftsman

# RIVETS AND RIVETING

will do a job of riveting which will give a strength conforming with the design requirements.

**RIVETING TECHNIQUE.** For the assembly of fittings, brackets, etc., a supply of bolts, sheet metal screws, or hole-aligning clamps are needed. Prior to riveting, all fittings should be bolted in place, and their positions and alignment (if any) carefully checked. Use of sheet metal screws of the proper size to prevent scarring of the hole usually is the quickest and easiest way of assembling fittings where only two metal thicknesses are being worked. Bolts are used with more than two metal thicknesses or with very thin material. Bolts are preferred to sheet metal screws in structural members where punching of holes is forbidden lest cracks appear later in service.

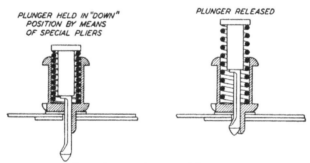

Fig. 5. Cleco Sheet Holder Used for Fastening Skin Preparatory to Riveting

*Courtesy of The Cleveland Pneumatic Tool Company, Cleveland, Ohio*

Bolt and screw sizes for corresponding size rivet holes are:

| Rivet Holes | Bolts | Sheet Metal Screws |
|---|---|---|
| 3/32" | 3–48 | No. 4 |
| 1/8" | 4–40 | No. 6 |
| 5/32" | 6–32 | Not to be used |
| 3/16" | 10–32 | Not to be used |

A better way to hold sheets together, where applicable, is with the Cleveland skin clamp shown in Fig. 5. These sheet holders are referred to by mechanics as Cleco fasteners, and the finished assembly of parts is said to be *Clecoed together.*

**Drilling.** One of the most exacting procedures in assembly is drilling, especially in the application of skin or of relatively thin sheets where buckling or distortion would be likely. Where a flat piece of skin is fitted to a curved surface, such as a fuselage, without being formed, the piece usually is removed after the holes have been drilled, in order to clear the dural shavings or chips, caused by drilling, from between the sheet and the frame. Sealing compound is then applied to the seams before the riveting begins.

The holes which are drilled in the formers, i.e., the longitudinal stiffeners (longerons or stringers), and in the circumferential stiffeners (frames) in a fuselage are drilled undersize. These holes are reamed to the proper size with the drill that is used to drill the skin. From this you would assume that the skin is drilled from the inside, which is true. In order to properly fit a sheet of skin to a fuselage frame, drilling and bolting, or *Clecoing* must begin in the center of the sheet and move out. Riveting also must proceed from the center, working out in both directions. If riveting starts from both ends of the material and moves to the center, an *oil can* or buckle is apt to result.

Drilling through thick sections or multiple sheets in aluminum alloy should be accomplished with a drill in which the flutes have a large spiral angle; that is, more twists per inch than the ordinary twist drill. Ordinary drills are suitable when the holes are not deep, as in skin, stringers, etc. Polished flutes improve the cutting action of the drill and also assist materially in chip removal. When the hole is deep in proportion to its diameter, it may be necessary to withdraw the drill from the hole occasionally to dispose of the cuttings. A high speed and light feed are also helpful, especially for drills of small size.

One point that cannot be overemphasized is the importance of care in drilling. To drill properly you must keep your mind on your work, remembering that once you drill a hole it is there to stay, that it cannot be removed, and if wrong may spoil the material plus many hours of labor. The following rules are important to remember:

1. Make sure that you have the hole locations marked in their proper places with the proper edge distance on all material being drilled.

2. Know what is behind the material you are drilling.

3. Never hold your finger on the other side of a piece of metal while drilling, expecting to drill through the metal only.

# RIVETS AND RIVETING

The drill will go through your finger before you can release the pressure. Never drill against any part of your body.

4. Use drills of proper size. On holes that are to be dimpled, obtain specific instructions from your foreman or leadman as to the drill size.

5. When drilling with an extension drill of any appreciable length of shank, use your free hand as a guide in which the drill turns. Never let go of the drill until it stops turning, as the centrifugal force will bend the drill at ninety degrees to its former position, making it dangerous.

6. Use a cutting lubricant when necessary. Keep your drill sharp, as explained in Chapter I.

7. Hold drill at proper angle to work.

8. Do not press on the work harder than necessary and hold the hand not operating the trigger or control under the body of the drill, to counteract the pressure and receive the weight when the drill goes through, thus preventing damage to the material by the chuck. Use drill stop (a small tube around drill) when advisable.

9. When drilling on a bench, drill against a block of wood to prevent making holes in the bench.

10. Use a chuck wrench to tighten drill in chuck and tighten sufficiently to prevent spinning, which would damage the shank of the drill.

A slightly smaller hole is drilled for the $1/16''$, $3/32''$, and $5/32''$ rivet size when a countersink is to be used. The following table is for countersunk holes.

| | |
|---|---|
| 1/16" Rivet | No. 52 Drill |
| 3/32" Rivet | No. 41 Drill |
| 1/8" Rivet | No. 30 Drill |
| 5/32" Rivet | No. 21 Drill |
| 3/16" Rivet | No. 10 Drill |
| 1/4" Rivet | 1/4" Drill |

For holes that are to be dimpled (counterpunched) through both sheets as at (A) in Fig. 6, the drill used is the same size as noted in the table for countersunk holes. The smaller hole is to allow the maximum amount of metal for drawing in the dimpling operation without danger of fracturing or splitting the edge of the hole as it would if too small a hole were drilled. However, with the 100° or 120° rivet, there is little danger of cracked dimpled holes because the greater angle requires a shallow dimple, whereas the 78° head, (B) in Fig. 6, is the deepest dimple, used only for repair work and replacement riveting.

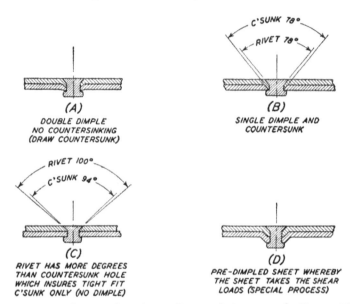

Fig. 6. Typical Examples of Installation of Countersunk Rivets. Note Differences in Angle between Rivet Head and Countersunk Hole at C, which Insures Tight Fit. At B, Rivet Head and Countersunk Hole in Sheet Have Same Angle, Depending on Extrusion of Dimple to Create a Tight Fit

**Countersinks.** High-performance airplanes are usually flush-riveted on the air-stream surfaces with countersunk rivets. The type of countersinking ordinarily depends on the thickness of the part on the outside surface. The thicker parts are frequently machine-countersunk; thin stock is usually dimpled or embossed. Another method is to machine-countersink the member being attached to the skin and then emboss the skin sheet to form a recess for the rivet head in the outer surface of the sheet, (B) in Fig. 6.

There are several types of countersinks (center reamers) with included angle of cutting point of 60°, 72°, 78°, 82°, 90°, 94°, 100°, or 115°; or any other angle may be made to order. Special countersinks are made by airplane tool manufacturers, as shown

# RIVETS AND RIVETING

in Fig. 7; these have removable cutters so that holes may be countersunk at any angle. Note the depth stop that is adjustable for countersinking to just the proper distance. The angle of the standard aircraft countersink rivet has been and is still used in the 78° size, but rivets of 90°, 100°, and 120° are used extensively for fastening skin where flush type rivets are desired. The 90° countersunk rivet is replacing the 78° rivet for new design.

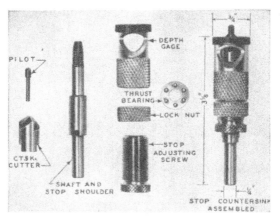

Fig. 7. Stop Countersink, Easily Adjustable to Any Desired Depth
*Courtesy of Aircraft Tools, Inc., Los Angeles, California*

The following table gives degrees of countersinks and the corresponding angles for rivet heads with which they are to be used.

| | |
|---|---|
| 72 Countersink | 78 Rivet |
| 78 Countersink | 78 Rivet |
| 82 Countersink | 90 Rivet |
| 90 Countersink | 90 Rivet |
| 94 Countersink | 100 Rivet |
| 100 Countersink | 100 Rivet |
| 115 Countersink | 120 Rivet |

The 82° countersink is used for the AN505 (Army and Navy flat head) machine screw, and the 100° is used for countersinking to fit the 100° Reed and Prince machine screws.

A countersink with less degrees than the rivet, as at (C) in Fig. 6, should always be used except where the skin is dimpled into a countersunk member as shown at (B).

The double dimple, (A) in Fig. 6, is used where the skin-support ribs or cellular structure are thin gage, which will allow the process to be carried out. A rivet is merely inserted in the hole and drawn with a draw set by the bucker before bucking the head, Fig. 8. This is the common method of riveting flush rivets in fuselage and wing skin. Where the skin is riveted to heavy thick sections, the section is countersunk first and the skin drawn into the countersunk hole as shown at (B) in Fig. 6. In this case the angle of the countersink is the same as that of the rivet. Pre-dimpled

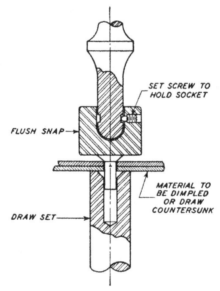

Fig. 8. Flush Rivet about to Be Drawn, Forcing a Dimple into Two Sheets of Metal

sheets, as at (D), Fig. 6, are press counterpunched and drilled afterward, or a special die is used which punches the hole and forms the dimple in one operation.

The stop on the countersink must be checked by trying it out on a piece of scrap metal before it is put into use on the airplane.

**Heading.** When a rivet is driven in any other way than as at 2, 3, 15, or 16, Fig. 9, it must be removed by drilling out the head, which is then removed with a chisel; the shank is driven out with a punch. Never leave a wrongly driven rivet intending to

# RIVETS AND RIVETING

come back later and change it, because you are likely to forget. The time for replacement is immediately after the error is made.

The following tabulation explains what happened, or what is wrong, with the incorrectly driven rivets shown in Fig. 9 by corresponding numbers.

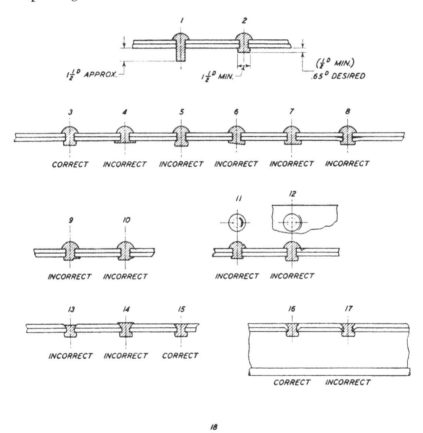

Fig. 9. Correct and Incorrect Practice, Showing Typical Rivet Imperfections to Be Avoided. Nos. 2, 3, 15, and 16 Illustrate Properly Driven Rivets

1. Shows the length of that part of a rivet which should protrude through the metal thickness before the upsetting of the head, approximately one and one-half times the rivet diameter.

2. Shows the diameter and thickness of a properly driven head. (Dimensions are comparative with shank diameter in all cases). Note the statement that the minimum thickness of the driven head should be one-half the shank diameter, and that a head $^{65}/_{100}$ths times the shank diameter is preferable. The minimum diameter of the driven head is one and one-half times the shank diameter. If the head is too small, use a longer rivet.

3. Correct. Uniform head completely filling hole.

4. Not correct. Poor timing; driven too flat.

5. Not correct. Rivet too hard, or bucking bar too light or not held solidly against the rivet. This rivet may have been out of the ice box too long and may need to be reheat-treated.

6. Not correct. Bucking bar held on rivet at an angle.

7. Not correct. Rivet was too long, improperly bucked, or caused by a sloppy hole.

8. Not correct. Rivet swelled between sheets, chips between sheets, or not properly bolted up.

9 and 10. Not correct. Bucking bar slipped off during riveting procedure.

11. Not correct. Rivet snap jumped on head of rivet, causing a disfiguration. Contributing factors for this fault are: either the buck was held too loosely on a flexible structure, or the gun was not held steady enough against the rivet head.

12. Not correct. Rivet snap indented the work because it was held at an angle or the cupping was too deep.

13. Not correct. Countersunk too deep.

14. Not correct. Not countersunk deep enough.

15. Correct. Countersunk to a depth which allows the head to drive flush.

16. Correct. The understructure was countersunk just right to allow a flush head, and the hole drilled in the dimpled sheet was small enough to allow stock for a close bearing around the rivet.

17. Not correct, the hole in the skin was drilled too large and stretched even larger through the dimpling operation, thus preventing a close bearing fit.

18. Not correct. Rivet apparently good, but sheet was disfigured by excess riveting because rivet was too hard.

# RIVETS AND RIVETING

**TECHNIQUE OF USING RIVET BUCKS.** Rivet bucking bars are made to fit the particular job at hand. Every mechanic will collect some, but a large assortment is kept in the tool crib.

One of the most common bucking bars is that used for bucking on bulb angles. Such a bar is milled out on one side for clearance for the bulb of the angle, as shown in Fig. 10. The milled slot may be cut on one end or along the side of a rectangular bar

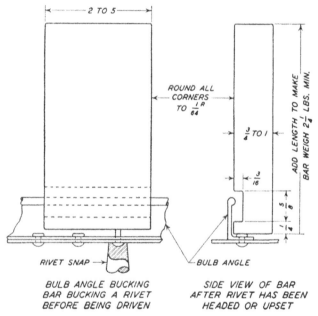

This bar may be made to your own specifications. Multiply volume in cubic inches by .2833 (weight of steel per cu. in.). The result must be over 2 lbs., especially when using slow-hitting riveters. Subtract or add to the length of the bar until the desired weight is obtained. If the bar is made 5" wide, the length will be reduced to less than the width.

Fig. 10. Bulb Angle Bucking Bar in Process of Bucking a Rivet

which afterwards should be case-hardened. Occasionally a rivet cannot be bucked by holding the buck squarely on the end of the rivet shank and letting the gun do the work in the customary way. In such places as under bolt heads, brackets, or other obstacles which prevent using an ordinary buck, rivets are bucked with the gun and snap, while the bucker forms the head by using a spar flattener, Fig. 11. The spar is hit with a hammer at W while one end of spar is held in the hand; the other end forces head on rivet.

When a rivet hole has become somewhat enlarged by several rivet removals, as happens in repair work, the head must be formed by a flat punch or spar flattener by the bucker, while the riveter bucks the rivet on the head side with the gun. Sometimes a hole becomes so large that the next size rivet must be used in order to save the work. If changing to a larger rivet is permissible except

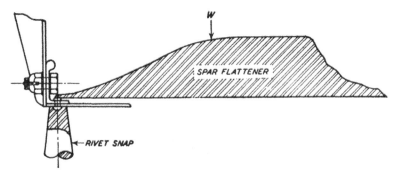

Fig. 11. Method of Upsetting a Rivet with Spar Flattener. (Used where Rivet Cannot Be Bucked with Regular Bucking Bar)

that, being in a conspicuous place, it would mar the looks of the work, a longer rivet should be used with a shank expander, a tool that looks like a draw but has a tapered hole which the rivet bot-

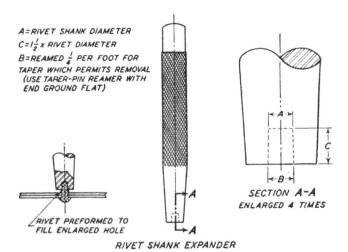

Fig. 12. Rivet Shank Expander which Enlarges Shank of Rivet for Use in Enlarged or Oversized Hole

# RIVETS AND RIVETING

toms in. When this expander is driven on the shank of a bucked-up rivet, as shown in Fig. 12, it increases the shank diameter, yet is easily removed because of the taper. This procedure is permissible only for relatively large holes; its use should not be overdone.

For extremely large holes, another method is to chuck a rivet of the next larger size in the drill press and turn the head down to

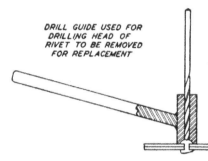

Fig. 13. Typical Example of Use of Drill Guide, Showing Drill Extending into Rivet Head

the size being used. This, however, should never be done without permission from your foreman, since it is not approved for all places.

The drill guide shown in Fig. 13 is for spotting the heads of a group of rivets prior to drilling the heads when they are being removed for replacement. A17S-T rivets are drilled without a guide or centerpunch mark because the identification mark serves as an indentation in which to start the drill.

A guide used for transferring holes from bottom sheet to top sheet, Fig. 14, is called a hole duplicator. It is used for locating rivet holes when replacing skin or in laying out de-icer cap strips when rivnuts are already installed in the wing. The peg in the

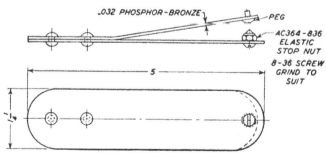

Fig. 14. Hole Duplicator Used for Locating Rivet and Rivnut Holes

upper section is placed in the rivet or rivnut hole, and the lower section is pulled back and released. The springing action causes the point to center-punch the exact center, so that when the hole is drilled it will be in perfect alignment. A drill bushing may be placed where the center punch is, and the tool is then used for a drill guide. A separate duplicator is required for each rivet size.

**Rivet Bucking in a Tube.** In riveting fittings or sheet to round or square tubing, we must rivet by a method called blind bucking. The bucking tool used, Fig. 15, is called a *mouse*. The mouse is

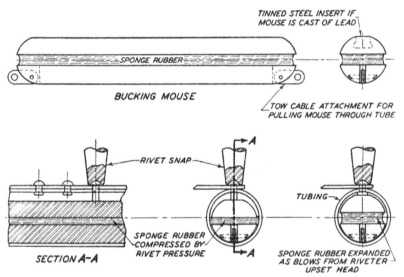

Fig. 15. Typical Method of Riveting Blind in a Tube Where a Bar Cannot Be Held by Hand

made to conform to the inside of the tube that is to be riveted, either straight or curved, with sponge rubber inserted between the two halves. Lugs are provided for attaching a cable on each end so that the mouse may be towed into position through the tube. Locating the approximate center of the mouse under the rivet to be driven is done by judging half the length of the mouse as drawn in either direction by the cable tow after the end of the mouse appears in the rivet hole. For riveting $3/32''$ rivets, steel is a satisfactory material for making a bucking mouse; in cases where adding length to the bar for weight is not desirable, it may be made of lead, which is cast around a tinned steel insert as shown by the

# RIVETS AND RIVETING

dotted lines of the end view in upper right in Fig. 15. Tinning the insert provides a bond between the lead and steel so that they will not become loose through continued riveting vibrations caused by the rivet gun.

The thickness of the layer of sponge rubber to be inserted and cemented to the two halves is determined by clearance of the mouse plus one and one-half times the rivet shank diameter (which protrudes before riveting) plus minimum space needed for the compressed sponge rubber. A flat area is milled or filed on the side which is used to form the rivet head. Good riveting with this type of mouse depends on the elasticity or resistance of the rubber.

A bucking mouse may have coil springs in place of sponge rubber; or it may be the wedge type controlled by rods fastened to each wedge and operated by pulling on one rod and pushing on the other. The sponge rubber type, Fig. 15, is preferred.

**HEAT TREATMENT OF RIVETS.** The method of heat-treating rivets is the same as for any aluminum and its alloys. For the treatment in the salt bath, rivets are placed in a suitable receptacle, usually a length (of 30" or more) of round tubing with a closed end. The container is placed vertically in the bath, which comes to within about four inches of the top. When the period of heating is finished, the tubes are lifted from the bath with asbestos gloves, and the rivets are dumped into a wire basket and immersed in cold water as quickly as possible. They scatter through the water, assuring a thorough quenching. In heat-treating, 17S rivets are held at a temperature of 930° to 950°F., 24S rivets at 910° to 930°F. Repair shops keep their salt bath barely above 930°F. when 24S rivets and material are being used, so that it will not take so long to reduce the temperature for heat treating the latter alloy. The temperature of the salt bath is controlled by a pyrometer, and it is very important that specified temperatures be maintained. The following tabulation gives time of heat treatment for common rivet sizes:

| Shank Diameter of Rivets | Minimum Period for Heat Treating |
| --- | --- |
| 3/32" | 20 minutes |
| 1/8" to 5/32" inclusive | 30 minutes |
| 3/16" to 1/4" inclusive | 45 minutes |
| 5/16" to 3/8" inclusive | 1 hour |

**SELECTION OF RIVET ALLOY.** The proper rivet to use in any aluminum alloy structure depends upon many factors; consequently each application must be considered in the light of its own requirements. The following paragraphs cover some of the more general points influencing the selection of an alloy and present certain combinations which have been found to be satisfactory.

If an attempt is made to drive a hard rivet such as 17S into a soft plate such as 3S–0, the plate will be unduly distorted and the resulting appearance poor. Also, the strength of the rivet is likely to be governed by its bearing value against the soft plate, so that the shear strength of the rivet may be largely wasted. For these reasons, it is poor practice to use a hard rivet in a soft alloy.

A soft rivet can sometimes be used to advantage in a hard alloy, especially if the joint is not highly stressed. Generally, however, it is advantageous to use a rivet having about the same properties as the material into which it is to be driven.

Steel rivets should be used only where the finished structure is to be painted and kept painted, for dust stains are likely to mar the appearance of the structure if bare steel is exposed. Properly protected steel rivets are satisfactory in many kinds of aluminum construction.

The following combinations of structure and rivets have been found to work together satisfactorily in ordinary applications:

| Structure | Rivets |
|---|---|
| 2S, any temper | 2S |
| 3S–O, 4S–O, 52S–O | 3S |
| 3S, 4S, 52S, quarter hard, or harder | 3S, 53S–W |
| A17S–T, 17S–T | A17S–T, 17S, Steel |
| 51S–W, 51S–T, 53S–W, 53S–T | 53S–W, 53S–T, Steel |

**STRENGTH OF RIVETED JOINTS.** In a properly designed riveted joint, rivet sizes and spacings are selected so that the force acting on any individual rivet does not exceed the safe design value of the rivet. Safe design values are given in shear, tension and bearing (crushing) strengths. They are determined so that an ample margin of safety is provided against each type of failure.

**Safe Shear Value of Rivets.** The first consideration in the selection of the proper rivet is its shear value, which depends upon:

# RIVETS AND RIVETING

1. Its cross-sectional area
2. The shear strength of the material
3. The number of planes along which the rivet tends to shear, as single shear, double shear, etc.

The cross-sectional area of a driven rivet is an uncertain quantity because hole sizes vary. In addition, the size of the hole may be increased by the swelling of the rivet in upsetting. Some engineers use the nominal diameter of the rivet for calculating areas and others use the diameter of the hole. However, it seems unnecessarily conservative, in most cases, to base calculations on the nominal diameter of the rivet; and it is recommended that the hole diameter be used up to a limiting value five per cent greater than the nominal diameter of the rivet.

In general, the nominal shear strength of the rivet material is a good basis for calculating the shear strength of driven rivets. Actual tests on driven rivets indicate that in some cases the driving operation increases the shear strength, a fact that is noted in Table 2.

The number of planes on which a rivet tends to shear depends entirely on the design of the joint. Usually rivets are used either in single or double shear, but occasionally they may be used in triple or even in quadruple shear. The shear value of a rivet increases directly as the number of planes upon which shear occurs, so that in double shear a rivet is twice as strong as in single shear.

Table 3 gives safe design values for various rivets, bolts and pins in single shear.

**Safe Tensile Values of Rivets.** Rivets are not well suited to transmitting loads in tension because a slight eccentricity of load exerts a prying action on the head which may result in early failure. This tendency is especially marked under repeated loads; consequently, it is generally accepted practice to avoid the use of any connection designed principally for transmitting loads by tension in the rivets. Tensile stresses in rivets cannot always be avoided because the racking of the framework and other secondary effects may produce appreciable tensile loads. In such cases, the safe tension value of a rivet may be taken as one-half the safe single shear value shown in Table 2. An effort should always be made to keep such secondary tensile stresses in rivets as low as possible.

# Table 3—Shear Strength of Rivets, Bolts, and Pins—Areas and Moments of Inertia

| Material | Aluminum Alloy | | | Low Carbon Steel | Heat-Treated Alloy Steel | Standard Aircraft, Heat-Treated Alloy Steel |
|---|---|---|---|---|---|---|
| | A17S Grade 4 | 17S-T Spec. QQ-A-351 | 24S-T Spec. 11071 | | | |
| Tensile strength, lbs. per sq. in. | 40,000 | 50,000 | 62,000 | 55,000 | 100,000 | 125,000 |
| Shear strength, lbs. per sq. in. | 25,000 | 30,000 | 35,000 | 35,000 | 65,000 | 75,000 |

| Size of Rivet, Pin or Bolt | Machine Screw Size | Area of Solid Section in Sq. In. | Moment of Inertia of Solid Section, Sq. In. | Allowable Single Shear Strength, Pounds (For double shear multiply by 2) | | | | | Ten. Str. at Root Dia., Lbs. | Yield Str. at Root Dia., Lbs. |
|---|---|---|---|---|---|---|---|---|---|---|
| 1/16 | .... | .003068 | .00000075 | 77 | 92 | 107 | 107 | 199 | 230 | .... | .... |
| 3/32 | .... | .006902 | .00000379 | 172 | 207 | 242 | 242 | 449 | 518 | .... | .... |
| .112 | No. 4 | .009852 | .00000772 | 246 | 296 | 345 | 345 | 640 | 739 | .... | .... |
| 1/8 | .... | .012272 | .00001198 | 314 | 368 | 430 | 430 | 798 | 920 | .... | .... |
| .138 | No. 6 | .014957 | .00001781 | 374 | 449 | 523 | 523 | 972 | 1122 | .... | .... |
| 5/32 | .... | .01918 | .00002926 | 480 | 575 | 671 | 671 | 1247 | 1438 | .... | .... |
| .164 | No. 8 | .02112 | .00003549 | 528 | 634 | 739 | 739 | 1372 | 1584 | .... | .... |
| 3/16 | .... | .02761 | .00006066 | 690 | 828 | 966 | 966 | 1794 | 2070 | 2136 | 1709 |
| .190 | No. 10 | .02835 | .00006399 | 709 | 850 | 992 | 992 | 1842 | 2126 | .... | .... |
| .216 | No. 12 | .03664 | .0001069 | 916 | 1099 | 1282 | 1282 | 2381 | 2748 | .... | .... |
| 7/32 | .... | .03758 | .0001125 | 940 | 1127 | 1315 | 1315 | 2442 | 2818 | .... | .... |
| 1/4 | .... | .04908 | .0001918 | 1227 | 1472 | 1717 | 1717 | 3190 | 3681 | 3982 | 3186 |
| 5/16 | .... | .07669 | .0004682 | 1917 | 2300 | 2684 | 2684 | 4984 | 5751 | 6429 | 5143 |
| 3/8 | .... | .1105 | .0009710 | 2762 | 3315 | 3868 | 3868 | 7183 | 8287 | 9953 | 7962 |
| 7/16 | .... | .1503 | .001797 | 3757 | 4509 | 5261 | 5261 | 9770 | 11272 | 13433 | 10746 |
| 1/2 | .... | .1963 | .003069 | 4907 | 5889 | 6871 | 6871 | 12760 | 14722 | 18356 | 14685 |
| 9/16 | .... | .2485 | .004914 | 6212 | 7455 | 8697 | 8697 | 16152 | 18637 | 23313 | 18650 |
| 5/8 | .... | .3068 | .007492 | 7670 | 9204 | 10738 | 10738 | 19942 | 23010 | 29676 | 23741 |
| 3/4 | .... | .4418 | .01553 | 11045 | 13254 | 15463 | 15463 | 28717 | 33135 | 43494 | 34795 |
| 7/8 | .... | .6013 | .02878 | 15032 | 18039 | 21046 | 21046 | 39085 | 45097 | 59515 | 47612 |
| 1 | .... | .7854 | .04908 | 19635 | 23562 | 27489 | 27489 | 51051 | 58905 | 80159 | 64127 |

Courtesy of Aluminum Company of America, Pittsburgh, Pa.

# RIVETS AND RIVETING

**Safe Bearing Value of Rivets.** Bearing value depends upon:
1. The area in bearing
2. The bearing strength of the metal in the rivet or plate, whichever is the lower
3. The edge distance in the direction in which the joint is stressed

The area in bearing is the thickness of metal times the diameter of hole. It is recommended that, for calculating bearing areas, the hole diameter be taken as not greater than 1.05 times the nominal rivet diameter.

The ultimate bearing strength of aluminum alloys in contact with driven rivets is about 1.8 times the nominal tensile strength of the metal, provided the edge distance in the direction of stressing is equal to at least twice the diameter of the hole, the edge distance being measured from the center of the hole. For smaller edge distances, the bearing strength drops off about in proportion to the decrease in edge distance; thus for an edge distance of one and one-half times the diameter, the bearing strength is about three-fourths of that for an edge distance of twice the diameter.

Bearing tests of joints show that the first appreciable permanent distortion of the hole occurs when the bearing stress is approximately equal to the nominal tensile strength of the material, and that this yielding is practically independent of the edge distance. The safe bearing design stresses for various aluminum alloys shown in Table 4 have an adequate factor of safety against both hole distortion and crushing failure.

The safe bearing stresses for driven rivets shown in Table 5 are to be used only when rivet is softer than the material through which it is driven. For instance, if $1/4''$ 4S–$1/2$H plates are joined by $1/2''$ 53S–W rivets driven in $17/32''$ diameter holes spaced $1\frac{1}{4}''$ from the edge of the plates, the safe bearing value is governed by the rivets and is approximately 1840 lbs. ($1.05 \times 1/2 \times 1/4 \times 14{,}000$). If the same rivets are used in 4S–O plates, the safe bearing value is governed by the plates and is approximately 1440 lbs. ($1.05 \times 1/2 \times 1/4 \times 11{,}000$). In the first case, if the edge distance had been only $7/8''$, the plate would have governed and the value would have been 1620 lbs. $\left( 1.05 \times 1/2 \times 1/4 \times 15{,}000 \times \dfrac{7/8}{2 \times 17/32} \right)$

### Table 4—Safe Bearing Design Stresses for Aluminum Alloy Plates and Shapes

These values should be used only when the edge distance measured from the center of rivet hole in the direction of stressing is equal to or greater than twice the diameter of the rivet hole. For smaller edge distances, values should be reduced proportionately.

These values have a factor of safety of about two against permanent distortion of the rivet hole and, when the foregoing restrictions are applied, they have a factor of safety of about four against crushing failure.

| Alloy | Safe Bearing Design Stress Lbs. per Square Inch | Alloy | Safe Bearing Design Stress Lbs. per Square Inch |
|---|---|---|---|
| 2S–O | 5,000 | A17S–T | 19,000 |
| 2S–½H | 7,000 | 17S–T | 26,000 |
| 2S–H | 10,000 | 24S–T | 29,000 |
| 3S–O | 7,000 | 27S–T | 26,000 |
| 3S–½H | 9,000 | 51S–W | 15,000 |
| 3S–H | 13,000 | 51S–T | 21,000 |
|  |  | 52S–O | 13,000 |
| 4S–O | 11,000 | 52S–¼H | 15,000 |
| 4S–¼H | 14,000 | 52S–½H | 17,000 |
| 4S–½H | 15,000 | 52S–H | 18,000 |
| 4S–H | 18,000 | 53S–W | 14,000 |
|  |  | 53S–T | 18,000 |

*Courtesy of Aluminum Company of America, Pittsburgh, Pa.*

### Table 5—Safe Bearing Design Stresses for Driven Rivets

These values to be used only when smaller than corresponding value for metal in contact with rivets as taken from Table 4.

| Rivet | Driving Procedure | Safe Bearing Design Stress Lbs. per Square Inch |
|---|---|---|
| 2S | Cold, as received | 7,000 |
| 3S | Cold, as received | 9,000 |
| A17S–T | Cold, as received | 19,000 |
| 17S | Cold, immediately after quenching | 26,000 |
| 53S–W | Cold, as received | 14,000 |
| 53S–T | Cold, as received | 18,000 |
| 17S | Hot, 930° to 950°F. | 26,000 |
| 53S | Hot, 960° to 980°F. | 14,000 |
| Steel | Hot, 1700° to 1900°F. | 30,000 |

*Courtesy of Aluminum Company of America, Pittsburgh, Pa.*

**PROPORTIONS OF RIVETED JOINTS.** The first requirement of riveted joints is that they be strong enough to transfer safely the forces acting on the parts joined. This requirement is responsible only in a general way for the design of the joint, because a number of joints can be designed for any given case, all sufficiently strong, though varying widely in size and spacing of rivets. In the following paragraphs, the factors influencing the proportions of riveted joints are discussed. Such factors will be found helpful in laying out highly stressed joints as well as those in which the rivets are used simply to fasten two or more parts together.

**Size of Rivet, Maximum and Minimum Limits.** If a large rivet is used in thin metal, the bearing strength usually governs and there is an excess of shear strength; moreover, the pressure required to drive the large rivet frequently causes an undesirable bulging of the thin material around the rivet head. For these reasons, the diameter of the rivet should rarely exceed two and one-half to three times the thickness of the sheet or plate.

On the other hand, if a small rivet is used in a thick plate, the shear strength is the determining factor and there is an excess of bearing strength. Small holes in thick plate usually make fabrication difficult. Experience indicates that the rivet diameter should be not less than the thickness of the thickest plate through which it is driven. The reason for this practice is that a hole made for inserting a rivet must be larger in diameter than the rivet; when the rivet is properly driven it completely fills the hole. A rivet too small in diameter will not fill the hole completely.

**Spacing of Rivets.** The spacing of rivets in any joint ordinarily depends upon the proportions of the members joined. The minimum spacing is determined by driving conditions; the space between the rivets must be sufficient to permit them to be driven without interference. Three times the nominal rivet diameter is the recommended minimum spacing. In tension members and joints the spacing must be such that the net stressed area is not too small. In some cases this becomes the controlling consideration.

The maximum spacing of rivets in compression members may be governed by the possibility of the buckling of the component parts between rivets. In other types of members, the maximum spacing is determined usually by the designer's sense of proportion, especially when the rivets are used simply to hold two or more plates or shapes together. As a general rule it is recommended that the maximum distance be not greater than twenty-four times the thickness of the sheet or plate.

**Edge Distance.** It already has been said that for maximum bearing strength, the edge distance measured from the center of the hole in the direction of stressing should be at least twice the diameter of the rivet hole and that the allowable bearing stress is decreased if the edge distance is less than this amount. Aside from considerations of strength, the edge distance is important because of the possibility of bulging the edge of the plate, which would cause an unsightly finished joint. Such bulging may be avoided by maintaining an edge distance of at least one and one-half times the diameter of the rivet and being careful to avoid overdriven rivets. When edge distances less than one and one-half diameters are used, the rivets generally must be underdriven to avoid bulging the plate.

**TYPES OF RIVETS.** The tremendous advances in aviation have been made possible in great degree through the development of all-metal design, pre-eminently employing the lighter metals such as aluminum and magnesium alloys. This style of construction requires some 40,000 to 500,000 rivets or more per plane, depending on the size. Riveting, therefore, is one of the most exacting and tedious jobs confronting airplane builders, and it grows more so as larger and larger planes are being built. For example, the B-19 Douglas Bomber, largest ship of its kind when completed, is said to have 3,000,000 rivets.

Gang-riveting machines, automatic hole-punching and rivet-driving devices, and the occasional replacement of rivets by high-amperage spot welding, all have tended to simplify the tremendous fastening problem. However, these methods, as well as the conventional driving of rivets by two-man crews—usually at the rate of two to three rivets a minute—are applicable only in assemblies which permit access to both sides.

**Blind Rivets.** There are many structures or portions of airplane structures which are totally or partially inaccessible from both sides. These require fasteners that are capable of being installed from one side; usually such fasteners are termed *blind rivets*. Many kinds of mechanical blind rivets have been developed, but in general they do not equal bucked rivets in physical and structural properties, besides being less economical; therefore, airplane design tends to avoid their use where possible.

One of the most popular of all blind rivets is the *rivnut* manufactured by Goodrich Rubber Co. Rivnuts, Fig. 16, are primarily

# RIVETS AND RIVETING

used for installing de-icer boots which hold the cap strip on the leading edge of wings and stabilizers. They are also used extensively for fastening fillets, cowling, fairings and patches, the installation of cabin accessories and other nonstructural applications.

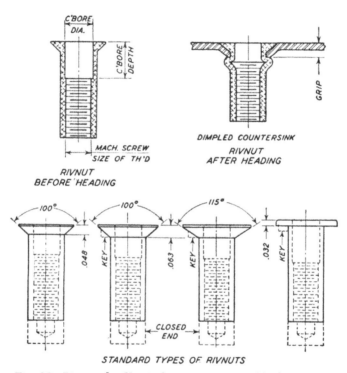

Fig. 16. Rivnuts for Use in Locations Inaccessible from Rear
*Reproduced by Permission of B. F. Goodrich Company, Akron, Ohio*

The rivnut actually is a hollow rivet, and the Civil Aeronautics Manual No. 18 states "Hollow rivets should not be substituted for solid rivets in load-carrying members without specific approval of the application by a representative of the Civil Aeronautics Administration."

The Cherry Blind rivet, Fig. 16A, is designed to do a first-class riveting job in the blind or hard-to-get-at places in aircraft structures. It is actually an assembly: a standard aircraft rivet, through which passes an aluminum alloy mandril or stem. Positive head formation and shank expansion are produced by the rivet stem. No bucking bar is needed.

Special guns are used. The rivet can be placed in the work first, or the projecting stem can be placed in the gun head and then inserted in the work. The gun may be either a hand-operated or power tool; in either case the gun pulls on the stem and pushes on the rivet head. As the stem is pulled through the rivet, it forms a tulip head on the blind side, expands the shank of the rivet, and permanently plugs the rivet. The pull continues until the stem breaks and frees the gun. The stem is then trimmed flush with flat-ground nippers. Steps in this process are shown in Fig. 16A.

Cherry rivets can be applied at the rate of 800 to 1,600 per hour, depending upon the type of rivet used. They give satisfac-

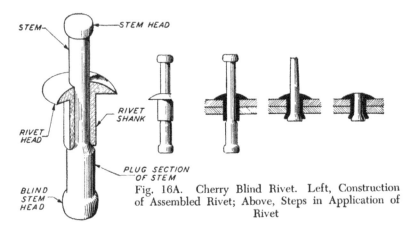

Fig. 16A. Cherry Blind Rivet. Left, Construction of Assembled Rivet; Above, Steps in Application of Rivet

tory performance with reasonable tolerances in grip length and hole size.

The Dill Lok-Skru is another widely used type of blind fastener in the same status as the rivnut. In Fig. 17 at left are shown the three parts of the Lok-Skru. At (*1*), Lok-Skru has been inserted through a drilled hole for blind attachment. At (*2*), Lok-Skru has been securely locked by drawing together with a special tool. (*3*) illustrates how attachment can be made to the fastener by means of the screw *C*. The Lok-Skru is a three-piece fastener, while the rivnut is two-piece.

**Explosive Rivets.** The explosive rivet made by Du Pont, Fig. 18, is a development that doubtless will play an important part in speeding aircraft production and simplifying design.

In 1921, Frank Allan, an American, patented several types of explosive rivets and bolts that he hoped might prove the solution of the blind rivet problem. His rivets did not prove commercially feasible. In 1937 a patent issued by the U. S. Patent Office to Karl and Otto Butter, two brothers in the employ of Ernst

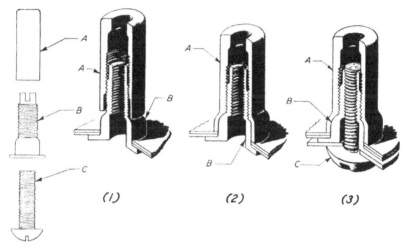

Fig. 17. Dill Lok-Skru; Blind Fastener Used for Installation of Cabin Furnishings, Fillets, etc.
*Courtesy of The Dill Manufacturing Company, Cleveland, Ohio*

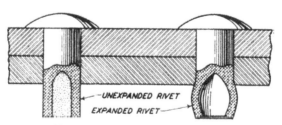

Fig. 18. Du Pont Explosive Rivets
*Courtesy of E. I. du Pont de Nemours and Co., Inc., Wilmington, Del.*

Heinkel, builder of the famous German plane bearing his name, revived the idea of setting blind rivets by expanding the shank with a minute explosives charge.

Early in 1939 the Du Pont explosives department became interested in the idea and began seriously to explore the possibilities. The Butter rivet seemed to offer advantages over existing types of blind rivets, providing it could be perfected and manufactured economically. North American rights to the invention were purchased by Du Pont, and an intensive research and development program was launched in their eastern laboratory at Repauno, New Jersey.

Fundamental details were disclosed by the patents, but many problems remained to be solved. First the design had to be developed to conform to American standards of precision. Then a new method of manufacture of the rivet blanks was necessary, since the small tolerances, or allowances for microscopic variations in size, could not be provided with existing equipment. The anodizing process, or treatment to prevent corrosion, had to be improved. Equipment for extremely accurate loading of almost infinitesimal charges of explosive had to be designed and built.

Finally, in the fall of 1940, after they had been tested and evaluated by the U. S. Army and Navy, the improved Du Pont rivets were shipped in limited numbers to a few aircraft manufacturers for further testing and actual shop installations. These rivets were made with experimental tools, but were loaded on production equipment. Since then, and in close cooperation with American airplane builders, details of manufacture and installation have been perfected, and they are now being made in commercial quantities.

This rivet has a high explosive secreted in a cavity at the end of the shank. Heat applied to the rivet head by an electric gun detonates the charge, Fig. 19. The explosion expands the charged end of the shank, thus forming a *blind* head and setting the rivet. The whole operation is performed from one side with greater ease and speed than is possible by any mechanical means now being used in aircraft factories.

Engineers estimate that from 800 fastening points (in an all-metal pursuit ship) to as many as 10,000 (in the largest all-metal bomber) are accessible only from one side, a fact responsible for one of the most troublesome bottlenecks in the mass production of fighting planes.

A skilled workman can set about two to four blind fasteners a minute, after they have been placed in the holes; in comparison, the Du Pont explosive rivets may be installed by one workman at a rate of 15 to 20 rivets a minute, once they are in place. The

# RIVETS AND RIVETING

riveting gun or iron weighs less than 5 pounds, and the rivets themselves weigh only about one-fourth as much as the ordinary mechanical blind fastener. Moreover, so well is the explosive charge controlled, that the expansion it causes may be held within limits of $21/1000$ths inch—an example of modern skill and precision.

Two years of experimental work, supplemented by extensive tests on airplane production lines in recent months, now stand behind this interesting development. Engineers expect the inven-

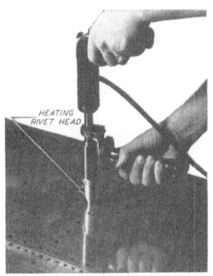

Fig. 19. Installing Du Pont Explosive Rivets with Electrically Heated Tool
*Courtesy of E. I. du Pont de Nemours and Co., Inc.
Wilmington, Del.*

tion to find wide application in industry in general, and to effect radical changes in riveting methods and structural designs.

### RIVETS AND RIVETING PRACTICE (See Fig. 20)

(Rules for rivets and riveting, and illustrations of riveting practice, have been taken in part from the Civil Aeronautics Manual No. 18, published by U.S. Dept. of Commerce, Civil Aeronautics Administration, Washington.)

**IDENTIFICATION OF RIVET MATERIAL.** The kinds of rivets in general use, listed in the order of their decreasing strength properties, are as follows and are shown in Fig. 2.

(a) 24S–T, identified by two small raised radial dashes at the ends of a diameter on the periphery of the head.

(b) 17S–T, identified by a small raised spot, or pimple, in the center of the head.

(c) A17S–T, identified by a small depression or dimple in the center of the head.

Fig. 20. Riveting Skin on Fuselage Section of Douglas Attack Bomber
*Courtesy of Douglas Aircraft Co., Inc., Burbank, California*

**REPLACEMENT OF RIVETS.** Rivets may be replaced by those of higher strength properties, but not vice versa, unless the lower strength is compensated for by an increase in diameter or a greater number of rivets. It is advisable to stock all rivets in heat-treated condition (S–T) in order to prevent unheat-treated rivets being used. The A17S–T rivets may be driven in the condition received, but the 17S–T rivets above 3/16", and all 24S–T rivets, should be reheat-treated just prior to driving, as they would otherwise be too hard for satisfactory riveting.

**Use of A17S–T Aluminum Alloy Replacement Rivets.** It will be considered acceptable to replace all 17S–T rivets of 3/16" diameter or less with A17S–T rivets for general repairs, provided the replacement rivets are 1/32" larger in diameter than the rivets they replace.

**Hollow Rivets.** Hollow rivets should not be substituted for solid rivets in load-carrying members without specific approval of the application by a representative of the Civil Aeronautics Administration.

# RIVETS AND RIVETING

**Rivet Size.** In replacing rivets, the original size should be used if this size will fit and fill the holes. If not, the holes should be drilled or reamed for the next larger size, care being taken, however, that the edge distance and spacings are not less than the minimums shown in Fig. 21. A general rule for selecting rivets of proper diameter to join aluminum alloy is approximately

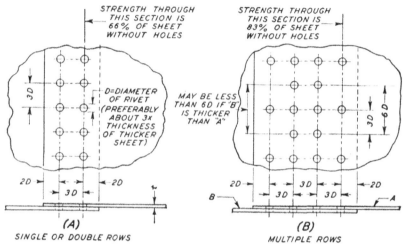

Fig. 21. Rivet Hole Spacing and Edge Distance

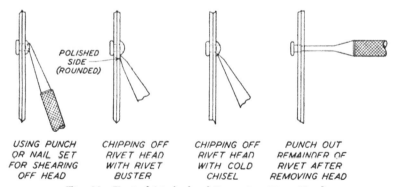

Fig. 22. Typical Methods of Removing Rivet Head

3 times the thickness of the heavier sheet, or somewhat larger for thin sheets. Dimensions for forming flat rivet heads are shown in Fig. 4.

**Rivet Spacing.** A new or revised rivet pattern should be designed for the strength required; but, in general practice, the spacing should not be closer than the minimums shown in Fig. 21. Rivets should not be used where they would be in tension tending to pull the heads off. A lap joint of thin sheets should be backed up by a stiffening section. Fig. 22 illustrates methods of rivet removal.

# 168   AIRCRAFT SHEET METAL WORK

**Precautions.** When adding or replacing rivets adjacent to previously installed 17S–T or 24S–T rivets, great care should be exercised so that the older rivets will not be loosened or made to fail due to the sharp vibrations in the structure caused by action of the rivet gun and bucking bar. In every case, all adjacent rivets should be carefully examined after a repair or alteration is finished, to be sure that they have not been damaged by the newer riveting operation.

Rivet holes should be drilled round, straight, and free from cracks. The snap used in driving the rivets should be cupped slightly flatter than the rivet head, as seen in Fig. 6, Chapter I.

## REPAIR METHODS

**SPLICING OF TUBES.** Round or streamlined tubular members may be repaired by splicing, as shown in Fig. 23. Splices in struts should be adjacent to the fittings.

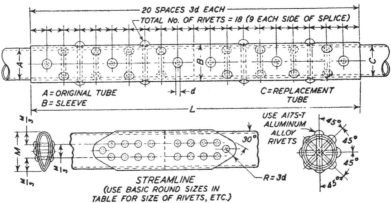

| | | | | | | | | | | |
|---|---|---|---|---|---|---|---|---|---|---|
| A, C° | 3/4 | 7/8 | 1 | 1 1/8 | 1 1/4 | 1 3/8 | 1 1/2 | 1 5/8 | 1 3/4 | 1 7/8 |
| | .065 | .065 | .065 | .065 | .058 | .058 | .058 | .058 | .058 | .058 |
| B | 7/8 | 1 | 1 1/8 | 1 1/4 | 1 3/8 | 1 1/2 | 1 5/8 | 1 3/4 | 1 7/8 | 2 |
| | All .058 thick | | | | | | | | | |
| Rivet Dia. | 5/32 | 5/32 | 3/16 | 3/16 | 3/16 | 3/16 | 1/4 | 1/4 | 1/4 | 1/4 |
| L | 9 3/8 | 9 3/8 | 11 1/4 | 11 1/4 | 11 1/4 | 11 1/4 | 15 | 15 | 15 | 15 |

°Includes all thicknesses up to and including maximum shown.

Fig. 23. Typical Repair Methods for Tubular Members of Aluminum Alloy

When solid rivets go completely through hollow tubes, their diameter should be at least 1/8 of the outside diameter of the outer tube. Rivets which are loaded in shear should be hammered only enough to form a small head.

# RIVETS AND RIVETING

No attempt should be made to form the standard round head, since the amount of hammering required to do so often causes the rivet to buckle inside the tube. Satisfactory rivet heads may be produced in such installations by spinning if the proper equipment is available. Examples of correct and incorrect rivet application to tubular members are included in Fig. 24.

*Note:* Spinning the heads of rivets is not recommended, because it tends to twist the shank in the hole, sometimes making a poor bearing surface.

Whenever practical in riveting through tubing, use the method illustrated at (D), Fig. 24. Fill the space inside the tube with an alloy that melts below

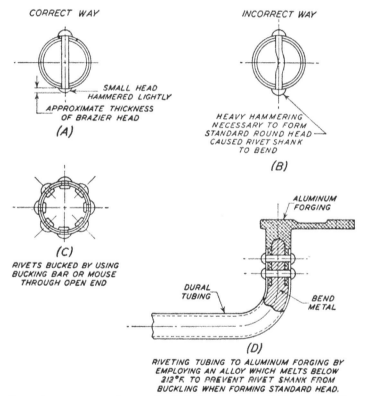

Fig. 24. Riveting Technique for Tubular Members of Aluminum Alloy

212°F. Such an alloy is *Wood's Metal*, with the following approximate composition:

| Amounts | Elements | Melting Points |
|---|---|---|
| 4 parts | Bismuth (Bi) | 550°F. |
| 2 parts | Lead (Pb) | 621°F. |
| 1 part | Tin (Sn) | 449°F. |
| 1 part | Cadmium (Cd) | 610°F. |

These elements, in the proportions specified, constitute the alloy known as Wood's Metal, which has a melting point of only 140°F. or 60°C. Note that the melting point drops to less than the boiling point of water, a temperature that may safely be applied to heat-treated aluminum alloys without changing their temper.

Ready-made alloys of this type, having various melting points between 150° and 200°F., are on the market as *Bend Metal, Bendalloy, Cerrobend*, and similarly designated products. As their names imply, they are used principally in the bending of sharp radii in tubing. This type of alloy is melted in a double boiler (heated over water) in order not to raise the temperature above 212°F. at sea level.

After the bend metal has been poured in the tube to be riveted, rivet holes are drilled through tubing and bend metal.* Riveting may then be accomplished without danger of bending the shank of the rivet inside the tube as shown at (B) in Fig. 24. The bend metal is removed by running hot water over the outside of the tube until the alloy remelts and runs out of the tubing into a container. Not many assemblies are so designed that this method of riveting can be used, but where it is applicable a very good job of riveting can be done, since it not only holds the two thicknesses of tubing together, but permits the forming of a standard head on the rivet. Under no circumstances should melted lead be used for such procedure, unless the finished part is to be heat treated afterward, because lead melts above 620°F., which will anneal aluminum alloys. Furthermore, it is considered poor practice to heat treat assembled units because of the difficulty afterward in washing the salts out from between cracks, seams and joints.

Where the end of the tube is open, the rivets are usually bucked on the inside of the tube. Rivets so bucked are shown at (C), Fig. 24.

**SPLICING OF SHEETS.** In some cases the method of copying the seams at the edges of a sheet may not be satisfactory; for example, when the sheet has cutouts or doubler plates at an edge seam, or when other members transmit loads into the sheet. In these cases the splice should be designed to carry the full allowable tension load for the sheet, as illustrated in the following example:

Material: 17S–T Alclad Sheet, .032" thickness. Width of sheet (i.e. length at splice) $= W = 10"$. To determine rivet size and pattern for a single lap joint, similar to that shown in Fig. 21.

*a*) Choose diameter the nearest size larger than three times the sheet thickness, as stated at (A), Fig. 21, $3 \times .032" = .096"$. Use 1/8" A17S–T rivets.

*b*) Determine the number of rivets required per inch of width, W, from Table 7. No. per inch = 5.9.

Total No. of rivets required $= 10 \times 5.9 = 59$ rivets.

*c*) Lay out rivet pattern with spacings not less than those shown in Fig. 21. Referring to Fig. 21, at (B), it is seen that a 4-row pattern with the center rows spaced 1/2" and the outer rows 1" will give a total of 60 rivets in the 10" splice, and is therefore satisfactory.

**SPLICING OF STRINGERS.** Typical splices for various shapes of sections are shown in Figs. 25 and 26. Stringers are designed to carry both tension and compression, and the splice shown in Fig. 26 will be used as an example illustrating the following general principles:

---

* Use a drill sharpened for soft metals for drilling through the bend metal at temperatures of 100° F. or less.

# RIVETS AND RIVETING

1. To avoid eccentric loading and consequent buckling in compression, splicing or reinforcing parts should be placed as symmetrically as possible about the centerline of the member, with attachment made to as many elements as necessary to prevent bending in any direction.

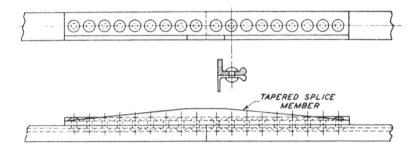

*NOTE: UNSHADED SECTIONS ARE ORIGINAL AND/OR REPLACEMENT SECTIONS. SHADED SECTIONS ARE CONNECTING OR REINFORCING SECTIONS.*

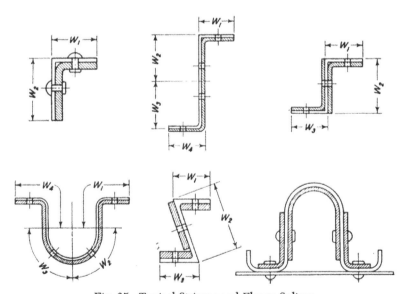

Fig. 25. Typical Stringer and Flange Splices

2. To avoid reducing the strength in tension of the original bulb angle, the rivet holes at the ends of the splice are made small (no larger than the original skin-attaching rivets), and holes in the second row (those through the bulb leg) are staggered back from the ends.

3. To avoid concentration of load on the end rivet and consequent tendency toward progressive rivet failure, the splice is tapered off at the ends,

in this case, Fig. 26, by tapering the backing angle and by making it shorter than the splice bar.

These principles are especially important in splicing stringers on the lower surface of stressed skin wings, where high tension stresses may exist. When several adjacent stringers are spliced, the splices should be staggered if possible.

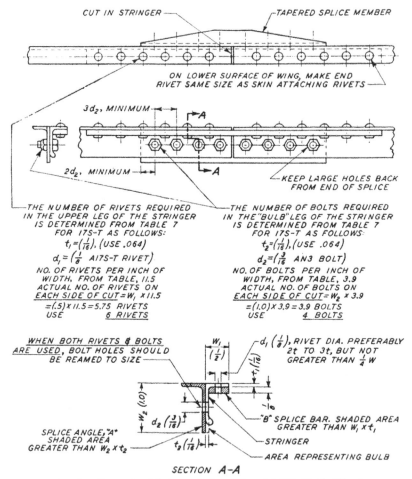

Fig. 26. Example of Stringer Splice (Material 17S–T Al. Alloy)

**1. Size of Splicing Members.** When the same material is used for the splicing member as for the original member, the net cross-sectional area (i.e., the shaded areas of the splicing member, Fig. 25) should be greater than the area of the section element which it splices. The area of a section element (e.g. each leg of an angle or channel) is equal to the width multiplied by the thickness. For example, in Fig. 26, the bar $B$ is assumed to splice the upper

# RIVETS AND RIVETING

leg of the stringer, and the angle $A$ to splice the bulbed leg of the stringer. Since the splice bar $B$ is not so wide as the adjacent leg, and since the rivet diameter is also subtracted from the width, the bar is made twice as thick in order to obtain sufficient net area.

**2. The Diameter of Rivets in Stringers.** The diameter of rivets in stringers should preferably be between 2 and 3 times the thickness, $t$, of the leg, but should not be more than 1/4 the width, $W$, of the leg. Thus, 1/8" rivets are chosen in the example shown in Fig. 26. If this splice were in the lower surface of a wing, the end rivets would be made the same size as the skin-attaching rivets, say 3/32".

**3. The Number of Rivets.** The number of rivets required on each side of the cut in the stringer may be found in Tables 6, 7 or 8, depending on the material. In determining the number of rivets required in Fig. 26 for attaching the splice bar $B$ to the upper leg, the thickness $t_1$ of the element of area being spliced is 1/16" (use .064), the rivet size is 1/8", and Table 7 shows that 11.5 rivets are required per inch of width. Since the width $W_1$ is 1/2", the actual number of rivets required to attach the splice bar to the upper leg, *on each side of the cut is*

11.5 (rivets per inch) $\times$ .5 (in. width) $= 5.75$, use 6 rivets.

For the bulbed leg of the stringer, $t_2 = 1/16$ (use .064), AN3 (3/16")

**Table 6—Number of Rivets Required for Splices (Single Lap Joint) in 24S–T, 24S–T Alclad, 24S–RT and 24S–RT Alclad Sheet**

| Thickness "t" in Inches | Number of A17S–T Rivets Required per Inch of Width, "W" | | | | | |
|---|---|---|---|---|---|---|
| | 3/32" | 1/8" | 5/32" | 3/16" | 1/4" | 3/16" bolts (AN3) |
| .016 | 7.4 | 5.6 | .... | .... | .... | .... |
| .020 | 7.4 | 5.6 | 4.5 | .... | .... | .... |
| .025 | 9.0 | 5.6 | 4.5 | .... | .... | .... |
| .032 | 11.5 | 6.5 | 4.5 | 3.7 | .... | .... |
| .036 | 13.0 | 7.3 | 4.7 | 3.7 | 2.8 | .... |
| .040 | 14.4 | 8.1 | 5.2 | 3.7 | 2.8 | .... |
| .051 | .... | 10.3 | 6.6 | 4.6 | 2.8 | .... |
| .064 | .... | 13.0 | 8.3 | 5.8 | 3.2 | .... |
| .081 | .... | .... | 10.5 | 7.3 | 4.1 | 3.7 |
| .091 | .... | .... | 11.8 | 8.2 | 4.6 | 3.7 |
| .102 | .... | .... | 13.2 | 9.2 | 5.2 | 3.7 |
| .128 | .... | .... | .... | 11.5 | 6.5 | 3.8 |

**Notes:** (*a*) For stringers in the upper surface of a wing, or in a fuselage, 80% of the number of rivets shown in the table may be used.

(*b*) For intermediate frames, 60% of the number shown may be used.

**Engineering Notes:** The above table was computed as follows:

1. The load per inch of width of material was calculated by assuming a strip 1" wide in tension at a stress of 62,000 p.s.i., which is slightly unconservative for 24S–RT and conservative for 24S–T Alclad and 24S–RT Alclad.

2. No. of rivets required was calculated for A17S–T rivets, using allowable stress values of 25,000 p.s.i. in shear, and a value in bearing calculated on the basis of a ratio between ultimate tensile strength and ultimate bearing strength of 1.43.

3. Combinations of sheet thickness and rivet size above the double line are critical in (i.e. will fail by) bearing on the sheet; those below are critical in shearing of the rivets.

Reprinted from *Civil Aeronautics Manual 18*, C.A.A.

bolts are chosen; the number of bolts required per inch of width = 3.9. The width $W_2$ for this leg, however, is 1", and the actual number of bolts required *on each side of the cut is* $1 \times 3.9 = 3.9$, use 4 bolts. When both rivets and bolts are used in the same splice, the bolt holes should be accurately reamed to size. It is preferable to use only one type of attachment, but in the above example the dimensions of the legs of the bulb angle indicated rivets for the upper leg and bolts for the bulb leg.

**SPLICING OF INTERMEDIATE FRAMES.** The same principles that are used for stringer splicing may be applied to intermediate frames, and the following points are also considered:

Conventional frames of channel or zee section are relatively deep and thin compared to stringers, and usually fail by twisting or by buckling of the free flange. The splice joint should be reinforced against this type of failure by using a splice plate heavier than the frame and by splicing the free flange of the frame with a flange on the splice plate, as illustrated in Fig. 27. The number of rivets required in each leg on each side of the cut is determined by the width W, thickness of frame material $t$, and rivet diameter $d$, using Table 7 (for 17S–T) in a manner similar to that for stringers, Fig. 26. Note (b), Table 7, indicates that only 60% of the number of rivets so calculated need be used in splices in intermediate frames. Since a frame is likely to be subjected to bending loads, the length of splice plate $L$ should be more than

**Table 7—Number of Rivets Required for Splices (Single Lap Joint) in 17S–T, 17S–T Alclad, 17S–RT and 17S–RT Alclad Sheet**

| Thickness "$t$" in Inches | Number of A17S–T Rivets Required per Inch of Width, "W" | | | | | $3/16"$ bolts (AN3) |
|---|---|---|---|---|---|---|
| | $3/32"$ | $1/8"$ | $5/32"$ | $3/16"$ | $1/4"$ | |
| .016 | 7.8 | .... | .... | .... | ... | ... |
| .020 | 7.8 | 5.9 | .... | .... | ... | ... |
| .025 | 8.0 | 5.9 | .... | .... | ... | ... |
| .032 | 10.2 | 5.9 | 4.7 | 3.9 | ... | ... |
| .036 | 11.5 | 6.5 | 4.7 | 3.9 | ... | ... |
| .040 | 12.8 | 7.2 | 4.7 | 3.9 | 2.9 | ... |
| .051 | .... | 9.2 | 5.9 | 4.1 | 2.9 | ... |
| .064 | .... | 11.5 | 7.4 | 5.1 | 2.9 | 3.9 |
| .081 | .... | 14.6 | 9.3 | 6.5 | 3.6 | 3.9 |
| .091 | .... | .... | 10.4 | 7.2 | 4.1 | 3.9 |
| .102 | .... | .... | 11.7 | 8.1 | 4.6 | 3.9 |
| .128 | .... | .... | 14.7 | 10.2 | 5.7 | 3.9 |

**Notes:** (a) For stringers in the upper surface of a wing, or in a fuselage, 80% of the number of rivets shown in the table may be used.

(b) For intermediate frames, 60% of the number shown may be used.

**Engineering Notes:** The above table was computed as follows:

1. The load per inch of width of material was calculated by assuming a strip 1" wide in tension at a stress of 55,000 p.s.i., which is slightly conservative for 17S–RT Alclad and 17S–T Alclad.

2. No. of rivets required was calculated for A17S–T rivets, using allowable stress values of 25,000 p.s.i. in shear, and a value in bearing calculated on the basis of a ratio between ultimate tensile strength and ultimate bearing strength of 1.36.

3. Combinations of sheet thickness and rivet size above the double line are critical in (i.e. will fail by) bearing on the sheet; those below are critical in shearing of the rivets.

Reprinted from *Civil Aeronautics Manual 18*, C.A.A.

# RIVETS AND RIVETING

**Table 8—Number of Rivets Required for Splices (Single Lap Joint) in 52S (All Hardnesses) Sheet**

| Thickness "t" in Inches | Number of A17S–T Rivets Required per Inch of Width, "W" | | | | | |
|---|---|---|---|---|---|---|
| | $3/32''$ | $1/8''$ | $5/32''$ | $3/16''$ | $1/4''$ | $3/16''$ bolts (AN3) |
| .016 | 6.0 | .... | .... | .... | ... | ... |
| .020 | 6.0 | 4.5 | .... | .... | ... | ... |
| .025 | 6.0 | 4.5 | .... | .... | ... | ... |
| .032 | 7.3 | 4.5 | 3.6 | .... | ... | ... |
| .036 | 8.2 | 4.6 | 3.6 | 3.0 | ... | ... |
| .040 | 9.1 | 5.1 | 3.6 | 3.0 | 2.2 | ... |
| .051 | 11.6 | 6.5 | 4.2 | 3.0 | 2.2 | ... |
| .064 | 14.5 | 8.2 | 5.2 | 3.6 | 2.2 | ... |
| .081 | .... | 10.3 | 6.6 | 4.6 | 2.6 | ... |
| .091 | .... | 11.6 | 7.4 | 5.1 | 2.9 | 3.0 |
| .102 | .... | 13.0 | 8.3 | 5.8 | 3.2 | 3.0 |
| .128 | .... | .... | 10.4 | 7.2 | 4.1 | 3.0 |

**Notes:** (*a*) For stringers in the upper surface of a wing, or in a fuselage, 80% of the number of rivets shown in the table may be used.

(*b*) For intermediate frames, 60% of the number shown may be used.

**Engineering Notes:** The above table was computed as follows:

1. The load per inch of width of material was calculated by assuming a strip 1" wide in tension at a stress of 39,000 p.s.i., which is slightly conservative for the ¾ hard, ½ hard and ¼ hard material.

2. No. of rivets required was calculated for A17S–T rivets, using allowable stress values of 25,000 p.s.i. in shear, and a value in bearing calculated on the basis of a ratio between ultimate tensile strength and ultimate bearing strength of 1.78.

3. Combinations of sheet thickness and rivet size above the double line are critical in (i.e. will fail by) bearing on the sheet; those below are critical in shearing of the rivets.

Reprinted from *Civil Aeronautics Manual 18*, C.A.A.

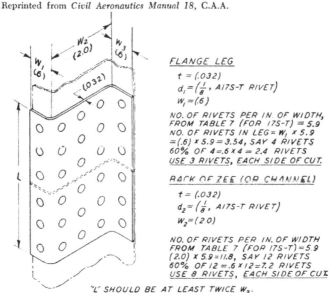

FLANGE LEG
$t = (.032)$
$d_1 = (\frac{1}{8}, A17S-T\ RIVET)$
$W_1 = (.6)$

NO. OF RIVETS PER IN. OF WIDTH, FROM TABLE 7 (FOR 17S-T) = 5.9
NO. OF RIVETS IN LEG = $W_1 \times 5.9$
= $(.6) \times 5.9 = 3.54$, SAY 4 RIVETS
60% OF 4 = .6 × 4 = 2.4 RIVETS
USE 3 RIVETS, EACH SIDE OF CUT.

BACK OF ZEE (OR CHANNEL)
$t = (.032)$
$d_2 = (\frac{1}{8}, A17S-T\ RIVET)$
$W_2 = (2.0)$

NO. OF RIVETS PER IN. OF WIDTH FROM TABLE 7 (FOR 17S-T) = 5.9
$(2.0) \times 5.9 = 11.8$, SAY 12 RIVETS
60% OF 12 = .6 × 12 = 7.2 RIVETS
USE 8 RIVETS, EACH SIDE OF CUT.

"L" SHOULD BE AT LEAST TWICE $W_2$.
THICKNESS OF SPLICE PLATE GREATER THAN THAT OF FRAME.

Fig. 27. Example of Splice of Intermediate Frame, 17S–T Al. Alloy

twice the width $W_2$, and the rivets spread out to cover the plate. Although only 8 rivets are necessary on each side of cut in $W_2$ width according to calculation, 9 are used for the sake of symmetry and to provide a margin of safety.

**INSPECTION OF DRIVEN RIVETS.** The standards to which driven rivets should conform frequently are somewhat uncertain. In addition to dimensions (thickness, diameter) and perfection of shape, inspection is concerned with whether the driven head is co-axial with the shank (not clinched) and whether the metal is in a sound condition (not cracked).

**SPOT WELDING VS. RIVETING OF ALUMINUM ALLOY.** Rivets probably will always be used in structural fabrication of airplanes, but spot welding has been growing more popular in fabrication of nonstructural parts.

Flush riveting is already taking the lead over brazier-head type. Nearly all leading edges of wings, stabilizers and the bottoms of flying boats are flush riveted, as is the entire outer surface of fast military equipment. A brazier head in the airstream will soon look as outmoded as exposed wires or struts and projecting arms on control surfaces would look today.

Where flush rivets are used for holding skin, the skin stiffeners are press-countersunk (counterpunched) to fit the form of the rivet head. If a stiffener

Table 9—Bearing and Shear Strength of A17S–T Rivets

| Size rivet, in. | 3/32 | 1/8 |
|---|---|---|
| Single shear | 173 | 307 |
| Double shear | 346 | 614 |
| Thickness of Sheet, In. | Bearing Strength of A17S–T Rivets, Lbs. | |
| .013 | 73 | 97 |
| .016 | 90 | 120 |
| .018 | 101 | 135 |
| .020 | 112 | 150 |
| .022 | 124 | 165 |
| .025 | 141 | 188 |
| .028 | 157 | 210 |
| .032 | 180 | 240 |
| .035 | 197 | 263 |
| .040 | 225 | 300 |
| .049 | 275 | 367 |
| .051 | 287 | 383 |
| .057 | 321 | 428 |
| .064 | 360 | 480 |
| .065 | 365 | 487 |
| .072 | 405 | 540 |
| .081 | 456 | 608 |
| .083 | 467 | 623 |

Values above the upper double line are less than single shear.
Values below the lower double line are greater than double shear.
The allowable shearing stress of A17S–T rivets = 25,000 lbs. per sq. in.
The allowable bearing stress of A17S–T rivets = 60,000 lbs. per sq. in.

# RIVETS AND RIVETING

### Table 10—Bearing and Shear Strength of 17S–T Rivets and Sheet

| Size rivet, in. | 1/16 | 3/32 | 1/8 | 5/32 | 3/16 | 1/4 | 5/16 | 3/8 | 7/16 | 1/2 |
|---|---|---|---|---|---|---|---|---|---|---|
| Single shear | 92 | 207 | 369 | 576 | 829 | 1473 | 2301 | 3315 | 4509 | 5892 |
| Double shear | 184 | 414 | 738 | 1152 | 1658 | 2946 | 4602 | 6630 | 9018 | 11784 |

| Thickness of Sheet, In. | Bearing Strength of Sheet, Lbs. (Aluminum Alloy 17S–T) | | | | | | | | | |
|---|---|---|---|---|---|---|---|---|---|---|
| .013 | 61 | 91 | 121 | 152 | 182 | 243 | 304 | 365 | 426 | 487 |
| .016 | 75 | 112 | 150 | 187 | 225 | 300 | 374 | 450 | 525 | 600 |
| .018 | 84 | 126 | 169 | 211 | 252 | 337 | 422 | 506 | 591 | 675 |
| .020 | 94 | 141 | 187 | 234 | 282 | 375 | 469 | 562 | 656 | 750 |
| .022 | 103 | 154 | 206 | 258 | 310 | 412 | 516 | 619 | 722 | 825 |
| .025 | 117 | 176 | 234 | 292 | 352 | 469 | 586 | 703 | 820 | 937 |
| .028 | 131 | 196 | 262 | 328 | 394 | 525 | 636 | 787 | 919 | 1050 |
| .032 | 150 | 225 | 300 | 375 | 451 | 600 | 750 | 900 | 1050 | 1200 |
| .035 | 164 | 246 | 328 | 410 | 492 | 656 | 820 | 984 | 1148 | 1312 |
| .040 | 188 | 281 | 375 | 469 | 563 | 750 | 938 | 1125 | 1313 | 1500 |
| .049 | 230 | 344 | 459 | 573 | 685 | 919 | 1148 | 1378 | 1608 | 1837 |
| .051 | 239 | 359 | 478 | 598 | 716 | 957 | 1196 | 1435 | 1674 | 1913 |
| .057 | 267 | 401 | 534 | 668 | 802 | 1069 | 1336 | 1603 | 1870 | 2138 |
| .064 | 300 | 450 | 600 | 750 | 900 | 1200 | 1500 | 1800 | 2100 | 2400 |
| .065 | 305 | 457 | 609 | 761 | 916 | 1219 | 1523 | 1828 | 2133 | 2437 |
| .072 | 338 | 506 | 675 | 844 | 1013 | 1350 | 1688 | 2025 | 2363 | 2700 |
| .081 | 380 | 570 | 759 | 949 | 1139 | 1519 | 1898 | 2278 | 2658 | 3038 |
| .083 | 389 | 583 | 778 | 971 | 1167 | 1556 | 1945 | 2334 | 2723 | 3112 |
| .091 | 427 | 640 | 853 | 1066 | 1280 | 1706 | 2133 | 2559 | 2986 | 3413 |
| .095 | 445 | 668 | 891 | 1112 | 1336 | 1781 | 2226 | 2672 | 3117 | 3563 |
| .102 | 478 | 717 | 956 | 1195 | 1434 | 1913 | 2391 | 2869 | 3347 | 3825 |
| .125 | 586 | 879 | 1172 | 1465 | 1758 | 2344 | 2930 | 3516 | 4102 | 4688 |
| .156 | 731 | 1097 | 1463 | 1828 | 2194 | 2925 | 3656 | 4388 | 5119 | 5850 |
| .188 | 879 | 1318 | 1758 | 2197 | 2637 | 3516 | 4395 | 5273 | 6152 | 7031 |
| .250 | 1172 | 1758 | 2344 | 2930 | 3516 | 4688 | 5859 | 7031 | 8203 | 9375 |

Values above the upper double line are less than single shear.
Values below the lower double line are greater than double shear.
The allowable shearing stress of alum. alloy 17S–T rivets = 30000 lbs. per sq. in.
The allowable bearing stress of 17S–T sheet = 75000 lbs. per sq. in.
The above values for single shear shall be used when the joint is formed with the two parts in direct contact without any intervening space or filler block. When the parts are not in contact, lower values shall be used, depending upon the width of the intervening space. The values shall be decreased 6.7% for each .010 inch, to a minimum of 67% of the above values.

is too thick to be counterpunched easily, it is drill-countersunk and the skin is drawn into it by the rivet head.

Spot welding may be said to compare favorably with riveting in shear, but is definitely inferior in tension. The low strength in tension is to be expected, because a spot weld is in reality comparable to a countersunk rivet as shown at (C) in Fig. 6. A truer comparison in favor of the flush rivet would be with that of the press-countersunk, counterdrawn, or drill-countersunk under-

### Table 11—Bearing and Shear Strength of 24S–T Rivets and Sheet

| Size rivet, in. | 3/32 | 1/8 | 5/32 | 3/16 | 1/4 | 5/16 | 3/8 | 7/16 | 1/2 |
|---|---|---|---|---|---|---|---|---|---|
| Single shear | 242 | 430 | 671 | 966 | 1717 | 2684 | 3868 | 5261 | 6871 |
| Double shear | 484 | 860 | 1342 | 1932 | 3434 | 5368 | 7736 | 10522 | 13742 |

| Thickness of Sheet, In. | Bearing Strength of Sheet, Lbs. (Aluminum Alloy 24S–T) | | | | | | | | |
|---|---|---|---|---|---|---|---|---|---|
| .013 | 110 | 146 | 183 | 219 | 293 | 366 | 439 | 512 | 585 |
| .016 | 135 | 180 | 225 | 270 | 360 | 450 | 540 | 630 | 720 |
| .018 | 152 | 202 | 253 | 304 | 405 | 506 | 608 | 709 | 810 |
| .020 | 168 | 224 | 280 | 336 | 448 | 560 | 672 | 784 | 896 |
| .022 | 186 | 248 | 309 | 371 | 495 | 619 | 743 | 866 | 990 |
| .025 | 211 | 282 | 352 | 423 | 564 | 705 | 846 | 987 | 1128 |
| .028 | 236 | 315 | 394 | 473 | 630 | 788 | 945 | 1103 | 1260 |
| .032 | 270 | 360 | 450 | 540 | 720 | 900 | 1080 | 1260 | 1440 |
| .035 | 295 | 394 | 492 | 591 | 788 | 985 | 1182 | 1379 | 1576 |
| .040 | 337 | 450 | 562 | 675 | 900 | 1125 | 1350 | 1575 | 1800 |
| .049 | 413 | 551 | 689 | 827 | 1103 | 1378 | 1654 | 1929 | 2205 |
| .051 | 430 | 574 | 717 | 861 | 1148 | 1435 | 1722 | 2009 | 2296 |
| .057 | 481 | 642 | 802 | 963 | 1284 | 1605 | 1926 | 2247 | 2568 |
| .064 | 540 | 720 | 900 | 1080 | 1440 | 1800 | 2160 | 2520 | 2880 |
| .065 | 548 | 731 | 914 | 1097 | 1463 | 1828 | 2194 | 2559 | 2925 |
| .072 | 607 | 810 | 1012 | 1215 | 1620 | 2025 | 2430 | 2835 | 3240 |
| .081 | 684 | 912 | 1140 | 1368 | 1824 | 2280 | 2736 | 3192 | 3648 |
| .083 | 700 | 934 | 1167 | 1401 | 1868 | 2334 | 2801 | 3268 | 3735 |
| .091 | 768 | 1024 | 1280 | 1536 | 2048 | 2560 | 3072 | 3584 | 4096 |
| .095 | 801 | 1069 | 1336 | 1603 | 2138 | 2672 | 3206 | 3741 | 4275 |
| .102 | 861 | 1148 | 1435 | 1722 | 2296 | 2870 | 3444 | 4018 | 4592 |
| .125 | 1055 | 1406 | 1758 | 2109 | 2813 | 3516 | 4219 | 4922 | 5625 |
| .156 | 1315 | 1754 | 2192 | 2631 | 3508 | 4385 | 5262 | 6139 | 7016 |
| .188 | 1586 | 2115 | 2644 | 3173 | 4230 | 5288 | 6345 | 7403 | 8460 |
| .250 | 2109 | 2812 | 3515 | 4218 | 5625 | 7030 | 8436 | 9842 | 11250 |

Values above the upper double line are less than single shear.
Values below the lower double line are greater than double shear.
The allowable shearing stress of 24S–T rivets = 35,000 lbs. per sq. in.
The allowable bearing stress of 24S–T sheet = 90,000 lbs. per sq. in.

structure, as shown at (A), (B), or (D), Fig. 6. In defense of the spot weld, however, it may be said that few joints are important in tension, and that spot welding does represent a definite saving both in weight and cost.

A marked advantage of flush rivets over spot welding is that the latter fails sooner in service, due to fatigue cracks which develop under exposure to considerable vibration, as in areas in the wake of the propeller, and high-frequency vibration points.

From the repair angle, riveted sections are much more easily disassembled for replacement of skin, patching, etc. If a spot weld gives way in the field, the only recourse is to repair by riveting, as it would not be possible to have the spot-welding equipment ready at every point where it might be needed. However, in the future there probably will be further development toward

# RIVETS AND RIVETING

**Table 12—Bearing and Shear Strength of 24S–T Rivets and 24S–T Alclad Sheet**

| Size rivet, in. | 3/32 | 1/8 | 5/32 | 3/16 | 1/4 | 5/16 | 3/8 | 7/16 | 1/2 |
|---|---|---|---|---|---|---|---|---|---|
| Single shear | 242 | 430 | 671 | 966 | 1717 | 2684 | 3868 | 5261 | 6871 |
| Double shear | 484 | 860 | 1342 | 1932 | 3434 | 5368 | 7736 | 10522 | 13742 |
| Thickness of Sheet, In. | Bearing Strength of Sheet, Lbs. (Aluminum Alloy 24S–T Alclad) | | | | | | | | |
| .013 | 100 | 133 | 167 | 200 | 267 | 333 | 400 | 466 | 533 |
| .016 | 123 | 164 | 205 | 246 | 328 | 410 | 492 | 574 | 656 |
| .018 | 138 | 185 | 231 | 277 | 369 | 461 | 554 | 646 | 738 |
| .020 | 153 | 204 | 255 | 306 | 408 | 510 | 612 | 714 | 816 |
| .022 | 169 | 226 | 282 | 338 | 451 | 564 | 677 | 789 | 902 |
| .025 | 192 | 256 | 320 | 384 | 512 | 640 | 768 | 896 | 1024 |
| .028 | 215 | 287 | 359 | 431 | 574 | 718 | 861 | 1005 | 1148 |
| .032 | 246 | 328 | 410 | 492 | 656 | 820 | 984 | 1148 | 1312 |
| .035 | 268 | 358 | 447 | 537 | 716 | 895 | 1074 | 1253 | 1432 |
| .040 | 307 | 410 | 512 | 615 | 820 | 1025 | 1230 | 1435 | 1640 |
| .049 | 377 | 502 | 628 | 753 | 1005 | 1256 | 1507 | 1758 | 2009 |
| .051 | 391 | 522 | 652 | 783 | 1044 | 1305 | 1566 | 1827 | 2088 |
| .057 | 438 | 584 | 730 | 876 | 1168 | 1460 | 1752 | 2044 | 2336 |
| .064 | 492 | 656 | 820 | 984 | 1312 | 1640 | 1968 | 2296 | 2624 |
| .065 | 500 | 666 | 833 | 999 | 1333 | 1666 | 1999 | 2332 | 2665 |
| .072 | 553 | 738 | 922 | 1107 | 1476 | 1845 | 2214 | 2583 | 2952 |
| .081 | 622 | 830 | 1037 | 1245 | 1660 | 2075 | 2490 | 2905 | 3320 |
| .083 | 638 | 851 | 1063 | 1276 | 1702 | 2127 | 2552 | 2978 | 3403 |
| .091 | 699 | 932 | 1165 | 1398 | 1864 | 2330 | 2796 | 3262 | 3728 |
| .095 | 730 | 974 | 1217 | 1461 | 1948 | 2434 | 2921 | 3408 | 3895 |
| .102 | 784 | 1046 | 1307 | 1569 | 2092 | 2615 | 3138 | 3661 | 4184 |
| .125 | 961 | 1281 | 1602 | 1922 | 2563 | 3203 | 3844 | 4484 | 5125 |
| .156 | 1198 | 1598 | 1997 | 2397 | 3196 | 3995 | 4794 | 5593 | 6392 |
| .188 | 1445 | 1927 | 2409 | 2891 | 3854 | 4818 | 5781 | 6745 | 7708 |
| .250 | 1921 | 2562 | 3202 | 3843 | 5125 | 6405 | 7686 | 8967 | 10250 |

Values above the upper double line are less than single shear.
Values below the lower double line are greater than double shear.
The allowable shearing stress of 24S–T rivets = 35,000 lbs. per sq. in.
The allowable bearing stress of 24S–T Alclad sheet is 82,000 lbs. per sq. in.

cheaper replacement parts, so that little repair work will be necessary except to isolated areas of structure damaged by accidents.

## QUESTIONS AND ANSWERS

**1. How is a proper edge distance determined?**

*Answer.* For maximum bearing strength, the edge distance measured from the center of the hole in the direction of stressing must not be less than twice the hole diameter. For a 1/8" rivet driven in a No. 30 hole, drill holes are centered at 17/64" or more from the edge.

**2. What is the deciding factor in choosing the size of rivets to be used on a job; that is, under what conditions would a 1/8" rivet be used rather than a 3/32" rivet?**

*Answer.* In aircraft construction and repair, the general practice is to

# 180 AIRCRAFT SHEET METAL WORK

design riveted joints in such a way that the rivet strength and strength of the sheet are approximately the same. The rivet size is usually selected with less shearing strength (see Tables 9 to 13) than the bearing strength of the hole when these values differ. However, the rivet shear value is increased by driving, because the drill used is slightly larger than the rivet shank. The bearing value of the hole in the sheet is also increased by use of the slightly larger drill, as can be seen from Table 10, which gives the bearing strength for holes of various sizes.

If a small rivet is used to hold heavy sheets together, the rivet will shear under load. If a large rivet is used to hold relatively thin sheets together, the rivets would tear out at low rivet stress. Therefore the proper size of rivet must be selected for a given thickness of plate.

*Example.* Choose the size of rivet required for use in a sheet of Alclad 24S–T that is 0.040″ thick.

As we choose rivets with less shear value than the bearing strength of their holes, they are the weakest factor.

The bearing strength for 0.040″ 24S–T Alclad, Table 12, is 307 lbs. for a 3/32″ rivet.

The single shearing strength for rivets is shown in Tables 9, 10, 11, or 12, and is as follows for a 3/32″ rivet:

$$3/32″ \; A17S–T \; \text{rivet} = 173 \text{ lbs.}$$
$$3/32″ \; 17S–T \; \text{rivet} = 207 \text{ lbs.}$$
$$3/32″ \; 24S–T \; \text{rivet} = 242 \text{ lbs.}$$

We see that the shear value of no 3/32″ aluminum alloy rivet will approach the bearing strength of 307 lbs. for 0.040″ 24S–T Alclad sheet. The next size rivet (1/8″) has a shear strength of:

$$1/8″ \; A17S–T \; \text{rivet} = 307 \text{ lbs.}$$
$$1/8″ \; 17S–T \; \text{rivet} = 369 \text{ lbs.}$$
$$1/8″ \; 24S–T \; \text{rivet} = 430 \text{ lbs.}$$

The bearing strength for 0.040″ 24S–T Alclad sheet for a 1/8″ rivet is found to be 410 lbs., which is less than single shear strength for a 1/8″ 24S–T rivet, according to Table 12, therefore that rivet is undesirable. A 1/8″ 17S–T

### Table 13—Shear Strength of Copper and Aluminum Rivets

| Rivet Size, Inch | Area, Sq. In. | Single Shear, Lbs. | | Rivet Size, Inch | Area, Sq. In. | Single Shear, Lbs. | |
|---|---|---|---|---|---|---|---|
| | | Copper | Aluminum | | | Copper | Aluminum |
| 1/16 | .0031 | 89 | 50 | 3/16 | .0276 | 795 | 441 |
| 3/32 | .0069 | 199 | 110 | 1/4 | .0491 | 1414 | 786 |
| 1/8 | .0123 | 353 | 197 | 5/16 | .0767 | 2210 | .... |
| 5/32 | .0192 | 553 | 307 | .... | .... | .... | .... |

The allowable shear stress of commercial copper rivets = 28,800 lbs. per sq. in.
The allowable shear stress of aluminum rivets = 16,000 lbs. per sq. in.

rivet would be chosen, because a check of tables will show that no rivet in 5/32″ size is suitable.

**3. Can A17S–T rivets be used in 17S–T, 24S–T, 24S–T Alclad, or 24S–RT Alclad sheet?**

*Answer.* Yes, providing a rivet is selected of proper size to give comparable shear and bearing strength.

# RIVETS AND RIVETING

### Table 14—Physical Properties of Aluminum Alloys

| Material | | Ultimate Tensile Strength, Lbs. per Sq. In. | °Yield Point, Lbs. per Sq. In. | Bearing Strength, Lbs. per Sq. In. | Shear Strength, Lbs. per Sq. In. | Modulus of Elasticity, Lbs. per Sq. In. |
|---|---|---|---|---|---|---|
| Sheet | 17S–T | 55,000 | 32,000 | 75,000 | 27,000 | 10,000,000 |
| | 24S–T | 62,000 | 38,000 | 90,000 | 30,000 | 10,300,000 |
| | 24S–RT | 65,000 | 50,000 | 94,000 | 31,000 | 10,300,000 |
| Alclad Sheet | 17S–T | 50,000 | 27,000 | 68,000 | 24,000 | 10,000,000 |
| | 24S–T | 56,000 | 37,000 | 82,000 | 27,000 | 10,300,000 |
| | 24S–RT | 58,000 | 46,000 | 85,000 | 28,000 | 10,300,000 |
| Tubing | 17S–T | 55,000 | 40,000† | 75,000 | 27,000 | 10,300,000 |
| | 24S–T | 62,000 | 42,000‡ | 90,000 | 30,000 | 10,300,000 |
| | 24S–RT | 65,000 | 58,000 | 94,000 | 31,000 | 10,300,000 |
| Castings | 195–T4 | 32,000 | | 45,000 | 25,000 | 10,000,000 |
| | 220–T4 | 40,000 | 23,000 | 55,000 | 33,000 | 10,000,000 |
| Bar More than ½" Thick | 17S–T | 50,000§ | 28,000 | 75,000 | 30,000‖ | 10,000,000 |
| Bar Less than ½" Thick | 17S–T | 55,000 | 30,000 | 75,000 | 30,000 | 10,000,000 |
| Small Forgings, Extrusions and Rolled Shapes | 17S–T | 50,000¶ | 30,000 | 75,000 | 30,000°° | 10,300,000 |
| | 24S–T | 57,000¶ | 40,000 | 90,000 | . . . . . . . . | . . . . . . . . |

°Corresponds to unit set of .002 inch per inch. (Tension yield point.)
†Applies only to 17S–T as received. For 17S–O HT 17S–T use 30,000 lbs. per sq. in.
‡Applies only to 24S–T as received. For 24S–O HT 24S–T use 37,000 lbs. per sq. in.
§Parallel to grain only. Perpendicular to grain, values may be smaller, depending on thickness and location in bar.
‖Across grain only.
¶Parallel to grain. For perpendicular to grain it is sometimes necessary to use smaller values depending upon the section shape.
°°Across grain. For parallel to grain it is sometimes necessary to use smaller values depending upon the section shape.

### 4. How close may rivets be spaced?

*Answer.* Ordinarily rivets are spaced three diameters apart (three times the diameter of the rivet) for inside rows, and six times the diameter of the rivet for outside rows, which leaves over 80% of the sheet for tensile strength. See Fig. 21. Each time a 1/8" hole is drilled in a sheet, the sheet loses one-

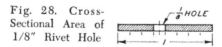

Fig. 28. Cross-Sectional Area of 1/8" Rivet Hole

eighth of its cross-sectional area per inch. This is illustrated by Fig. 28, where the 1/8" hole is one-eighth of the cross-sectional area of the sheet.

# 182    AIRCRAFT SHEET METAL WORK

**5. How are we to place the required number of rivets in a stringer splice?**

*Answer.* The splice must be long enough for the spacing of the rivets required, calculated and shown in Fig. 26.

**6. Is it necessary to form a hat section channel splice exactly like the channel being spliced?**

*Answer.* No. It may be spliced as shown in Fig. 25; that is, it may be built up in sections.

**7. Where stress is unimportant, how are brackets fastened when it is impossible to buck rivets?**

*Answer.* By employing blind rivets or fasteners.

**8. How is the bearing strength found in 0.032", 24S–T Alclad without the use of the tables in this book?**

*Answer.* Cross-sectional area removed for a given rivet size is multiplied by bearing strength of sheet.

*Example.* For a 1/8" rivet in 0.032" gage: $(0.032 \times 1/8) \times 82,000$ p.s.i. $= 328$ pounds. See Table 14.

**9. How many rivets would be necessary for a patch to cover a 2" round hole on the under side of a DC–3 metal wing about 8' from the root and between two stringers?**

*Answer.* Cross-sectional area removed × stress in skin, divided by allowable bearing strength of sheet or shear value of rivet, whichever is smaller.

Assume that: The skin is 0.032" 24S–T Alclad. Skin ultimate stress is 56,000 p.s.i. Diameter of hole to be patched is 2". Cross-sectional area removed is $2 \times 0.032 = 0.064$ sq. in. Load for reinforcement $= 0.064 \times 56,000 = 3,584$ lbs.

If a 1/8" A17S–T rivet is to be used, its single shear strength $= 307$ lbs. Bearing strength for 0.032" 24S–T Alclad $= 328$ lbs. As the shear strength is less than the bearing strength, the rivets are the critical item on which we must base our figures. Then

$$\frac{3584}{307} = 11.67, \text{ or } 12 \text{ rivets per } 2" \text{ width.}$$

This number of rivets would be required on each side of the patch. Any patch, of whatever shape, should be visualized in two halves, or as having two sides. Draw an imaginary line through the center, bisecting the patch across the maximum portion of the hole and at 90° to the line of stress. Generally, the rule of using the rivet pattern at the next joint inboard is followed. In that case, the patch would be square, which is satisfactory, providing there are no reinforcements or doubler plates in the area.

The hole in question is a 2" round hole. If a round patch is used, it should conform in rivet design to the one shown at $H$ in Fig. 4, Chapter 11. There only 8 rivets in 5/32" size were required on each side, or 16 in all. Our question concerns skin of different weight and our calculations showed 12 rivets in 1/8" size necessary on each side, or 24 in all. These may be placed as shown in Fig. 29.

A repair of this kind is considered a major repair and must be inspected by the C.A.A.

**10. Can 24S–T rivets be replaced with A17S–T or 17S–T rivets?**

# RIVETS AND RIVETING

*Answer.* Yes, providing there is enough room to install the size and quantity needed. 24S–T rivets cannot be replaced with A17S–T of the same size.

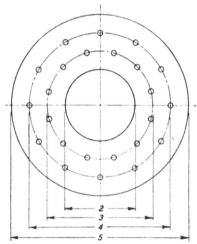

Fig. 29. Rivet Pattern for Round Patch over a 2" Round Hole in 0.032" 24S–T Alclad Using 1/8" A17S–T Rivets

### QUESTIONS

1. After heat treatment, how are rivets preserved for driving?
2. What is the *cupping* in a rivet snap?
3. What is a *rivet snap*?
4. Name the common types of rivets.
5. What sized drill is used for clearance for a 1/8" rivet?
6. At what temperature are 24S rivets heat-treated? 17S rivets?
7. Show three different methods of installing countersunk rivets.
8. How far should the shank of a rivet protrude through the metal for heading?
9. Is the A17S rivet heat-treated?
10. What means is used for holding skin together?
11. What must be removed from between sheets before riveting?
12. What are the characteristics of a drill for drilling through thick sections of aluminum?
13. In riveting on a large sheet, where is the place to start riveting?
14. What is a *dimple*?
15. How do you properly fasten a drill in a chuck?
16. What is meant by *drawing* a rivet?
17. What determines the degrees in a countersink used for installing countersunk rivets?
18. What is the first thing to do when a mistake is made?
19. Tell how to remove a bad rivet.

20. What is a *stop countersink?*
21. When heat-treating rivets, how much time should be allowed from the time when they are removed from the heat until they are quenched in water?
22. How are heads of rivets of different alloys drilled off for removal of the rivet?
23. How would you buck up a rivet in a tube?
24. What kind of head would you put on a rivet that extends through a hollow tube?
25. How could you put a standard head on a rivet that extends through a tube without buckling the rivet?
26. What is a *rivnut?*
27. What is a *Dill Lok-Skru?*
28. What governs the choice of the size of a rivet?
29. Are rivets used extensively in tension?
30. Is a rivet as strong in double shear as it is in single shear?
31. What is the meaning of *bearing value?*
32. How thick should a bucked rivet head be?
33. What is the diameter of a properly bucked rivet head?
34. Is it better to use a large rivet for riveting together thin sheets?
35. What is meant by *edge distance?*
36. What is a blind rivet?

## PROJECT 1

Repeat Project 3, Chapter 1, to learn more of the technique of operating the rivet gun or riveter and practice making properly driven heads by timing your shots.

## PROJECT 2. REMOVING RIVETS BY DRILLING

**Objective**

To teach the proper procedure in removing rivets without damage to original rivet holes or surrounding material.

**Tools and Equipment**

1. Center punch
2. Ball-peen hammer
3. Hand drill and assortment of twist drills
4. "Rivet buster" (specially-ground cold chisel)
5. Rivet punch (pin punch)

**Material**

Material riveted in Project 1

**Procedure**

1. Center punch the exact center of the rivet head which is to be removed.

*Note.* Since you have A17S–T rivets in the patch you may skip procedure No. 1 unless you have substituted other rivets. The identification mark is ideal for starting the drill.

2. Select a twist drill slightly smaller than the shank of the rivet and mount in drill chuck.

# RIVETS AND RIVETING

3. Drill only through the head of the rivet.
4. Remove the rivet head with a rivet buster (a chisel ground mostly on one side and polished on the other). Care must be taken to prevent damage to surrounding material by gouging.

*Note.* If you have no rivet buster, use an ordinary small cold chisel, but place the cutting edge well up on the rivet head, rather than under the head. See Fig. 22.

5. Punch the sheared rivet out with the pin punch.
6. Show your work to the instructor for inspection, and for next operation replace all rivets removed.

### PROJECT 3. DRILLING RIVET HEADS WITH ELECTRIC DRILL TO REMOVE THEM

**Objective**

To develop the ability to drill to the base and to the exact center of a rivet head.

**Tools**

1. Electric hand drill
2. Twist drill, same size as used in drilling clearance hole for rivet
See *Note,* under Procedure 3.

**Procedure**

1. Place point of twist drill on the center of the rivet head.
2. Applying a little pressure, rotate the twist drill to the right by turning the chuck of the drill by hand. When the appearance of shavings indicates that the drill point has bitten into the metal, turn on the switch.
3. Drill to a shallow depth; remove drill, and see if the hole is apparently in the center. If not, return the drill to center by pointing it the way you want it to go, then straightening up until the drilling is in dead center. *Caution:* Never try to force a drill to center by applying pressure sideways, as chances are that the drill will break, run into the work and ruin it.

*Note.* The judgment of an expert mechanic decides whether the rivets in a group are to be drilled before removing heads, center punched before drilling, or drilled by the method just outlined. The different methods follow; their choice should depend upon the skill of the operator.

1. Center punch the rivets before drilling. (A17S rivets have a punch mark as identification, so they need not be center punched.)
2. Drill the head by skill, without center punching.
3. Drill the center of the rivet by aid of a centering guide, Fig. 13.
4. Experts drill the heads off, allowing them to accumulate on the twist drill, which must be cleared of them at intervals. In this method the drill pushes the shank of the rivet out of the hole. This method is applied only to thin skin sections.
5. From heavy gage skin or from a solid structure that is to be replaced, rivets are removed without drilling by use of the rivet buster. This is poor practice on new structure, as where a rivet must be removed because improperly driven. Rivets should be removed from thin skin, or structure that is not being replaced, by drilling the heads and pushing out the shank.

## PROJECT 4. INSTALLING A REINFORCING TYPE PATCH ON STRESSED SKIN

**Objective**

To teach the proper way of riveting a flat patch over a hole punched in the skin, such as might occur in the bottom of a stabilizer if a rock were thrown up by the propeller blast on takeoff.

**Tools and Equipment**

1. Aircraft snips (left and right)
2. Half round file (smooth), 8" size
3. Hand drill
4. Rivet snap to fit gun
5. Riveter
6. Bucking bar, 2-lb.
7. Scale, 6"
8. Combination square
9. Dividers, 4"
10. Hermaphrodite dividers (with pencil)
11. Ball-peen hammer, 1-lb.
12. Cold chisel, 1/2" (rivet buster)

**Material**

1. .025" 24S–T, Al. 6" × 8" (piece to be patched)
2. .032" 24S–T, Al. 4" × 4" (patch material, next heavier gage)
3. Supply of A17S–T rivets, 1/8" × 1/4", brazier head (AN455AD4–4)
4. Supply of 4–48 or 4–40 machine screws

**Procedure**

1. Place the larger of the two metal sheets on a sand bag and give it a heavy blow, as close to the center as possible, with the ball peen of the 1-lb. hammer. Then, to simulate a punctured dent, with the aid of a 1/2" cold chisel, cut through the metal in the deepest part of the dent, backing it up with a block of wood in which a one-inch hole has been drilled. This piece of metal is to represent a section of stressed skin between two stringers, near the tip on the bottom side of a stabilizer on a Douglas DC-3 airplane.

2. Since a crack in this location (near the tip) is not under as much stress as it would be near the root of the stabilizer, it is permissible to cover the hole and dent with a patch; however, the dent must not be cut out because it acts as a reinforcement at that point. With a rat-tail file, smooth the hole, removing all ragged edges so that cracks will not start to travel out under the patch later, as often occurs from vibration.

3. Cut a neat square or rectangular patch from the same kind of material but in the gage next heavier than the piece which is to be patched. Be sure that the patch covers sufficient area so that the two thicknesses of metal are in contact where the rivets are to be driven. All rivet holes must be one-fourth inch out from the crest or edge of the dent. We assume the original skin with the hole punched in it will bear half the required strength and, since the repair is near the tip where the loads are not as heavy as they are toward the root, a single row of rivets will be sufficient.[*]

4. We have learned that the diameter of a rivet must be three times the thickness of the heaviest sheet, and a 3/32" 24S–T rivet with a single shear value of 242 pounds is comparable to the bearing strength of .032" 24S–T of 246 pounds as shown in Table 12.

However, the only rivets we have on hand that are heat treated are A17S–T and, as shown in Table 9, a 3/32" rivet of this alloy in single shear

---

[*] Providing, however, it is approved by the manufacturer's repair specifications and by a Civil Aeronautics Inspector.

# RIVETS AND RIVETING

is only good for 173 pounds. What are we going to do? Would it be good judgment to use a rivet with a shearing strength of 173 pounds in sheets having 246 pounds bearing strength? No. We try to build an airplane like "The One-Hoss Shay."

The next step is to increase the rivet size. Table 9 shows a single shear for a 1/8" rivet at 307 pounds. Table 12 shows 328 pounds bearing strength for a 1/8" hole in .032" 24S–T Alclad. A 1/8" rivet is selected because the rivet shear strength and sheet bearing strength are more nearly equal for this thickness of sheet.

5. Lay out the rivet holes in the patch to dimensions no less than those shown at (B) in Fig. 21 for the outside row: six times the rivet diameter, or, for this project, 3/4" spacings; edge distance two times rivet diameter, or 1/4"; use 17/64" edge distance to allow for oversize in the No. 30 drill. Center punch and drill holes.

6. Clamp pieces of material in vise, bolt on patch with 4–40 machine screws, drilling one hole at a time until two or more holes are bolted, then drill remainder of holes and rivet.

7. Present for inspection.

## PROJECT 5

Riveting a patch over a small hole on an engine nacelle fillet, resulting when a mechanic's automatic screw driver slipped off a screw due to use of a dull worn-out bit in screw driver.

### Objective

1. To teach the proper way to rivet and patch a small hole in a non-structural part.

2. To develop the student's technique in riveting by simulating a mechanic's movements in the student.

### Tools and Equipment

Same as used in Project 3.

### Material

1. Two pieces, each 4" × 4", .025" 17S–T
2. Supply of 3/32" × 3/16" A17S–T rivets

### Procedure

1. Punch a 1/2" hole in one sheet of dural.

2. Run both pieces of dural through the rolls, shaping them to a 6" radius (approx.) to simulate a curved fillet.

3. Cut patch from the piece of dural that has no hole in it, allowing proper amount of material for edge distance in all directions.

**Question: How large should the patch be?**

*Answer.* Since weight is always to be considered in airplanes, you do not want the patch larger than necessary, yet it should be large enough to cover the hole and have proper edge distance from hole to rivet and from rivet to edge of patch. Try laying out a square 1⅛" × 1⅛" and inscribe a half-inch circle in the center. You will find this size patch is large enough and will require only four rivets.

4. File 1/32" radii on corners, install, and submit work to instructor for inspection and comments.

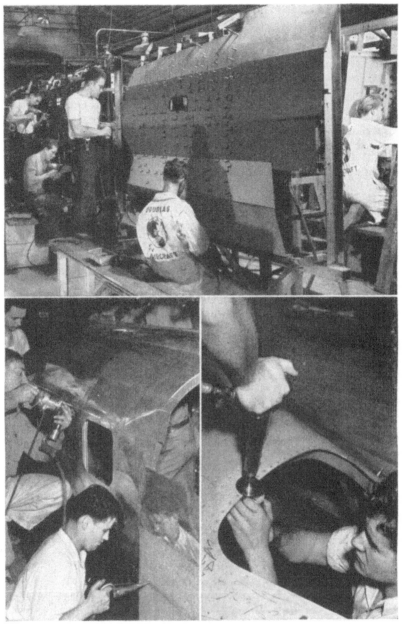

Above. These workmen are performing a skinning operation on the fuselage section of an attack bomber.

*Courtesy of Douglas Aircraft Co., Inc., Santa Monica, Cal.*

Lower Left. Workers riveting fuselage. Note the bucker inside. Lower Right. Installation of screws and flush riveting doubler plate surrounding hatch of fuselage.

*Courtesy of Lockheed Aircraft Corp., Burbank, Cal.*

*Chapter VIII*

# Skin Fitting and General Fabrication

**SKIN FITTING.** The operation of applying metal plates to wings, control surfaces, and fuselage shells of airplanes takes place in the skin fitting department, usually called the skinning department. Thickness of skin varies from 0.012" in nonstressed areas in smaller planes to 0.125" (1/8") for highly stressed areas in larger craft.

In factories where wings are of multicellular construction they are made in sections, and the skin is a part of each section, being riveted to the cellular structure. When these sections are assembled, the wing is complete. Wing structures of the rib and spar type are built on vertical jigs which are wheeled to the skinning department, except in the case of very large wings, when the skin is fitted to the frame on a stationary jig.

**Grain of Sheet Metal.** The grain in sheet metal runs in the direction in which the sheet was rolled at the factory. The sheet stock of various aluminum alloys, steel, and other materials has the grain running lengthwise of the sheet. Whenever practical, the skin is cut from the sheets so that the grain runs laterally on the wing; that is, with the span of the wing. The skin of the fuselage is placed so that the grain runs fore and aft. This practice is followed to provide the greatest strength where needed. For example, values secured on rolled stock in the direction of rolling are referred to as longitudinal or *with grain* properties, and these usually are noticeably greater than those secured in a direction at right angles to the direction of rolling, or the transverse or *cross grain* properties.

**Cutting and Blanking.** Cutting and blanking of the skin is one of the first operations in skin fitting. It involves the cutting, from rectangular sheets of metal, of the sizes and shapes called for. The sizes of sections are taken from blueprints which show where to make all joints and which give the rivet spacings.

**Fitting Skin Sections.** On large wing and fuselage frames where the profiles are of large radii, the skin is laid on by springing flat sheets on to the frame of the structure and riveting by the divisional method explained in the preceding chapter. Where the curvature is considerable, the sheet is held in place with shock cords or straps until the bolts or Cleco sheet holders are applied; these hold it in place until riveted.

On small sized wings and fuselages, skin fitting around small radii must be preformed; that is, the sheet is formed in a hydraulic press, power brake, or drop hammer before it is applied to the frame. When the skin is preformed, skinning a wing usually begins by laying on the nose (leading edge). However, each airplane plant has developed its own special techniques of applying skin, using whatever is best for each particular plane or job.

The sheets of skin must be applied to the frame and fastened by whatever mechanical means gives greatest accessibility to the riveting crews, especially on those wings where much of the bucking up is done by a man working from the outside and reaching in with his bucking tool. In wing skinning, the sheets are usually placed alternately on top and bottom until all skinning is finished except one strip. This strip is riveted at both edges simultaneously as the strip is gradually unrolled or peeled back.

The wing jig is built so as to allow skinning and riveting crews to work on both sides at one time. Fuselage jigs are built of heavy angles and channel iron that hold the fuselage frame, circumferential rings or bulkheads, and longitudinal stringers in place.

A considerable change in room temperature and consequent expansion of the steel jig may throw rivet holes out of alignment on large wings or fuselages where long runs of holes have been drilled previous to riveting.

Seaplane hulls are built upside down for convenience in applying the skin or shell plating to bulkheads and stiffeners. Here, also, the sheets are applied as cut and are sprung on, except in the case of sharp bends or small radii, when they are preformed.

When using heat-treated rivets with brazier heads that have been treated against corrosion, a small piece of masking tape stuck to the cupping of the rivet snap will prevent marring the rivet head.

All sections of skin must be marked to correspond with the

# SKIN FITTING AND FABRICATION

part numbers on the blueprint. A protective coat of primer is applied so they will not easily be scratched in handling. Suitable stock bins are provided and the sections are stored resting on edge, as it is not good practice to stock sheet metal lying flat.

**Joining Sheets.** The skin is joined together by lap joints, each joint being lapped in such a way that it is not open toward the slip stream. In other words, the forward sheet is always lapped over the rear sheet, and the upper sheet over the lower sheet.

In areas where no particular stresses are present, longitudinal seams are single-row riveted, but all circumferential joints are riveted as outlined in the chapter on rivets and riveting. All seams must be waterproofed with zinc oxide primer paste, or its equivalent. Joints of skin on wings, floats, or hull bottoms of seaplanes must be made thoroughly watertight by use of one of the various compounds or cements. Gasoline and oil tanks integral with the structure require some type of sealing compound that is resistant to gasoline and oil. Good watertight or gastight seams are dependent upon workmanship, proper sealing material, and low rivet stresses. Material of heavier gage is used for skin where seams must hold pressure, because bulging between rivets will not permit a tight joint where thin gages are used.

*Neoprene,*\* a synthetic rubber, is used for sealing gas and oil tanks. It is available in sheets or strips as well as in the form of cement. A thin layer of Neoprene cement is used to hold the strip to the plates, which are afterward riveted at close pitch. Neoprene is also used in thick liquid form. This is applied to the area of the joint and allowed to air-dry before riveting; a thin second coating is applied after the first coat dries. The rivets are dipped into the cement before driving.

On boat hulls, fabrics impregnated with sealing compound are widely used. Some of the compounds used are: synthetic resin base, spar varnish, bakelite varnishes, marine glue, bituminous paint, and synthetic rubber compounds.

**ABUTTING PARTS.** Regardless of how close abutting parts appear on a drawing, they need not fit tightly or precisely unless a special fit is specified. This means that in fitting parts it is not

---

\*Neoprene synthetic rubber, E. I. du Pont de Nemours & Co., Inc., Rubber Chemicals Division, Wilmington, Delaware.

necessary to do an extreme amount of filing and precision work. Draftsmen and designers usually are required to allow for about $3/32''$ variation on all such parts. Sometimes as high as $3/16''$ is allowed in fitting larger parts that are difficult to assemble precisely, such as major units of wing or body work.

The clearance of parts such as circumferential or diagonal stiffener ends at longitudinal stringers, or longerons, longeron ends at splices or bulkheads, rib chords, or spars, may be allowed to fall short in length up to .050" or more. This is specified on the face of the blueprint by dimensions or notes such as: ABUTTING PARTS MAY CLEAR BY 0 to .060. There are some places, in fact, where there must be a clearance to prevent wear or chafing. Clearance of at least .020" to .030" usually is required between longerons and stiffeners and the bulkheads or circumferentials through which they pass. The drawing will specify these instances, as well as indicate when parts are to be butt fitted; otherwise the mechanic will make only reasonable fits, not necessarily tight or uniform, even on machined parts.

**OFFSETS OR JOGGLES.** As a general rule, joggles are avoided by designers and are omitted where the offset is .032" or less. This holds true for wing and fuselage skin, as well as for all internal structures such as bulkheads, except in specific places to prevent dimpling of the skin when being riveted. The offset on joggled extruded sections, or formed sheet sections, should be made through a distance of not less than six times the depth of the offset, unless otherwise specified.

**RIVET HOLES.** Punching of holes is permitted on Army and commercial work except where drawings specify: DRILL. Holes which are to be drilled or reamed may be either punched or drilled undersize.

On Navy work where punched holes are used for structural bolts and pins, they must be punched $1/32''$ under size and reamed to fit, and a mechanic will not go wrong by following this procedure on Army and commercial work, also.

Providing that tools for punching are kept sharp, it is permissible to punch holes, without drilling or reaming, for structural rivets in sheets under .093" in thickness. Unless there is an order to do so, no holes are to be plugged with rivets where they have been

## SKIN FITTING AND FABRICATION

drilled through a mistake of the mechanic, or where they were previously drilled in certain parts and are no longer required because of design changes.

**INSULATION OF DISSIMILAR METALS.** On flying boats and Navy airplanes, when steel bolts and nuts are used in conjunction with aluminum alloy, they must be insulated by dipping in zinc chromate primer, and installed while the primer is still wet. Where steel, copper, and aluminum alloy surfaces are in contact, each surface must be insulated by receiving one additional coat of zinc chromate primer, or the assembly must be filled with zinc chromate plastic compound in such a way that it will be forced out at all boundaries; the excess must be removed in such a way as to leave a complete fillet all around the boundary. This insulation process must be carried out for stainless steels, and even for self-tapping screws when in contact with aluminum alloys, because of the corrosive action between them, especially over or near salt water.

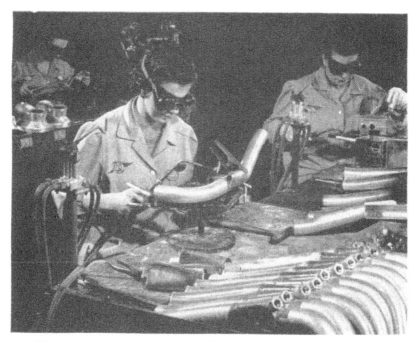

Women shown here are oxyacetylene welders and are working on P-39 Army Airacobra parts. Note that all are protected by goggles.
*Courtesy of Bell Aircraft Corp., Buffalo, N. Y.*

Upper Left. This workman is operating a Taylor-Winfield Spot Welding machine.
Upper Right. These men are using oxyacetylene flame to weld parts. Note gang welding jig.
*Courtesy of Vought-Sikorsky Aircraft Division, United Aircraft Corp., East Hartford, Conn.*

Below. These experts are arc welding the upper truss of a Mainliner landing gear.
*Courtesy of United Air Lines, Cheyenne, Wyo.*

## CHAPTER IX
# Soldering, Brazing, and Welding

### PART I

To cover the all-important subject of welding in one chapter is of course impossible. Entire and very good books have been published in recent years on the newer welding applications and techniques.* However, since this book applies to aviation sheet metal work as a whole, it would not be complete without giving the student at least a brief outline of practical methods of soldering, brazing, and welding.

Practical experience is the quickest approach to soldering, yet the student can do a good job, once he understands the principles involved; these are simple and few. Two methods are used for soldering: one is by use of the soldering copper (often called the soldering iron) or electrical soldering iron, for soft soldering; the other is by use of a welding torch for silver soldering (hard soldering) and brazing.

Brazing is more difficult than soft soldering; it is similar to silver soldering in that the torch is used. Furnace brazing is done on certain kinds of production jobs.

Welding usually is classed as a trade by itself; as applied to aircraft, it must include a study of the chemical and physical properties of metals, besides the actual welding operation. Yet every sheet metal craftsman in aviation should be familiar with welding practices and techniques. In this chapter, only oxyacetylene welding technique will be considered. Actual practice at welding is the only way to become an expert. Also, in welding, as in many other trades, constant progress is being made in laboratories and in the field, and materials, equipment, and technique are being modified rapidly. Any text, therefore, can only hope to be an outline guide, and the sheet metal craftsman must keep his eyes and mind open to improvements and even quite drastic changes.

---
* See *Electric Welding* and *Oxyacetylene Welding*, American Technical Society, Chicago, Ill.

The various methods of welding might be classified in four groups, as follows:
1. Gas
    a) oxyacetylene
    b) oxyhydrogen
2. Electric
    a) arc welding
    b) carbon arc
    c) resistance—spot, butt, flash, shot, etc.
3. Gas and Electric
    a) atomic
4. Chemical
    a) thermit

The last two methods are only occasionally used in aircraft work.

**SOFT SOLDERING.** Soldering in this form is the simple process of uniting two pieces of metal by means of any alloy having a relatively low melting point. Common soft solder is composed of lead and tin in varying proportions. Table 1 gives the percentages of various lead-tin solders with their melting points.

Table 1—Composition and Melting Point of Soft Solder

| Percentage of Lead | Percentage of Tin | Melting Point in Degrees Fahrenheit |
|---|---|---|
| 100 | 0 | 619 |
| 90 | 10 | 563 |
| 80 | 20 | 529 |
| 70 | 30 | 504 |
| 60 | 40 | 464 |
| 50 | 50 | 428 |
| 40 | 60 | 374 |
| 30 | 70 | 365 |
| 20 | 80 | 392 |
| 10 | 90 | 421 |
| 0 | 100 | 450 |

The solder of lead-tin-bismuth, which has a much lower melting point, is seldom used in aircraft. For soldering the Navy wrapped and soldered splice, the solder should be half-and-half tin and lead to conform to Navy specifications. The melting point of this solder varies from 400 to 450° F., and the tensile strength is approximately 5,700 pounds per square inch.

Two methods are widely used in soft soldering. The first is known as *sweating:* a thin coating of solder is applied to the two surfaces to be joined; they are then brought together, usually under pressure, and the whole assembly is heated until the solder is thoroughly melted. The second method involves the flowing in of solder between the metals to be joined, by means of a heated soldering copper or an electrical soldering iron, the latter being more popular in the aircraft industry.

Of first importance in soldering equipment are the shape and condition of the copper point, which should be filed sharp, with clean square edges. The point must be well tinned; that is, coated with solder so that it will pick up the metal and carry it to the work.

Next in importance is having the iron hot enough to heat the part to be soldered; overheating the copper point must be avoided also. Turn your iron off before the tinned point takes on a dull appearance. If the point is overheated, the surface corrodes, which will necessitate filing the point and retinning.

There are many methods of tinning, but in aircraft common rosin, one of the oldest fluxes, is used for tinning as well as for a flux. An oxidized metal surface cannot be soldered, and rosin used as a flux retards oxidation until the solder has taken hold.

The choice of flux will depend on the type of material being soldered; the rules and regulations of the Civil Aeronautics Manual, and Army and Navy specifications must be observed. The object of the flux is to maintain the chemical cleanliness of the metal while the solder is being heated and before it adheres.

For illustration, clean a piece of sheet copper by polishing with fine emery cloth or with a rag buffing wheel. Place the copper over a gas flame and it will immediately dull; this is caused by oxidation. Now drop a small piece of rosin on the dulled surface and watch the original brightness of the copper appear where the rosin melts. The rosin, used as a flux, has broken up the oxide on the copper sheet so that solder will adhere. Table 2 gives soft soldering fluxes and solders.

Solder is sold in bar and wire form, the latter being more generally used in aircraft. Wire solder comes in solid, acid core, and rosin core. Acid core is used to some extent in aircraft construction, especially by sheet metal craftsmen. Acid core solder is defi-

### Table 2—Fluxes and Solders for Soft Soldering

| Metal to be Soldered | Flux | Soldering Mixture | | |
|---|---|---|---|---|
| | | Lead | Tin | Bismuth |
| Brass | °Zinc chloride, rosin or chloride of ammonia .................... | 34 | 66 | 0 |
| Copper | Zinc chloride, rosin or chloride of ammonia .................... | 40 | 60 | 0 |
| Gold | Zinc chloride .................. | 33 | 67 | 0 |
| Lead | Rosin or Tallow ................ | 67 | 33 | 0 |
| Pewter | Gallipoli oil .................... | 25 | 25 | 50 |
| Silver | Zinc chloride .................. | 33 | 67 | 0 |
| Steel or Iron | Chloride of ammonia ............ | 50 | 50 | 0 |
| Steel (tinned) | Zinc chloride or rosin ........... | 36 | 64 | 0 |
| Steel (galv.) | Hydrochloric acid (muriatic) ...... | 42 | 58 | 0 |
| Zinc | Hydrochloric acid .............. | 25 | 25 | 50 |

° To make zinc chloride, drop a few pieces of scrap zinc into a wide-mouthed jar of hydrochloric (muriatic) acid. The acid will fume and boil as the zinc seems to disintegrate. When it will absorb no more zinc, and the fuming has subsided, it is ready for use.

nitely NOT to be used for soldering electrical wire, nonflexible and flexible minor control cables, or any assembly or repair work on an airplane. Also, a soldering iron that has been used with acid core solder should not be used with rosin core solder in airplane work unless it has been thoroughly cleaned, filed, and retinned. The acid will corrode electrical wires, causing current loss or complete disconnection, and where used on steel it will cause oxidation (rust). A solder flux containing stearic acid (there shall be no *mineral* acid present) and rosin, in proportions of 25 to 50% stearic acid and 75 to 50% rosin, is used in soldering steel cable, etc., on airplanes. Rosin core wire is generally used.

The soldering of aluminum is possible but not practical, and very little aluminum soldering is done. There is no way to determine the strength of a soldered aluminum joint since, up to this writing, all methods require scratching the solder into the aluminum surface with a wire brush, because of the rapid oxidation of this metal. This procedure makes it almost impossible to know how much of the solder has adhered to the surface.

**LEAD BURNING.** Lead burning, which requires more skill than soldering, consists in melting the lead, which has been scraped bright at the joint, and causing the metal to flow together and join without the aid of solder or filler rod. Usually this is done with a lead burning torch. The flame must be relatively small, sharp-pointed, and intense. Lead burning is practical in making linings for tanks to hold chemical solutions, in plating equipment, battery work, etc.

**BRAZING (HARD SOLDERING).** Brazing is done where joints must withstand considerable mechanical strain. Silver soldering, or more correctly, brazing with a silver alloy, gives one of the strongest joints that can be obtained short of bronze brazing or welding.

Brazing with silver is becoming very important in aircraft construction, since by this method the joining of metals may be accomplished at so low a temperature that alloy steel is not impaired. Its use with steel tubing fabricated with sleeve joints is especially desirable. The silver solder for this purpose contains 50% silver, 15.5% copper, 16.5% zinc, and 18% cadmium. Cadmium is the alloying agent essential in producing the high fluidity and the low melting point of the solder.

The brazing operation itself is similar to soft soldering in that the hard solder or spelters (copper and zinc) flow, as in soft soldering, around and between the work, but the torch is used in place of the soldering copper or iron. While the range of heat for melting soft solder is from below 400° F. to above 600° F., brazing is accomplished with heats of 790° F. to 1980° F., the melting points of zinc and copper respectively. Silver solders are available with melting points within the range of brazing alloys or spelters.

Table 3 shows the melting point of copper and zinc alloys.

Generally, brazing is accomplished where the work is brought to a red heat at the joint being brazed. The parts to be joined must be thoroughly clean, and as they are brought up to red heat, borax in powdered form is sprinkled on them to act as a flux. The flux melts and covers the surface; the brazing spelter is melted from the rod by the torch and the molten drop is applied to the joint, where it will run into the smallest crevices.

Next in importance to clean surfaces in brazing or silver soldering is to make the space between the surfaces of the work as small as possible. The thinner the film of brass or silver solder uniting the

### Table 3—Melting Point of Copper and Zinc Alloys

| Percentage of Copper | Percentage of Zinc | Melting Point in Degrees Fahrenheit |
|---|---|---|
| 100 | 0 | 1983 |
| 90 | 10 | 1904 |
| 80 | 20 | 1823 |
| 70 | 30 | 1706 |
| 60 | 40 | 1652 |
| 50 | 50 | 1616 |
| 40 | 60 | 1508 |
| 30 | 70 | 1436 |
| 20 | 80 | 1292 |
| 10 | 90 | 1076 |
| 0 | 100 | 786 |

surfaces, the better and stronger the work. Either brass or silver alloy when properly melted is very penetrating, and will seek out and fill the thinnest crevices.

Silver solder is used in preference to brass in places where the work is subject to severe shocks or vibration, as in gas, oil, and instrument lines of copper or brass which are connected to an engine. With either brazing or silver soldering, it is considered wasteful to build up fillets for strength, but in some cases it is practiced for appearance, to make a neat job. In either case, the parts are usually cadmium plated after the flux and scale have been removed with a wire brush, or by sandblasting or pickling in a weak solution of sulphuric acid and water, about 5 to 10%.

A new joining process, also termed *brazing*, has been applied in airplane production with some of the aluminum alloys. The process differs from welding in that only sufficient heat is applied to flow filler material into the joint, with little or no melting of the parent part. Afterward, the brazing heat may be applied to a batch of parts in a furnace, or by dipping in molten flux, or with a torch in a manner resembling torch welding.

The filler material is an aluminum base alloy, of low melting point, which comes in the form of wire. It is used with a special flux that permits the filler, on melting, to wet the surface of the joint. The wire filler is wrapped around or laid along the edges of the joint, see Fig. 1, after the flux has been used. The assemblies are then run into the brazing furnace. The fillets formed have a

# SOLDERING, BRAZING, WELDING

neat, meniscus shape, with a color approximating that of the parent metal.

This method of brazing for use on aluminum alloys has several advantages over torch and arc welding; it can be applied to thinner material, costs are lower, and the surface appearance is usually better. Since the brazing metals are all aluminum alloys, the difference in electrolytic potential between the joint and the main body of the work is relatively small. Service experience over the

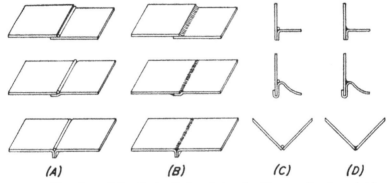

Fig. 1. How Wire Filler Is Applied to Joints for Furnace Brazing of Aluminum; (A) and (C) Before Brazing, (B) and (D) After Brazing

months in which the process has been in use indicates a resistance to corrosion more nearly comparable to welded than to soldered joints. In strength the joint is strictly comparable to a welded joint.

**WELDING.** *Welding* is the joining together of two pieces of similar metal by heating the surfaces to be joined and allowing them to fuse together, with or without the addition of molten metal of like character, or by hammering and compressing, with or without previous softening by heat.

*Welding* as defined by the American Welding Society is: "The localized intimate union of metal parts in the plastic, or plastic and molten states with the application of mechanical pressure or blows, and in the molten states without the application of mechanical pressure or blows."

**OXYACETYLENE WELDING.** The oxyacetylene process is still considered the most flexible type, and, in general, the one most suitable for repair work on aircraft structural elements. Unless the repair agency holds an approval for other methods of welding (electric arc or spot welding) in accordance with *CAR 04.4012* as interpreted in *CAM 04.4012*, or unless a particular application of a different welding process is specifically approved by the Civil Aeronautics Administration, only oxyacetylene torch welding will be considered generally acceptable.

In the aircraft industry oxyacetylene welding is confined to the welding of steel tubing in the motor mounts, landing gear, fuselages of some planes, fixtures, aluminum gas tanks, exhaust stacks, collector rings of stainless steel, and steel fittings. Welded joints are strong, and their use eliminates many of the gussets, rivets, and fittings that must otherwise be applied in aluminum alloy structures and which cannot be welded successfully and yet maintain their ultimate strength.

Some requirements for successful oxyacetylene welding are: selection of the proper type welding rod, the proper size tips for the torch, a torch capable of complete control in respect to size of flame, and the use of the right proportion of gases in the mixture for the given thickness of metal being welded.

The equipment necessary for oxyacetylene welding, in addition to a set of sheet metal mechanic's tools and shop facilities, are:

Oxygen tank (called bottle or cylinder)
Acetylene pressure generator or tank (cylinder)
Welding torch, with tips, graduated sizes
Cutting torch, with tips, graduated sizes
Oxygen and acetylene hose of proper length
Acetylene regulators with pressure indicators
Oxygen regulators with high and low-pressure indicators
Colored goggles
Tank wrenches, hose connections and gas lighter.

Acetylene is a hydrocarbon gas obtained by decomposition of calcium carbide and water, and having a flame temperature, with oxygen, of 6300° F. The water liberates the acetylene gas, leaving a residue of slaked lime. The gas is colorless, has a rather agreeable odor, and, compared with other gases is nearly harmless.

Oxygen is one of the most abundant elements, forming about one-fifth of air, by volume, 88.88% of water, and about 45% of the rocks of the earth's crust, chiefly as silicates. Each part of acetylene

---

* CAR stands for Civil Air Regulations; CAM stands for Civil Aeronautics Manual.

# SOLDERING, BRAZING, WELDING

requires two and one-half parts of oxygen for complete combustion. One part oxygen should be supplied through the torch, and one and one-half parts from the atmosphere surrounding the flame. This reaction deoxidizes the air coming in contact with the metal that is being melted by the oxyacetylene flame, thus forming the invisible blanket of protection to the molten metal so vital in welding. No other fuel gas approaches the absorptive power of acetylene, which accounts for its wide use in welding. To maintain this invisible protection, the torch must always be directed upon the work; never remove the flame from the molten metal.

The first consideration in welding is the protection of your eyes. *Never start welding without first adjusting colored goggles.*

The second consideration is to secure the correct flame adjustment, see Fig. 2. (A), (B), (C), and (D) are oxyacetylene flames, and (E), (F), and (G) are oxyhydrogen flames.

Third, you must select the proper type and size of filler rod for the weld, see Fig. 3. (Rods are alloyed for special jobs.)

Fourth, comes preparation of the joint to be welded—such procedures as cleaning, beveling edges, fitting, etc.

Fifth, you must have some knowledge of the properties of the metal to be welded; i. e., you must know whether you need to use a flux, the kind of flux, type of filler rod, and whether a carbonizing, oxidizing, or neutral torch flame is needed.

Sixth, you must have some mechanical knowledge of how to allow and compensate for the expansion or contraction of the metals, etc.

**Welding Torch.** Welding and cutting torches are tools which, used with the proper mixture of gases, develop a tremendous power and can be applied literally to melt together or cut apart huge masses of metal. The torch has a two-fold purpose: first, that of mixing the acetylene and oxygen or hydrogen and oxygen, then carrying the correct proportions to the tip to feed the flame; second, as a tool for directing the flame on the weld. The tips are interchangeable and come in different sizes to supply the correct amount of heat needed for any weld. The choice of tip size depends upon the gage of metal being welded. Table 4 shows the approximate size of hole in the tip for welding certain thicknesses of metal, and the size of drill to use in cleaning tips. See Fig. 3 and Table 5.

## AIRCRAFT SHEET METAL WORK

**Table 4—Welding-Torch Tips, Sizes, and Recommended Drills for Cleaning**

| Thickness of Steel | | Diameter of Hole in Tip | Proper Drill Size for Cleaning |
|---|---|---|---|
| Gage | Inches | | |
| 28 to 22 | .015 to .031 | .026 | 71 |
| 22 to 16 | .031 to .0625 | .031 | 68 |
| 16 to 11 | .0625 to .125 | .037 | 63 |
| 11 to 7 | .125 to .188 | .042 | 58 |
| 7 to 3 | .188 to .250 | .055 | 54 |
| 3 to 000 | .250 to .375 | .067 | 51 |

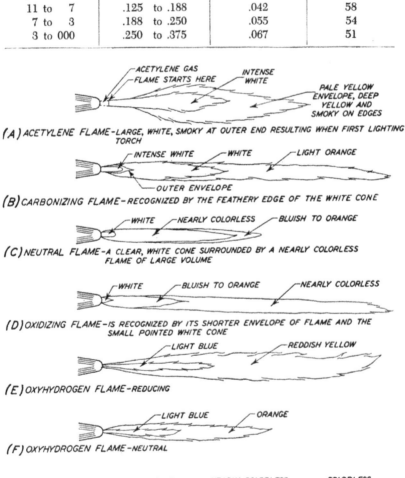

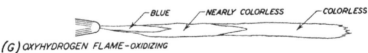

Fig. 2. Structure of Typical Welding Flames. (A), (B), (C), and (D) Secured by Adjusting Oxyacetylene Flame; (E), (F), and (G) Secured by Adjusting Oxyhydrogen Welding-Torch Flame

# SOLDERING, BRAZING, WELDING

**Table 5—Tip Sizes, Size of Hole, and Size of Drill Recommended for Cleaning***

| \multicolumn{3}{c|}{Smith No. 2 Torch} | \multicolumn{3}{c|}{Prest-O-Weld W-109 Torch} | \multicolumn{3}{c}{Oxweld} |

| Smith No. 2 Torch | | | Prest-O-Weld W-109 Torch | | | Oxweld | | |
|---|---|---|---|---|---|---|---|---|
| Tip No. | Hole Size, In. | Cleaning Drill No. | Tip No. | Hole Size | Cleaning Drill No. | Tip No. | Hole Size, In. | Cleaning Drill No. |
| 20 | .0225 | 74 | 1 | .021 | 75 | 1 | .0225 | 74 |
| 21 | .0255 | 72 | 2 | .031 | 68 | 2 | .031 | 68 |
| 22 | .031 | 68 | 3 | .040 | 60 | 3 | .039 | 61 |
| 23 | .0345 | 65 | 4 | .0465 | 56 | 4 | .0465 | 56 |
| 24 | .0375 | 63 | 5 | .0595 | 53 | 5 | .055 | 54 |
| 25 | .0415 | 58 | 6 | .070 | 50 | 6 | .0635 | 52 |
| 26 | .046 | 56 | 7 | .0785 | 47 | 7 | .070 | 50 |
| 27 | .055 | 54 | 8 | .086 | 44 | .. | ..... | .. |
| 28 | .0625 | 52 | .. | ..... | .. | .. | ..... | .. |
| 29 | .0695 | 50 | .. | ..... | .. | .. | ..... | .. |

* Sizes of tips made by various manufacturers are not standardized. The above table is merely typical of tip sizes, diameter of tip orifices, and recommended size of cleaning drills for a few among the many acceptable brands.

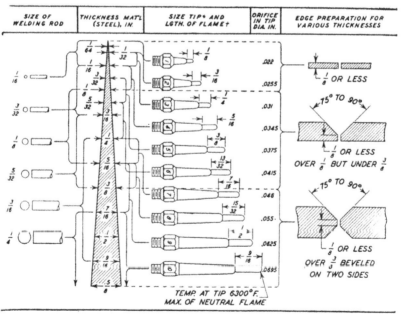

* Actually the size of the tip depends entirely upon the skill of the welder. This chart was compiled to help the beginner choose filler rods and tips of the proper size for various thicknesses of metals.
† Usual length of the inner flame when the torch is adjusted for a neutral flame.

Fig. 3. Approximate Size of Tip Orifice and Welding Rod, Edge Preparation, and Usual Length of Inner Cone of Neutral Flame for Welding Various Thicknesses of Steel

**Flame Control.** Securing the correct flame adjustment is most important in welding. The neutral flame, (*C*) of Fig. 2, ordinarily is the best for steel. The neutral flame must be maintained throughout the welding operation because the slightest deviation in gas flow will introduce one of the gases into the molten metal. A clear, well-defined white cone indicates the correct flame. Welds made with it will be thoroughly fused and free from burned metal or hard spots. If an excess of oxygen is used, see oxidizing flame (*D*) in Fig. 2, too much heat will be generated, resulting in burned metal. The oxidizing flame can be recognized readily by its shorter envelope of flame and the small pointed white cone; its use is limited to sheet brass, cast brass, and certain bronze welding. On the other hand, if too little oxygen, or an excess of acetylene, is used in the mixture, a carbonizing (reducing) flame will result, see (*B*) of Fig. 2. This flame would add carbon to the steel, or in other words, carbonize it. In the handle of the torch are two needle valves for adjusting the amount of gas flow to the mixing chamber and thence to the torch tip. Regulators on the storage tanks serve to control the working pressure of both gases.

**Welding Rods.** Choice of size of welding rods depends on the thickness of metal being welded. The composition of the rod depends on the chemical composition of the metal being welded. The general rule is to select rod with a diameter approximately equal to the thickness of the material being welded, see Fig. 3. Commercial sizes are: $\frac{1}{16}$, $\frac{3}{32}$, $\frac{1}{8}$, $\frac{5}{32}$, $\frac{3}{16}$, $\frac{1}{4}$, $\frac{5}{16}$, and $\frac{3}{8}''$ Dia.

**Preheating.** The procedure of preheating parts before welding may be followed for three reasons: First, it counteracts or compensates for stresses (internal strains) and warpage caused by the welding operation. Second, a saving in oxygen and acetylene results when natural gas, oil, or charcoal is used to heat some part that is too large for heating by torch. Third, compensation in preheating for the welding operation in wheels, gears, sprockets, etc., may prevent breakage from contraction through cooling in areas close to the weld or in spokes, rim, or hub.

Preheating is employed usually on the larger castings of steel, bronze, aluminum, and magnesium alloys, and may in some cases be followed by normalizing, which helps to remove all strains and improve the quality of the weld.

# SOLDERING, BRAZING, WELDING

Preheating may be applied with a torch for small parts; a furnace may be used for larger parts. Another method is to build bricks around the part and cover with asbestos sheets which are left around the major portion, only the place to be welded being uncovered. Slow cooling upon completion of the welding is essential.

**WELDING ALUMINUM.** Due to the low melting point of aluminum and the fact that it has no color range under the torch flame, it is often considered difficult to weld. Actually, aluminum is one of the easiest of all metals to weld, and the procedure is carried on throughout the airplane, in fabricating such structures as the aluminum tanks which carry oil, gasoline, hot and cold water, and propeller de-icing fluid. Engine sumps, aluminum tubing used for conveying air under pressure to the wing de-icers, cabin furnishings, and food boxes, all are fabricated by aluminum welding.

Because of the oxide film which forms on an exposed surface, it is necessary to use a flux in torch or arc welding of aluminum. For arc welding, a flux-coated rod is used to advantage. When welding the nonheat-treatable wrought alloys by either of these processes, a welding rod of 2S, or of the same composition as the alloy being welded, is often used. However, an aluminum alloy rod containing 5% of silicon is more readily handled and gives better results in complicated welds. This latter rod is recommended for most applications in the welding of heat-treatable wrought alloys.

Butt, lap, and fillet joints are made in aluminum by torch welding, using either the oxyhydrogen or the oxyacetylene flame, as easily as corresponding joints in steel.

There are some limitations to the applications of arc welding; however, butt joints are readily made on material thicker than about $5/64''$. This process has the advantage of greater speed and less distortion of the part, as compared with torch welding; also, the effect on structure and temper of the parent metal does not extend so far.

**Design of Welded Joints.** A few general considerations should be taken into account in designing parts containing welded joints. Strain-hardened alloys, after welding, are annealed for a short distance from the weld. Consequently, the design stresses for annealed alloy should be applied. The metal in the weld has a cast structure of about the same strength as the annealed metal, but it has less ductility. If the weld head is left as welded, the joint is usually stronger than the adjacent metal. Grinding of the welds, however, will somewhat reduce the strength of the joint; hammering them will generally accomplish the same purpose as grinding, without the sacrifice in strength.

Welding heat-treated alloys tends to destroy the effect of prior heat treatment. The annealing temperature range is exceeded in the metal adjacent to the weld, but the rate of cooling in the air is fairly rapid; consequently, the strength usually is intermediate between that of fully annealed alloy and that which would result from the solution heat treatment.

The heating of the metal also has an adverse effect upon its resistance to corrosion. The loss in these properties can be recovered partially by heat treatment or by performing the welding operation before heat treatment where this is feasible.

Where a joint is depended on for maximum efficiency, one that is torch welded, and in many cases arc welded, joints cannot be considered equal to a well-designed mechanical type of joint.

**Torch Welding of Aluminum.** The use of torch welding in aircraft is limited to applications where high unit stresses are not involved. Its chief merit is that it is the simplest way to obtain the gastight or liquidtight seams required in fuel and other tanks.

The heat of welding anneals a strip of metal on either side of the joint in the case of strain-hardened alloys, and seriously impairs the effect of heat treatment in the case of heat-treated alloys. Torch welding, therefore, is almost entirely confined to those alloys which are not heat-treated. These alloys, especially 2S and 3S, are well adapted to welded construction.

The standard welding equipment is suitable for aluminum also. Special flux (such as Alcoa No. 22 Torch Welding Flux) and a suitable filler material (either 2S wire or 43S wire containing 5% silicon) are necessary. Acetylene gas is most commonly used for welding aluminum, especially if the operators are accustomed to using it with other metals. However, in some cases hydrogen gas would be more satisfactory.

Since the welding flux is corrosive, it must be removed completely after the weld is made. Warm water and a brush or cloth will do a fair job if there is easy access to both sides of the weld. It is better practice, however, to follow this cleaning with a dip in

## SOLDERING, BRAZING, WELDING

a 5% solution of sulphuric acid at a temperature of about 150° F. for 10 minutes. If it is more convenient to use the dip at room temperature, increase acid content to about 10% and time to 30 minutes.

**Learning to Weld Aluminum.** After you have completed your course in welding steel, you need only practice on pieces of scrap sheet aluminum until you become proficient and skilled in the manipulation of the torch. The following hints will be found helpful:

1. Be sure that the work is clean in the area to be welded, and that all joints fit closely before tack-welding.

2. Apply flux to rod and seam with a glue brush.

3. Use a neutral flame with oxyhydrogen and oxyacetylene.

4. Divide the flame equally on both sides of the joint; hold it steady and move ahead as fast as the puddle melts and is filled.

5. To avoid blowholes or running of the metal out of control, the beginner should use a light volume flame (a neutral flame adjusted below the capacity of the tip). This will slow down the welding speed somewhat, while you are acquiring the technique; afterward the full capacity of the tip may be used. The flame should slant ahead at an angle which will keep the direct blast out of the puddle. The filler rod is kept in the flame, thus somewhat retarding the blast of the flame.

6. Whenever the metal starts to run apart or melt away, twist the handle, flicking the flame up, then return it to the weld as the puddle begins to freeze. Add filler rod in starting it again.

7. The flux must melt at a temperature just below the melting point of aluminum; it quickly removes all aluminum oxide when molten. The flux, having a specific gravity less than that of aluminum, will rise to surface of molten metal and remove the oxide.

8. When welding heavy sections of aluminum, or welding bosses, filler necks, or flanges of cast aluminum to relatively thin tanks, the heavier section should be preheated to 700° F. or 800° F. to avoid heat strains and reduce gas consumption. In aircraft work, preheating is accomplished with the torch because parts are not large. Welding temperature can be determined by rubbing a pine stick on the casting; this will leave a char mark at the proper heat.

9. Mix the flux with water to the consistency of a thin paste (two parts of flux to one part of water). Keep the mixture in a stone or glass jar; *do not use an iron container*. Make only enough flux

for a day's supply. Keep store of unmixed flux well covered to prevent it from absorbing moisture from the air.

10. As soon as the weld has cooled, be sure to wash off all traces of the flux, as previously described.

**Electric Resistance Welding of Aluminum.** Electric resistance spot and seam welding is economical and often results in smoother surfaces than riveting. Where a machine can be operated a large part of the time, the higher cost of the equipment, compared to that required for riveting, is often offset by the lower unit cost of spot welds. For spot welding, the design of the parts and their sequence of assembly must allow the necessary access for the welder's arms and the welding electrodes. Special electrodes and electrode holders often are used for joints that cannot be reached with standard equipment.

Leading manufacturers of spot-welding equipment are familiar with the requirements for handling aluminum and are prepared to give mechanical and performance details on their products. Several types of excellent machines are available.

Spot welding is ideal for the fabrication of subassemblies small enough to be handled easily by one or two operators. It is in this field that spot welding is finding favor in the aircraft industry.

Both the United States Army and Navy approve the use of spot welding, but require a qualifying test on each welding machine for each gage and alloy that is used. In setting up a machine for a job, it is customary to make several test welds in scrap strips of the same gage, alloy, and temper as the work itself. These set-up welds are tested by peeling or rolling the sheets apart. A good spot weld will pull a *button* out of one of the sheets. These buttons should be nearly round and about equal in diameter to twice the sheet thickness, plus 0.060".

Resistance welding may be used in the fabrication of aluminum in a manner similar to that employed on other materials. Because of the entirely different physical characteristics of aluminum, the technique and equipment employed will differ considerably from that used for steel. In some cases, however, equipment used for steel may be modified or added to, in order to provide excellent results when used with aluminum or aluminum alloys.

In addition to the required changes in equipment, several times

# SOLDERING, BRAZING, WELDING

the electrical capacity is required for aluminum, compared with similar resistance-weld applications in steel.

*Electric Spot Welding of Aluminum.* Spot welding is an automatic process performed by a machine which allows a heavy electrical current to pass through a point to be welded; the resistance of the two oxide films at the joining surfaces of the sheet or parts, results in the melting of the steel or aluminum alloy at the joining surfaces. Two types of spot-welding machines are used in aircraft

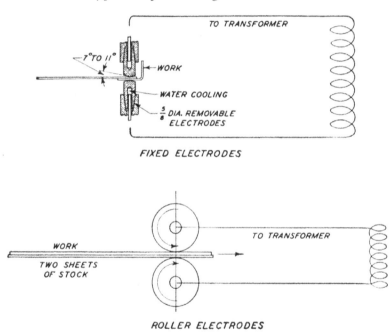

Fig. 4. Fixed and Roller Electrodes of Electric Spot Welders, Showing How Electrodes Grip Work

construction, one with a fixed electrode and one with roller-type electrodes, see Fig. 4. The electrodes are made of a copper alloy because pure copper is too soft to withstand the high pressure used on the two thicknesses of metal being spot-welded. The electrodes of both the fixed and the roll spot welders are water cooled. In the roll spot welder, a pair of water-cooled copper alloy wheels takes the place of the fixed electrodes of the ordinary spot welder; the upper of these wheels is power driven while the lower one idles. The sheets to be welded are fed between these wheels and the automatic electrical control passes current at fixed intervals. Rolls or electrodes must be kept clean and dressed down.

Roll welders can turn out over 100 spot welds per minute, or from 25,000 to 40,000 spots in an 8-hour shift, according to the design of the work being fabricated. The pressure is applied, in modern machines, either by pneumatic, hydraulic, or mechanical means. The operator of the machine is usually a man specially trained in the procedure.

Tables 6 and 7 give approximate figures on machine settings for various gages of sheet made from aluminum alloys. As the

Table 6—Machine Settings for Spot Welding Aluminum Alloys

| Gage | | Time Cycles | Current Amperes | Electrode Pressure | |
|---|---|---|---|---|---|
| B & S No. | Inch | | | Min. Lb. | Max. Lb. |
| 26 | .016 | 4 | 14,000 | 200 | 400 |
| 24 | .020 | 6 | 16,000 | 300 | 500 |
| 22 | .025 | 6 | 17,000 | 300 | 500 |
| 20 | .032 | 8 | 18,000 | 400 | 600 |
| 18 | .040 | 8 | 20,000 | 400 | 600 |
| 16 | .051 | 10 | 22,000 | 500 | 700 |
| 14 | .064 | 10 | 24,000 | 500 | 700 |
| 12 | .081 | 12 | 28,000 | 600 | 800 |
| 10 | .102 | 12 | 32,000 | 800 | 1000 |
| 8 | .128 | 15 | 35,000 | 800 | 1200 |

*Courtesy of Aluminum Co. of America, Pittsburgh, Pa.*

Table 7—Approximate Machine Settings for Seam Welding Aluminum Alloys

| Alloy* | Thickness Inch | Pressure Lb. | Cycles | | Spots per Inch | Approx. "On" RMS Amperes |
|---|---|---|---|---|---|---|
| | | | On | Off | | |
| 52S-½H | .025 | 600 | 1 | 6½ | 18.0 | 26,000 |
| 52S-½H | .032 | 680 | 1 | 6½ | 16.0 | 29,000 |
| 52S-½H | .040 | 760 | 1 | 6½ | 14.3 | 32,000 |
| 52S-½H | .051 | 855 | 1½ | 6 | 12.6 | 36,000 |
| 52S-½H | .064 | 960 | 1½ | 6 | 11.3 | 37,500 |
| 52S-½H | .072 | 1015 | 1½ | 6 | 10.6 | 39,000 |
| 52S-½H | .081 | 1080 | 2 | 11½ | 10.0 | 40,000 |
| 52S-½H | .102 | 1210 | 2 | 11½ | 9.0 | 42,500 |

*For 52S-¼H, reduce pressure 10%; for 52S-O, reduce pressure 25%; for 3S-½H, reduce pressure 25%.

*Courtesy of Aluminum Co. of America, Pittsburgh, Pa.*

tables imply, a seam weld is a series of spot welds. Both pressures and currents are approximate and vary with the tempers and kinds of alloys. The harder materials require greater electrode pressure than the softer materials.

This method of welding is applied on secondary structural parts, such as window and door frames, fixtures, furnishings, fillets, firewalls and cowling, whether of aluminum, aluminum alloy, magnesium alloy or stainless steel.

*Butt Welding of Aluminum.* Two methods ordinarily are used for making butt welds with aluminum alloys, one called *push welding* and the other *flash welding.* When push welding is employed, the pieces to be joined are clamped in dies, one movable and the other stationary. Pressure is applied to the movable die to force the two pieces together. Current is then turned on, and after a certain number of cycles, or seconds, melting will occur at the juncture. Since pressure is maintained, coalescence occurs at the interface, and some of the molten metal is forced out of the joint as flash. An adjustable contact device opens the main contactor when the movable die has traveled a predetermined fraction of an inch.

Success of push welding depends largely on, first, excellent mechanical fit between the two parts to be welded together and, second, a die design that will give uniform current distribution. This method is therefore most applicable for butt welding such simple sections as round or rectangular bars. It is difficult to design dies that will provide uniform current distribution in complicated extruded shape sections; consequently, flash welding is recommended as the better method for welding this type of section.

Flash welding methods for aluminum are similar to those used for other material. The metal is clamped in position with a small gap between the parts to be joined. When the main contactor is closed, a motor and cam mechanism bring the two pieces in contact at low initial speed, the cam continually accelerating this movement during the push up. Continuous burning will occur during the flashing period, the heat from which provides the molten condition desired at the interface of the parts. The cam is so designed as to provide a push up fast enough to stop the flashing after a certain period; immediately afterward the contactor is opened, and the weld is completed.

Since most of the heat is generated during the flashing period, current distribution is more or less automatically taken care of and is only slightly affected by the design of the dies. This permits the flash welding of complicated extruded sections such as those used in the manufacture of window frames.

Because of the difference in machine settings for various sections, it is not practical to provide figures showing suitable current densities, pressures, amount of burnoff, etc., for different types of jobs. It is essential, therefore, before obtaining equipment for butt welding aluminum alloys, that sufficient tests be made on the particular jobs in question to determine the electrical and mechanical specifications for a suitable machine.

**ARC WELDING.** Arc welding is accomplished by the successful effort of an electric current to jump across an air gap. In electric arc welding, the arc is struck between (a) a carbon or metallic electrode (welding rod) connected to one terminal of a direct-current electric generator, and (b) the metal to be welded, which is connected to the other terminal. The size of the arc can be varied by placing a varying resistance in the circuit, by increasing or decreasing the distance across the gap, which is controlled by the operator's skill, or by increasing or decreasing the amount of current (amperage) on the control panel.

The welding rod for electric arc welding is called an *electrode,* and it is covered with a thick coating of flux. The molten metal, passing from the electrode to the base metal weld through the arc, is in the form of small liquid globules from $\frac{1}{100}''$ to $\frac{1}{1000}''$ in diameter; these are moving at speeds varying up to more than one hundred miles an hour.

Gas expelled from burning flux on the electrode envelops the stream of tiny globules and rushes outward from around the pool of the weld, thus preventing penetration of air.

Much arc welding is done in both airplane manufacture and repair; however, the operator is a highly skilled specialist, knowing all the tricks of his trade, and with a special knack for handling the electrode. Not everyone possesses the ability to become a good arc welder; but those who do, seem to find the work relatively simple. It takes considerable practice to get the knack of (1) holding the arc at exactly the right distance to prevent the electrode from stick-

# SOLDERING, BRAZING, WELDING

ing (freezing) to the work, (2) feeding it downward at a uniform rate of speed as it shortens, and (3) at the same time continually advancing the arc so that a uniform deposit of metal will be made.

## PART II. OXYACETYLENE WELDING COURSE FOR BEGINNERS

To make a weld, an area at the edges of two adjacent parts is heated until a small pool of molten metal is formed. This pool is made to progress along the junction of the two parts, adding additional metal from a welding rod.

For the beginner there are certain important things to learn and to remember, whether you want to become a specialized welder or an all-around welder. The following rules should be mastered before any attempt at welding is made.

### CAUTIONS AND PROCEDURES

1. Always be sure to have enough gases on hand to complete a job.
2. Always use the wrenches provided for cylinder valves; leave wrench on acetylene cylinder valve when in use.
3. Under no circumstances use oil or grease on oxygen cylinder valves or regulators. This is very dangerous, as oil and oxygen mixed together will explode. Keep the hose away from any oil that would penetrate it.
4. Always clamp oxygen and acetylene hose securely at gages and torch handle. When checking for leaks, use soapy water applied to fittings or hose with a brush, and look for bubbles.
5. Before connecting the torch, blow out both hose lines with gas pressure; i.e., *crack* the valves.
6. Before connecting a regulator to a cylinder, open the cylinder valve just enough to blow out any possible dirt, to prevent its being carried into the regulator, as this may cause leakage or pressure creepage.
7. After attaching a regulator on a cylinder, turn pressure-adjusting handle on the regulator to the left before opening the cylinder valve.
8. The cylinder valve must be closed before attempting to tighten a leaky regulator to the cylinder connection.
9. Never try to decrease the pressure by turning the pressure-adjusting handle when pressure is indicated on the low-pressure gage, unless the torch valve is open.
10. Always shut off the gas at the cylinder valves when work is done.
11. Never leave pressure in the regulators (indicated by the pressure gages) when not in use.
12. Always use the red hose for acetylene and the green (or sometimes black, or blue) hose for the oxygen.
13. Do not attempt to disconnect a regulator from a cylinder before

releasing the pressure from the hose, low-pressure gage, and high-pressure gage by turning valve handle to the right (with torch open) after the pressure has been shut off at the tank. Then turn the pressure adjusting valve to the left, thus releasing the spring tension on the diaphragm.

14. When releasing gas pressure from cylinders to regulators, barely *crack* the cylinder valves at first, to let the high-pressure gage hand move up slowly, then open up. If gas is released suddenly, gage may be damaged.

15. The oxygen valve has two seats, one for full open and one for full closed. When open, it should be tight against the open seat.

16. Never open the acetylene cylinder valve more than one and one-half turns.

17. Consider the acetylene tank empty after pressure has reduced to 50 pounds. Acetone will begin to feed from the tank at pressures below this, causing loss of flame control. Also, there is danger of flame burning inside of the tank at low pressures.

18. Always wear goggles when working with the lighted torch.

19. Do not handle gas tanks roughly.

20. When working in a confined place or on an airplane, always have a helper near, to shut off the gas at the cylinders in case of necessity, and to man the fire-fighting equipment that should be handy in case of fire.

21. Never do any welding or cutting on containers that have held inflammable materials until they have been thoroughly cleaned or steamed with all vent holes or other openings open.

22. Never do any welding on gasoline or oil tanks until they have been steamed at least 8 hours.

23. Do not use torches, regulators, or any other welding equipment that is in need of repair.

24. Have equipment inspected at frequent intervals.

25. Use a carbonizing flame when welding 18–8 stainless steels.

26. The connection on the oxygen regulators has right-hand threads, while that of the acetylene or hydrogen line has left-hand threads. This feature is included in the equipment to prevent the gages from being crossed accidentally. Remember this when you use the wrench to remove regulators. The same applies in the connections of the hose to regulators and torch handle.

27. If the gas pressure is released from the hose by opening the torch valves after closing regulator valve, always see that the torch valves are closed afterward. If the top orifice should be clogged by dried flux, wet asbestos, etc., and the gas again turned on at the cylinders, a mixture of gases might be forced into either hose, and this might explode upon lighting the torch.

28. Cylinders should always stand upright, especially the acetylene cylinder, to prevent acetone from leaking into the gas streams and spoiling the physical properties of the weld.

**WELDING ROD.** A welding rod that has been, and still is, used quite extensively in aircraft is a low-carbon steel of the following analysis: 0.06% carbon, max.; 0.25% manganese, max.; and not over 0.05% silicon. This rod is sold in all sizes, and comes copper coated. Its tensile strength is considerably lower than that of SAE

# SOLDERING, BRAZING, WELDING

X4130 base metal, but because of the greater thickness of the weld in cross section and the alloying elements picked up by the weld metal from adjacent base metal, welds can be made of greater strength than the base metal, especially in the lighter gages. However, it is also true that the slight loss of alloying elements stolen from the base by the weld metal has somewhat weakened the base metal adjacent to the weld. For welding SAE 1025, this type of welding rod is nearly always used.

For landing-gear parts, motor mounts, and parts requiring heat treatment, a rod of higher carbon content, called *high test*, is used. This contains 0.13 to 0.18% carbon, about 1.10% manganese, and 0.25% silicon. The manganese and silicon in this welding rod have a fluxing action which aids in the operation.

Welding rods available in many different compositions for specific jobs will not be discussed here. The two rods mentioned are used most often for aircraft welding. The welders' guides or catalogs published by various welding equipment manufacturers are furnished to every welding shop and may be referred to for instruction on specialized welding, as for castings, various special alloys, and unique conditions.

**DESIGN OF WELDS.** Welded X4130 (chrome-molly) tubes depend upon the type of joint employed to obtain an ultimate strength from the weld which cannot be obtained from a plain 90° beveled butt weld. The specifications given in the Civil Aeronautics Manual 18 must be followed very closely.

Especially in repair work, the welder must make all splices and repairs in accordance with these specifications where tensile strengths of joints must be 90,000 p.s.i. Many different designs are illustrated in the Civil Aeronautics Manual 18.

### PROJECT 1

**Objective**

An introduction to the oxyacetylene equipment; to learn the procedure involved in changing empty gas cylinders for full ones, turning on gas, adjusting pressure regulators, and adjusting torch for proper flame.

**Procedure**

1. With the help of your instructor, remove and replace oxygen and acetylene regulators, hose, and torch. Tell the instructor as you go along exactly what you intend doing next. Do not proceed until he approves. You

will probably attempt to do some part of the job wrong, so listen carefully to the instructions given. After going through the whole procedure once, reread the text and try it again, still under the guidance of your instructor. When he is satisfied with your technique, go through the procedure by yourself.

2. The procedure of lighting and adjusting the torch may be accomplished by following these instructions. Ask for comments from your instructor.

*a*) Turn regulators out (to the left) so there is no tension on the spring and the valves are closed.

*b*) Turn oxygen tank valve full open until the valve tightens against the *open* seat.

*c*) Turn acetylene tank valve open three-fourths, or less, of a turn.

*d*) Open both valves on the torch one full turn or more.

*e*) Adjust the pressure regulator valve handle so that the low-pressure gages show about 8 to 10 pounds of oxygen and 5 to 8 pounds of acetylene.

*f*) Turn off both valves at the torch.

*g*) With the gas lighter in the left hand and the torch in the right, use the forefinger and thumb to open the acetylene needle valve on the torch slightly, and to just crack the oxygen valve, enough to relieve tightness without admitting any appreciable amount of oxygen. Then ignite the gas at the tip. Put on goggles.

*h*) Open the acetylene valve on the torch a little more, until the blaze has left the tip, as at (*A*), Fig. 2. A slight, sharp whip of the torch will help to make the flame separate from the tip. Try adjusting the valve if the flame does not separate to the space desired, which should be from 1/8" to 3/16" from the tip.

*i*) Turn the oxygen on until a small white inner cone is formed in the flame. If the flame pops and goes out, there is either too much oxygen or not enough acetylene being used. Adjust the flame until it looks like (*C*), Fig. 2. The small white inner cone in the flame should be 1/8" to 1/2" or more in length, depending on the size of the tip.

*j*) If the inner flame or cone gives a feathery edge or is encased with an outer envelope, you have a carbonizing flame, as shown at (*B*), Fig. 2, and should either increase the oxygen or reduce the acetylene. If you continue to reduce the flow of acetylene or increase the flow of oxygen after the outer envelope has moved into the white cone, the white cone will assume a pointed appearance, as at (*D*), Fig. 2, and the flame will be oxidizing.

As already explained, there are three definite types of oxyacetylene flames. First, the neutral flame used for most welding. It has two *distinct* parts, an intense white inner cone, and a nearly colorless outer envelope with a bluish to orange tip. Adjust your flame as at (*C*), Fig. 2.

The second type, (*B*) in Fig. 2, is the carbonizing flame. A third cone appears to envelope the inner cones; adjust your flame as shown at (*B*). This is the carbonizing flame used for welding 18–8 stainless steel.

In the third type, see (*D*) of Fig. 2, the inner cone is shorter, is

# SOLDERING, BRAZING, WELDING

nicked in, and is white with a purplish hue; the outer envelope is bluish to orange, tapering off to nearly colorless toward the end. Adjust your flame to the third type. This flame should not be used in welding on aircraft as, except on certain copper alloys, it oxidizes and burns the metal.

Before going to the next project where you will begin to use the flame, you should understand just how this flame works.

Acetylene gas is a compound of two parts carbon and two parts hydrogen, and the complete oxidation of acetylene takes place in two distinct steps. As you learned from the introductory material on welding, it requires about $2\frac{1}{2}$ volumes of oxygen to burn one volume of acetylene completely. Since the neutral flame burns approximately equal parts, by volume, of acetylene and oxygen from the tanks, much of the oxygen is supplied from the air. It is the oxygen supplied from the air that causes the secondary combustion, or the outer enveloping flame which makes it possible to do this type of welding successfully. This outer flame absorbs the oxygen surrounding the area of the weld, thus preventing oxidization of the metal.

After careful study of the following suggestions, you will be ready to begin your actual practice in welding.

### GENERAL SUGGESTIONS

1. Remember that no welded structure is as strong as the original base metal unless subsequently annealed or normalized. If 100% strength is desired, provision must be made for proper heat treatment. The procedure for heat treatment will depend on:

   *a*) Whether the structure is left in an annealed or a heat-treated state
   *b*) Whether the weld is made with heat-treatable filler rod
   *c*) The design of the weld.

2. Complete fusion must start at the bottom of the joint. Do not attempt to secure perfect penetration on any metal thicker than 14 gage or 1/8" without beveling. Perfect fusion and penetration are most important in good welding.

3. Speed is a secondary consideration in welding; excess speed is a contributing factor in welding failures, either by just plain hurrying, or by use of excess oxygen to pool the metal faster.

4. Keep the end of the filler rod in the puddle while adding the metal.

5. Melt the metal only once and keep in a molten state as short a time as possible.

6. Use just enough heat to get thorough penetration and no more; do this by using a tip of the proper size.

7. Avoid reheating a joint to produce a smooth effect. This practice is

called *wash weld*. Wash weld also refers to the practice of turning a plate over after welding a seam and welding the reverse side at the seam to cover up a poor weld—one that is either burned or has poor penetration. These are not good welding practices.

    8. Never dig the inner cone deep into the puddle. Keep it on top.
    9. Follow your shop procedure.
    10. Never remove the flame from a weld until you have finished.
    11. Always weld on clean metal. Use a wire brush to remove scale, rust and dirt.

## PROJECT 2

**Objective**

To learn how to fuse metal with an oxyacetylene torch without the use of welding rod

**Tools and Equipment**

    1. One set of oxyacetylene tanks, regulators, gages, hose, torch, and assortment of tips
    2. Gas lighter
    3. Goggles
    4. Gas pliers
    5. Bench and vise
    6. Fire brick
    7. Wire brush

**Material**

Mild steel (SAE 1010, 1020, or 1025) 5" long by 1/8" thick by 1" wide; (A) in Fig. 5. Six pieces should be plenty for this exercise.

**Procedure**

    1. Place one of the pieces of steel on edge on the brick, propped between two of the other pieces laid flat; or clamp one edge in a vise.

    2. Select welding tip of proper size, about No. 5, Fig. 3, with an orifice of, say, .037" for this metal thickness, as shown in Table 4. Set both regulators at correct working pressure with the torch valves open, oxygen at about 5 pounds, and acetylene at 3 pounds. Place goggles over eyes, light torch, and adjust flame to neutral.

    3. Hold the torch lightly but firmly in hand, with tip pointed in the direction of weld progression so that the metal ahead of the weld is preheated.

    4. Start at the right-hand corner of the piece of steel; hold the flame so that it divides on each side of the edge and the tip of the inner cone is about 1/16" away from the metal. As the edge melts at the point of the inner flame, move the torch forward along the edge. This practice of puddling the metal along the edge is called *fusion welding*. Watch how the metal melts into a pool at the tip of the inner cone and how it solidifies in the wake of the flame. Since in this exercise you do not have to think of your manipulation of the welding rod, you have a good chance to study the heat reactions and observe the metal in the molten state. Try to judge the depth of your penetration; i.e., how deep is the molten pool?

When you have fused the top edge all the way across, place upright in the vise and fuse across the end, and so on around all the edges. Repeat the practice on the other pieces of metal until your instructor judges you are ready

# SOLDERING, BRAZING, WELDING

for the next project. From 30 to 60 running inches, depending on the individual, usually is sufficient practice on this project.

## PROJECT 3

### Objective

To learn how to fuse two pieces of metal together (edge welding) without the use of welding rod. This practice project is very important because it introduces the first principles of welding—the fusing together of two or more metals.

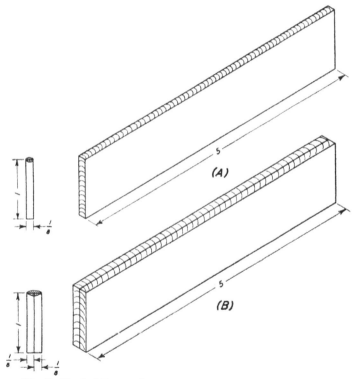

Fig. 5. Typical Project for Fusing Metal without Welding Rod

### Tools and Equipment

Same as for Project 2

### Materials

Same as for Project 2

### Procedure

1. Place two pieces of 1/8" material flat together, (B) in Fig. 5, and clamp in vise, leaving as much as possible protruding above the vise jaws.

2. Use a No. 6 tip, the next size larger than used in Project 2, put on goggles, light torch and adjust to a neutral flame.

# AIRCRAFT SHEET METAL WORK

3. Apply the flame to the end of the seam, keeping the tip of the inner white cone about 1/16" away from the metal. Watch the metal closely, and when the edges at the seam melt and fuse together, move the flame along slowly. If the flame is made to progress along the seam too quickly, the edges will not be properly fused. There must be good penetration, so watch the depth of your molten pools, trying to reach a depth of about 3/32" to 1/8" before moving on. On the other hand, if the torch is moved too slowly, too much metal will be melted, making an uneven weld.

4. Try about 24 to 48 running inches of fusing before going on to the next project. You will soon find yourself lowering and elevating the flame, alternately, describing a series of overlapping circles in controlling the pool of molten metal under the flame. At first, however, it is best to hold the flame steady until you can make the metal flow in an even bead of uniform depth.

## PROJECT 4

**Objective**

To learn how to flange-weld or fuse two pieces of metal without welding rod, in order to acquire the knack of penetration.

**Tools and Equipment**

Same as used in Project No. 3

**Material**

Several pieces, in pairs, of mild steel or black iron, 1" by 5", with thicknesses of .064", .049" and .035"

**Procedure**

1. Brake a 3/16" flange at 90° along the edge of two pieces of .064" material, Fig. 6 at (A). Brake 1/8" and 1/16" flange respectively at 90° along the edges of the pieces of .049" and .035" material.

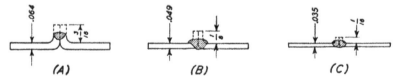

Fig. 6. Typical Project for Fusing Flanged Metal without Welding Rod

2. Place two pieces of .064" steel flat on a fire brick, with the flanges up and together, as shown in Fig. 6 at (A).

3. Select a proper size tip (orifice of .031 for metal .064 thick, see Fig. 3), light torch, and adjust flame to neutral. *Wear goggles.*

4. Begin by tack welding each end of flange, to hold the two pieces together.

*Note:* A tack weld is a short weld to hold two pieces together at the edge. No particular design of weld is necessary. This method, also called tacking, is used for holding material temporarily in place when it is to be welded solidly, to insure proper alignment and position. Tack welds should not interfere with

# SOLDERING, BRAZING, WELDING

the welding operation; that is, when approaching a tack weld, continue as though it were not there.

5. Practice fusing the two edges together without use of filler rod.

6. Consult your instructor as to your workmanship. It may be necessary to repeat this step several times, as it introduces the beginner's next step—penetration.

7. When your instructor is satisfied with your technique, repeat the operation on the .049" material with 1/8" flange. You might change the tip size here to one with an orifice of about .0255, a Smith No. 21, Oxweld No. 1, etc., see Table 5. A larger size tip is used after more experience.

8. Next try your skill on the pieces of .035" with 1/16" flange. Although you are not using a welding rod to add metal to the weld, you have actually been welding with surplus material, that contained in the flange. In this step try to make a smooth ripple weld, using up all of the flange and gaining thorough penetration to the reverse side of the metal.

**WELDING TECHNIQUES.** Now that you have tried welding without filler material, before you go on to the projects where filler rod will be used, it would be well to consider some additional welding techniques.

There is a range in the neutral flame where it is hard to detect whether it is slightly on the carbonizing or slightly on the oxidizing side. Always choose an adjustment leaning toward the carbonizing side, just at the point where the small envelope surrounding the white cone disappears. In aircraft work, a slight excess of acetylene in the flame adjustment is recommended, rather than an oxidizing flame, for welding with either low-carbon or high-test rod and X4130 base metal. The slight amount of carbon in the flame has a fluxing action which helps to reduce surface oxides, and yet does not cause carbon pick-up by the base metal. Even if there is a slight excess acetylene feather in the flame, it is better than to get a fluctuating flame adjustment unknowingly, as this would cause surface oxide.

In aircraft welding there are three different ways to apply or deposit weld metal in the weld: forehand ripple, backhand, and scale welding.

**Forehand Ripple Welding.** This method calls for the torch to be held in the right hand (for right-handed operator) and the rod in the left. The direction of welding is from right to left, placing the flame between the completed portion of the weld and the rod, see (A) in Fig. 7. The puddle is maintained continually with a meniscus front by moving the rod and flame alternately from side to side to melt and prepare the metal edges ahead of the welding action. The rod is added intermittently in the center of the puddle and must follow close to the inner cone to maintain the near-molten state and help prevent oxidation. If the puddle is moved forward by adding rod without first melting the edges of the base metal, it will cool without being welded. In butt welds, the edges must be melted all the way through to the opposite side.

**Backhand Welding.** In this welding technique, the direction of welding is from left to right, although rod and torch are held as in forehand welding: see (B), Fig. 7.

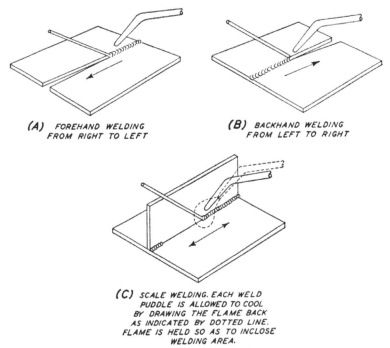

(A) FOREHAND WELDING FROM RIGHT TO LEFT

(B) BACKHAND WELDING FROM LEFT TO RIGHT

(C) SCALE WELDING. EACH WELD PUDDLE IS ALLOWED TO COOL BY DRAWING THE FLAME BACK AS INDICATED BY DOTTED LINE. FLAME IS HELD SO AS TO INCLOSE WELDING AREA.

Fig. 7. Direction of Progress in Making Forehand, Backhand, and Scale Weldings

There is some advantage in playing the flame on the edges of the base metal without interference from the rod. The hand holding the torch must back away as the torch progresses; the ripples are caused by the rod flow instead of the flame force. These ripples are somewhat coarser than those resulting in forehand welding, but some welders produce just as smooth a weld by this process.

# SOLDERING, BRAZING, WELDING

**Scale Welding.** In this procedure the welding puddle is not maintained continuously. After a puddle is created, welding rod, which enlarges the puddle, is added, and is left to solidify while the flame is concentrated farther along the weld, as shown at (C) in Fig. 7. The new puddle, when filled with rod, overlaps the former puddle, and the flame is not removed entirely from this overlapping section until the lap is completed. Never flick the flame off to one side.

This type of welding may be done by either forehand or backhand technique and is used in work on thinner sections where fillet welds are needed and where warpage would result from overheating.

In this course only the forehand ripple welding is taught. When you have mastered this type of welding, you can more easily acquire other techniques.

### PROJECT 5

**Objective**

To learn welding with the addition of filler rod

**Tools and Equipment**

Same as used in Project 2

**Material**

Same as used in Project 2, plus two lengths 1/16" low-carbon, copper-coated filler rod (Oxweld No. 7 or equivalent).

**Procedure**

1. Lay two pieces of 1/8" steel flat together with the edges up.
2. Light torch; adjust flame to neutral; adjust goggles.
3. Hold welding rod lightly between first two fingers and thumb of left hand.
4. Tack weld at each end and once in center.
5. Begin to melt a puddle at the right-hand corner.
6. Hold the welding rod in the outer flame, bringing it close to the inner flame when the puddle begins to melt.
7. When the puddle has melted and fused both pieces together, and the penetration seems to be from 3/32" to 1/8" deep, drop end of heated rod into center of puddle. Do not push the rod into puddle; just hold it there until it melts away. As the torch moves forward, creating an advancing center of the puddle, dip rod into it again, melting off another supply of steel. Practice this welding by holding torch still, except for the forward movement, until you have done from 60" to 100". Then you may proceed to the next step.
8. As soon as the molten puddle begins to form at the seam, insert heated end of the rod and move it slightly from side to side. As rod is added

at each side, the flame is moved to the opposite side while the rod is left hanging at the spot where the metal was melted from it. When the rod is needed on the opposite side, where the flame was moved, it is moved over to be melted off, and the flame is moved to the other side again. This action takes place so fast it appears as if the motion of the rod is opposite that of the torch; actually this is somewhat the case, except that it trails a little.

9. Continue with this welding until you finish the first two pieces. Then check your work with instructor, who should tell you approximately how many inches or feet of welding you should finish before going on to the next project.

## PROJECT 6

**Objectives**

1. To learn how to make a butt weld
2. To learn technique of applying filler rod
3. To increase your ability to get thorough penetration in welding

**Tools and Equipment**

Same as used in Project 5

**Material**

Same as in Project 5

**Procedure**

1. In this exercise the pieces of metal are to be laid out flat on the surface of the fire brick and will be welded as shown at (A) in Fig. 7.

2. We are now ready to make an actual butt weld, so clean the edges to be welded with a wire brush if there is any rust or scale present. Space the right-hand ends where welding is to begin at least 1/32" apart. This space should taper out on the far end to between 3/32" and 1/8" for pieces 5" long.

*Note:* The reason for this spacing is that as the metal is being welded it melts so fast that most of the expansion is taken care of; as the molten metal cools, however, it contracts and has a tendency to pull the two edges of the seam together. This action is resisted by the part previously welded and now solidified, proving that, in the process of welding, the stresses in a welded seam have different direction and amplitude than they have after the weld has cooled.

Welders compensate for this effect by setting up the two pieces to be butt welded so that the edges are nearly in contact at one end of the seam and separated at the other end by approximately 1/4" per lineal foot of seam. The spacing should vary somewhat according to the metal and its thickness; welders acquire a sort of instinct for the correct amount, through practice. The general rule is to have the far end of a seam spread about 1/4" per foot for steel and 3/8" for aluminum, to prevent locked-up stresses.

3. Since the material is just 1/8" thick, no edge preparation is necessary, so proceed to weld.

4. Hold the rod close to the white cone as the puddle develops, and maintain penetration down through the thickness of the base metal; melt the base metal ahead of the front of the puddle, maintaining a crescent advance as rod metal is added.

# SOLDERING, BRAZING, WELDING

5. Let the weld cool, then clamp in a vise and bend back and forth until the piece breaks through the weld. Present to instructor for inspection and comments on defects or methods of improvement or correction.

### PROJECT 7

**Objective**

1. To learn how to weld two pieces without use of filler rod
2. To learn how to make a beveled corner weld with the use of filler rod

**Tools and Equipment**

Complete oxyacetylene outfit as used in previous projects

**Material**

1. Four pieces of steel such as used in previous projects, 1/8" by 1" by 5"
2. Two lengths of 3/32" low-carbon filler rod

**Procedure**

1. Place two pieces of steel on the fire brick; lay one piece flat, butt the other on the first at 90° as shown in Fig. 8 at (A) and tack weld the ends.

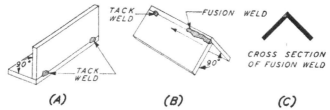

Fig. 8. Practice in Making Horizontal Butt Corner Weld without Filler Rod; (A) Position for Tack Welding; (B) Position for Making Fusion Weld; (C) Cross Section of Weld

2. Turn over and fuse the metal through, using no filler rod, see Fig. 8 at (B). Note the cross-sectional view of the weld.

3. Using proper support, tack weld the corners of the other two pieces so that they are in position to make a 90° beveled weld as shown in Fig. 9.

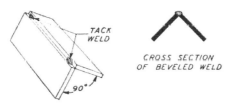

Fig. 9. Position for Making a Horizontal Beveled Weld on an Outside Corner, and Cross Section of Beveled Weld

4. Using filler rod, build the weld up as shown in the cross-sectional view, and carry out this weld thickness throughout the length of the weld.

*Note:* Spacing of the two pieces of steel is not necessary in either of these welds owing to the angle at which they are welded.

**WELD CROSS SECTION.** Perhaps we should stop here to consider what the cross section and contour of a weld should be like. Possibly your instructor has already pointed out that one of your welds was *starved* (i. e., not enough filler rod was added) or that the filler rod was not fed evenly to the weld along the entire length of the seam. You should get acquainted, then, with the way the profile of the cross section of your weld should look, and learn how to prepare the edges for welding.

**Edge Preparation.** For welds like those made in our projects, on material only $\frac{1}{8}''$ thick or less, no special preparation is needed except to clean off any rust, scale, grease, or other foreign matter, such as the paint used in the color code for marking different steels.

Materials thicker than $\frac{1}{8}''$ must have the edges beveled for a butt weld, to provide an included angle of at least 70°; 90° where possible. See Fig. 3. For sheets heavier than $\frac{3}{8}''$, bevel both sides, as shown at (*B*) Fig. 10.

**Weld Penetration.** At all points in the weld, fusion welding must show thorough penetration and fusion between the base metal and metal added from the rod, see Fig. 10.

**Weld Reinforcement.** Reinforcement in welding means the amount built up above the top surface of a sheet, or that portion in excess of the sheet thickness of part being joined together, see (*H*) in Fig. 10. This reinforcement must merge smoothly into the top surface without undercutting (starving) or excessive build-up at any point. For butt welds, the filler rod is added so as to build up the total thickness of the weld to $1\frac{1}{4}$ times the thickness of the base metal, see (*C*) in Fig. 10.

Fillet welds can be built up as a reinforcement to almost any specified size.

**PROJECT 8**

**Objective**
To learn how to make a fillet weld in a tee joint with filler rod, as shown in Fig. 11

**Tools and Equipment**
Same as used in previous projects

**Materials**
1. One piece of mild steel 1" by 6" by 1/8"

# SOLDERING, BRAZING, WELDING

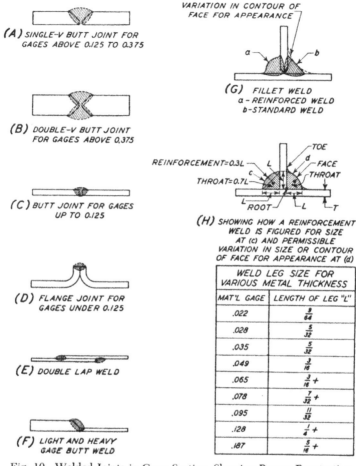

Fig. 10. Welded Joints in Cross Section, Showing Proper Penetration

2. One piece of mild steel 2" by 6" by 1/8"
3. Low-carbon welding rod, size 3/32"

**Procedure**

1. Lay the piece of steel which is 2" wide on the fire brick, and center the 1" piece upon it, on edge, as shown in Fig. 11 at (A); tack weld the ends in place.

2. Light torch; adjust flame to neutral; adjust goggles.

3. With welding rod in left hand, start welding at the right-hand end of one side of the joining point, Fig. 11 at (B).

4. When finished with one side, repeat the operation on the other side, (C) Fig. 11.

### Note

This project calls for making a fillet weld, one in which some fixture or member is welded to the face of a plate. The welding material is applied in both corners formed by the two pieces, and is finished at an angle of 45° to the plate. The reinforced fillet weld is shown at (G), in Fig. 10. Fillet welds must have equal penetration in both the vertical and the horizontal faces of the joint. The heat must be applied in such a way as to offset the pull of gravity and prevent sagging of the welding metal toward the horizontal plate.

5. Present your finished project to the instructor for comments.

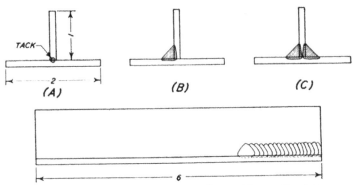

Fig. 11. Typical Project, Making Fillet Welds in a Tee Joint with Filler Rod

### PROJECT 9

**Objective**

1. To learn how to make a vertical weld, with filler rod, on an outside corner, welding up or down.

2. To learn how to make a vertical outside fillet or bevel-weld on the outside of a corner, using filler rod and welding both up and down.

3. To learn how to make a vertical inside fillet weld, using filler rod, welding both up and down.

**Tools and Equipment**

Choose those necessary, from past projects.

**Material**

1. Pieces of 1" x 5" x 1/8" mild steel
2. Low-carbon welding rod, size 3/32" or 1/8"

**Procedure**

1. Arrange pieces of steel as shown in Fig. 8 at (A), tack weld them together, and stand them upright so that the welding will have to be done vertically. Start at the bottom and weld up, using filler rod. See Fig. 12 at (A). Note the cross-sectional view of the weld.

2. Arrange two more pieces in the same way and repeat the welding operation, this time starting at the top and welding down.

# SOLDERING, BRAZING, WELDING

3. Probably the two welds are not alike when compared; the one welded from the bottom up is likely to be too heavy. Try several specimens until you have made some improvement in this respect.

4. Next, tack weld pieces together as shown in Fig. 9, stand them upright as shown at (B), Fig. 12, and weld, using filler rod. Try several specimens, some from the bottom up and some from the top down.

5. After you have completed the outside welding of a considerable number of specimens, make a corner weld on the inside of all the angles, welding half of them up and half down. Fig. 12 at A shows the inside corner weld completed.

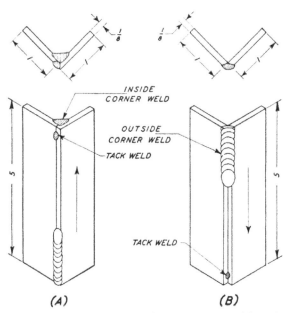

Fig. 12. Vertical Corner Welds. (A) Butt Weld Made with Filler Rod; (B) Vee Weld Made with Filler Rod

## PROJECT 10

### Objective

To learn to weld on thinner materials, thus acquiring the ability necessary for attacking the next project, the welding of tubing.

### Procedure

Select several pieces of steel in about the dimensions used in the previous projects, but in gages of .064", .049", .035", and .028". Practice welding these pieces in all the ways in which you did the heavier pieces. This work should take several periods. Your instructor must be satisfied with your progress before you start the next project.

## PROJECT 11

**Objective**

To learn to weld aircraft tubing

**Tools and Equipment**

As used in other projects

**Material**

For this and subsequent projects using tubing, the materials are:
1. Mild steel tubing, SAE 1025
2. Chrome molybdenum tubing, SAE X4130

Tubing sizes should vary from 1/2" to 1½", including wall thickness from .028" to .065". Use 4" lengths.

3. 3/32" low-carbon welding rod
4. 3/32" high-test welding rod

**Procedure**

Saw the selected tubing in two, lengthwise; then tack weld together again. Make longitudinal welds, using rod and joining the split tube. Cut tube again

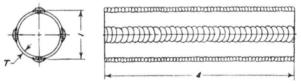

Fig. 13. Typical Welding Operation Performed Lengthwise on Steel Tubing

and repeat. Fig. 13 shows the tubing with the completed welds. Now cut the tube in two, crosswise of the weld; file smooth; present the cross-sectional area for inspection, and save it for future comparisons.

*Note:* Tables 8, 9, and 10 give properties of tubes for splices.

## PROJECT 12

**Procedure**

Cut several 4" lengths of different size tubing in two, crosswise, and rejoin with open butt welds, Fig. 14, by the *rotating and position* process; that is, rotate the tubing and weld around it, while the flame remains in one position.

To conserve materials, saw each piece of tubing in two several times, at least once after the first weld and weld together.

## PROJECT 13

**Procedure**

Fit a 2" length of tubing on a 4" length of tubing by grinding and filing, and weld as shown in Fig. 15. It is important for the beginner to make a close fit on this type of work because it is considered a difficult fillet weld, espe-

# SOLDERING, BRAZING, WELDING

**Table 8—Properties of Tubes for Splices Using Inside Sleeves**

A = Original Tube, B = Replacement Tube, C = Inside Sleeve

| Material A | Diameter, Inches A,B | Diameter, Inches C | A = .028 |  |  |  | A = .035 |  |  |  | A = .049 |  |  |  | A = .058 |  |  |  |
|---|---|---|---|---|---|---|---|---|---|---|---|---|---|---|---|---|---|---|
|  |  |  | 1025 |  | 4130 |  | 1025 |  | 4130 |  | 1025 |  | 4130 |  | 1025 |  | 4130 |  |
|  |  |  | B | C | B | C | B | C | B | C | B | C | B | C | B | C | B | C |
| 1025 | 5/8 | 1/2 | .028 | .049/.083 | .028 | .028/.049 | .035 | .049/.083 | .028 | .035/.049 | .049 | .065/.120 | .035 | .049/.065 | .058 | .083 | .049 | .058/.083 |
| 4130 |  |  | .028 | .058 | .028 | .028/.035 |  |  |  |  |  |  |  |  |  |  |  |  |
| 1025 | 3/4 | 5/8 | .028 | .035/.065 | .028 | .028/.035 | .035 | .049/.083 | .028 | .035/.049 | .049 | .065/.120 | .035 | .049/.065 | .058 | .083 | .049 | .058/.083 |
| 4130 |  |  | .028 | .058 | .028 | .028/.035 |  |  |  |  |  |  |  |  |  |  |  |  |
| 1025 | 7/8 | 3/4 | .028 | .035/.058 | .028 | .028/.035 | .035 | .049/.083 | .028 | .035/.049 | .049 | .058/.120 | .035 | .049/.058 | .058 | .083 | .049 | .058/.083 |
| 4130 |  |  | .028 | .058 | .028 | .028/.035 |  |  |  |  |  |  |  |  |  |  |  |  |
| 1025 | 1 | 7/8 | .028 | .035/.058 | .028 | .028/.035 | .035 | .049/.083 | .028 | .035/.049 | .049 | .058/.120 | .035 | .049/.058 | .058 | .083 | .049 | .058/.083 |
| 4130 |  |  | .028 | .058 | .028 | .028/.035 |  |  |  |  |  |  |  |  |  |  |  |  |
| 1025 | 1 1/8 | 1 | .028 | .035/.058 | .028 | .028/.035 | .035 | .049/.083 | .028 | .035/.049 | .049 | .058/.120 | .035 | .049/.058 | .058 | .083/.120 | .049 | .058/.083 |
| 4130 |  |  | .028 | .049/.058 | .028 | .028/.035 |  |  |  |  |  |  |  |  |  |  |  |  |
| 1025 | 1 1/4 | 1 1/8 | .028 | .035/.058 | .028 | .028/.035 | .035 | .049/.083 | .028 | .035/.049 | .049 | .058/.120 | .035 | .049/.058 | .058 | .083/.120 | .049 | .058/.083 |
| 4130 |  |  | .028 | .049/.058 | .028 | .028/.035 |  |  |  |  |  |  |  |  |  |  |  |  |
| 1025 | 1 3/8 | 1 1/4 | .028 | .035/.058 | .028 | .028/.035 | .035 | .049/.083 | .028 | .035/.049 | .049 | .058/.120 | .035 | .049/.058 | .058 | .065/.120 | .049 | .058/.065 |
| 4130 |  |  | .028 | .049/.058 | .028 | .028/.035 |  |  |  |  |  |  |  |  |  |  |  |  |
| 1025 | 1 1/2 | 1 3/8 | .028 | .035/.058 | .028 | .028/.035 | .035 | .049/.083 | .028 | .035/.049 | .049 | .058/.095 | .035 | .049/.058 | .058 | .065/.120 | .049 | .058/.065 |
| 4130 |  |  | .028 | .049/.058 | .028 | .028/.035 |  |  |  |  |  |  |  |  |  |  |  |  |
| 1025 | 1 5/8 | 1 1/2 | .028 | .035/.058 | .028 | .028/.035 | .035 | .049/.083 | .028 | .035/.049 | .049 | .058/.095 | .035 | .049/.058 | .058 | .065/.120 | .049 | .058/.065 |
| 4130 |  |  | .028 | .049/.058 | .028 | .028/.035 |  |  |  |  |  |  |  |  |  |  |  |  |
| 1025 | 1 3/4 | 1 5/8 | .028 | .035/.058 | .028 | .028/.035 | .035 | .049/.083 | .028 | .035/.049 | .049 | .058/.093 | .035 | .049/.058 | .058 | .065/.120 | .049 | .058/.065 |
| 4130 |  |  | .028 | .049/.058 | .028 | .028/.035 |  |  |  |  |  |  |  |  |  |  |  |  |
| 1025 | 1 7/8 | 1 3/4 | .028 | .035/.058 | .028 | .028/.035 | .035 | .049/.083 | .028 | .035/.049 | .049 | .058/.095 | .035 | .049/.058 | .058 | .065/.120 | .049 | .058/.065 |
| 4130 |  |  | .028 | .049/.058 | .028 | .028/.035 |  |  |  |  |  |  |  |  |  |  |  |  |
| 1025 | 2 | 1 7/8 | .028 | .035/.058 | .028 | .028/.035 | .035 | .049/.083 | .028 | .035/.049 | .049 | .058/.095 | .035 | .049/.058 | .058 | .065/.120 | .049 | .058/.065 |
| 4130 |  |  | .028 | .049/.058 | .028 | .028/.035 |  |  |  |  |  |  |  |  |  |  |  |  |

Wall Thickness—Inches

*From Civil Aeronautics Manual 18.*

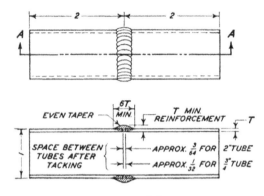

SECTION A-A

Fig. 14. Typical Welding Operation Performed Crosswise on Steel Tubing, with Specifications for Weld

**Table 9—Properties of Tubes for Splices Using Outside Sleeves for Conditions Not Covered by Tables 8 and 10**

(A, B, and C Either SAE 1025 or 4130 Steel)

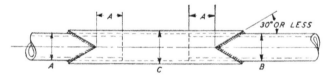

| Diameter | | Wall Thickness | |
|---|---|---|---|
| A and B | C | A and B | C |
| 1⅞″ or less | Diameter A + ¼″ | .095″ or less | .095″ |
| | | .120″ | .120″ |
| | Diameter A + ⅜″ | 5/32″ | 5/32″ |
| | Diameter A + ⅜″ | 3/16″ | 3/16″ * |
| | Diameter A + ¼″ | ¼″ | ¼″ * |
| 2″ or more | Diameter A + ¼″ | .095″ or less | .095″ |
| | | .120″ | .120″ |
| | Diameter A + ½″ | 5/32″ | ¼″ * |
| | | 3/16″ | |
| | | ¼″ | |

* Sleeve C must be reamed to sliding fit. *From Civil Aeronautics Manual 18.*

# SOLDERING, BRAZING, WELDING

**Table 10—Properties of Tubes for Splices Using Outside Sleeves**

A = Original Tube, B = Replacement Tube, C = Sleeve

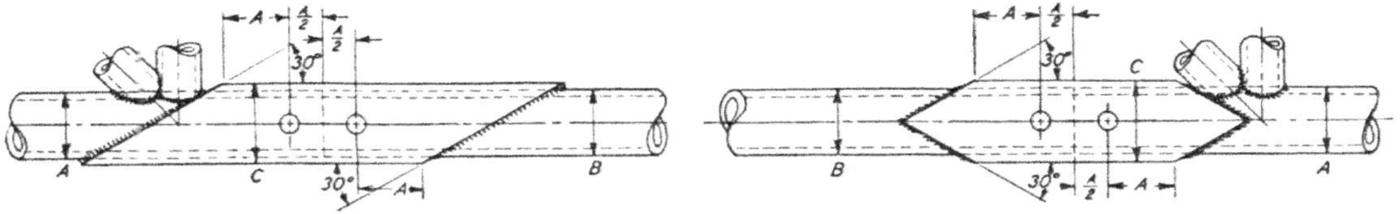

| Diameter, Inches | | Material | Wall Thickness—Inches | | | | | | | | | | | | | | | | | |
|---|---|---|---|---|---|---|---|---|---|---|---|---|---|---|---|---|---|---|---|---|
| | | | A = .028 | | | | A = .035 | | | | A = .049 | | | | A = .058 | | | | A = .065 | |
| | | | 1025 | | 4130 | | 1025 | | 4130 | | 1025 | | 4130 | | 1025 | | 4130 | | 1025 | 4130 |
| A,B | C | A | B | C | B | C | B | C | B | C | B | C | B | C | B | C | B | C | B/C | B/C |
| ½ | ⅝ | 1025 | .028 | .058 | .028 | .028 | .035 | .058 | .028 | .035 | .049 | .049 | .028 | .049 | .058 | .049 | .028 | .049 | .065 .058 | .028 .058 |
| | | 4130 | .028 | .049 | .028 | .028 | .035 | .065 | .028 | .035 | .049 | .095 | .028 | .049 | .058 | .120 | .028 | .058 | .065 | .065 .058 |
| ⅝ | ¾ | 1025 | .028 | .058 | .028 | .028 | .035 | .058 | .028 | .035 | .049 | .049 | .028 | .049 | .058 | .049 | .035 | .049 | .065 .058 | .035 .058 |
| | | 4130 | .028 | .049 | .028 | .028 | .035 | .065 | .028 | .035 | .049 | .095 | .028 | .049 | .058 | .120 | .028 | .058 | .065 | .065 |
| ¾ | ⅞ | 1025 | .028 | .058 | .028 | .028 | .035 | .058 | .028 | .035 | .049 | .049 | .028 | .049 | .058 | .049 | .035 | .049 | .065 .058 | .035 .058 |
| | | 4130 | .028 | .049 | .028 | .028 | .035 | .065 | .028 | .035 | .049 | .095 | .028 | .049 | .058 | .120 | .028 | .058 | .065 | .065 |
| ⅞ | 1 | 1025 | .028 | .058 | .028 | .028 | .035 | .058 | .028 | .035 | .049 | .049 | .028 | .049 | .058 | .058 | .035 | .049 | .065 .058 | .049 .058 |
| | | 4130 | .028 | .049 | .028 | .028 | .035 | .065 | .028 | .035 | .049 | .095 | .049 | .058 | .120 | .058 | .058 | .065 | .065 | |
| 1 | 1⅛ | 1025 | .028 | .058 | .028 | .028 | .035 | .058 | .028 | .035 | .049 | .049 | .035 | .049 | .058 | .058 | .035 | .049 | .065 .058 | .049 .058 |
| | | 4130 | .028 | .049 | .028 | .028 | .035 | .065 | .028 | .035 | .049 | .095 | .049 | .058 | .120 | .058 | .058 | .065 | .065 | |
| 1⅛ | 1¼ | 1025 | .028 | .058 | .028 | .028 | .035 | .058 | .028 | .035 | .049 | .049 | .035 | .049 | .058 | .058 | .049 | .049 | .065 .058 | .049 .058 |
| | | 4130 | .028 | .049 | .028 | .028 | .035 | .065 | .028 | .035 | .049 | .095 | .049 | .058 | .120 | .058 | .058 | .065 | .065 | |
| 1¼ | 1⅜ | 1025 | .028 | .058 | .028 | .028 | .035 | .058 | .028 | .035 | .049 | .049 | .035 | .049 | .058 | .058 | .049 | .049 | .065 | .049 |
| | | 4130 | .028 | .049 | .028 | .028 | .035 | .065 | .028 | .035 | .049 | .095 | .049 | .058 | .120 | .058 | .058 | .065 | .065 | |
| 1⅜ | 1½ | 1025 | .028 | .058 | .028 | .028 | .035 | .058 | .028 | .035 | .049 | .049 | .049 | .049 | .058 | .058 | .049 | .049 | .065 | .049 |
| | | 4130 | .028 | .049 | .028 | .028 | .035 | .065 | .028 | .035 | .049 | .095 | .049 | .058 | .120 | .058 | .058 | .065 | .065 | |
| 1½ | 1⅝ | 1025 | .028 | .049 | .028 | .028 | .035 | .058 | .028 | .035 | .049 | .049 | .049 | .049 | .058 | .058 | .049 | .049 | .065 | .049 |
| | | 4130 | .028 | .049 | .028 | .028 | .035 | .065 | .035 | .035 | .049 | .095 | .049 | .058 | .120 | .058 | .058 | .065 | .065 | |
| 1⅝ | 1¾ | 1025 | .028 | .049 | .028 | .028 | .035 | .058 | .035 | .035 | .049 | .049 | .049 | .049 | .058 | .058 | .049 | .049 | .065 | .058 |
| | | 4130 | .028 | .049 | .028 | .028 | .035 | .065 | .035 | .035 | .049 | .095 | .049 | .058 | .120 | .058 | .058 | .065 | .065 | |
| 1¾ | 1⅞ | 1025 | .028 | .049 | .028 | .028 | .035 | .058 | .035 | .035 | .049 | .049 | .049 | .049 | .058 | .058 | .049 | .049 | .065 | .058 |
| | | 4130 | .028 | .049 | .028 | .028 | .035 | .065 | .035 | .035 | .049 | .095 | .049 | .058 | .120 | .058 | .058 | .065 | .065 | |
| 1⅞ | 2 | 1025 | .028 | .049 | .028 | .028 | .035 | .058 | .035 | .035 | .049 | .049 | .049 | .049 | .058 | .058 | .049 | .049 | .065 | .058 |
| | | 4130 | .028 | .049 | .028 | .028 | .035 | .058 | .035 | .035 | .049 | .095 | .049 | .058 | .120 | .058 | .058 | .065 | .065 | |

*From Civil Aeronautics Manual 18.*

cially where thin-wall tubing is involved. Save your finished job for a future project (Project 16) where additional tubes will be welded to them.

Refer to Fig. 10 in checking your work for cross-sectional dimensions of the weld.

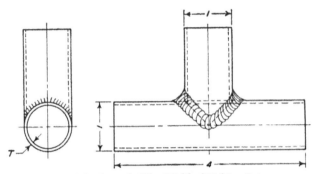

Fig. 15. Typical Fillet Weld of Tubing Joint

## PROJECT 14

**Procedure**

Select several sizes of the 4" lengths of tubing and prepare, see Figs. 16 and 17. Fig. 16 shows proper spacing. First, cut plate from .065 SAE 1025

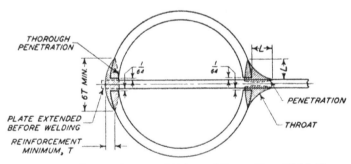

Fig. 16. Typical Cross Section of Tube and Insert Plate Weld, Showing Both Reinforced and Fillet Welds with Specifications

mild steel to dimensions shown, and tack weld into place. Refer also to Fig. 10 for weld size and spacing of parts before tack welding. In this project which is welding for practice, the tube merely has to be sawed in two through the center line and then spaced on each side of the plate as shown in Fig. 16. However, ordinarily, tubing joined to a plate in this manner would be milled or sawed out and filed smooth to a width of 1/32" greater than the insert plate to preserve its true shape. Note that your project will not look exactly like Fig. 17 because there milling and filing has been done as on an actual job where tubing must be kept round.

# SOLDERING, BRAZING, WELDING

## PROJECT 15

**Procedure**

Prepare work as in Project 14, and arrange as shown in Fig. 18. Refer to Fig. 16 for weld data.

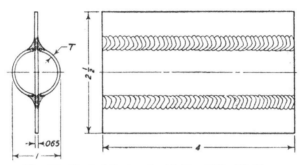

Fig. 17. Typical Project for Making Fillet Welds on Tubing with Insert Plate

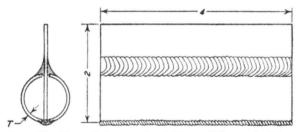

Fig. 18. Typical Project for Making Fillet and Reinforced Welds on Tubing with Insert Plate, to Specifications Shown in Fig. 16

## PROJECT 16

**Procedure**

Prepare two-inch lengths of tubing by grinding and filing to fit at an angle of 45°, as shown in Fig. 19, over the welded pieces saved from Project 13.

When a student has finished his school training and is hired as an apprentice in the aircraft industry, he will, of course, be supervised to see that his work is done in accordance with Army and Navy standards. However, a welder is expected to know how to follow the rules and procedures set forth by the Civil Aeronautics Authority for repairing or rebuilding welded structures of commercial and private airplanes. Space does not allow us to give all of these rules and regulations which are explained and illustrated in Civil Aeronautics Manual 18.

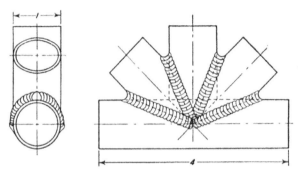

Fig. 19. Composite Joint Made by Fitting Additional Tubing to Welded Joint Shown in Fig. 15

**QUESTIONS**

1. Explain the method of attaching regulators to full cylinders. What is the procedure of attaching hose and torch, and bringing the gas to the tip?
2. Describe procedure for removing all equipment preparatory to exchanging gas cylinders.
3. In lighting torch, which gas is turned on first and why?
4. What is the temperature of the oxyacetylene flame?
5. What is the ratio of acetylene and oxygen passing through the tip for feeding a neutral flame?
6. How many parts of oxygen are necessary to burn one part of acetylene by volume?
7. What is meant by thorough fusion?
8. What is wash welding?
9. Is wash welding considered good practice?
10. Can a weld be judged by external appearance?
11. Explain forehand, backhand, and scale welding.
12. What precautions must be taken before welding on a gasoline, oil, or fuel tank which has been in use?
13. How would you check for leaks in welders' hose, fittings, or gas line?
14. Is an oxidizing flame ever used on steel?
15. Is it necessary to keep the weld covered by the flame while it is being welded? Why?
16. Should the white cone come in contact with weld?
17. What substance must be kept away from oxygen valves, gages, fittings, etc.?
18. What protection must be given the eyes in welding?
19. Why should acetylene tanks be kept upright before and while being used?
20. To what pressure reading of gage may acetylene gas be used?
21. Why is a **V** ground out at ends of pieces to be welded?
22. Above what gage of metal must joining ends be ground before welding at the splice?
23. Explain preheating. Why is it done?

# SOLDERING, BRAZING, WELDING

24. Should you attempt to weld parts of an airplane which have been subjected to careful heat treatment?
25. Why are there two pressure gages on the regulators?
26. Draw a diagram of a tubing joint, using an insert plate. Give dimensions of welds.
27. What is the first procedure after welding aluminum?
28. Explain the spot welding process.
29. Explain the difference between spot welding and arc welding.
30. What is resistance welding?

Workman inspecting deep drawn stainless steel part which he has formed on the double-acting hydro-press.
*Courtesy of Boeing Aircraft Co., Seattle, Wash.*

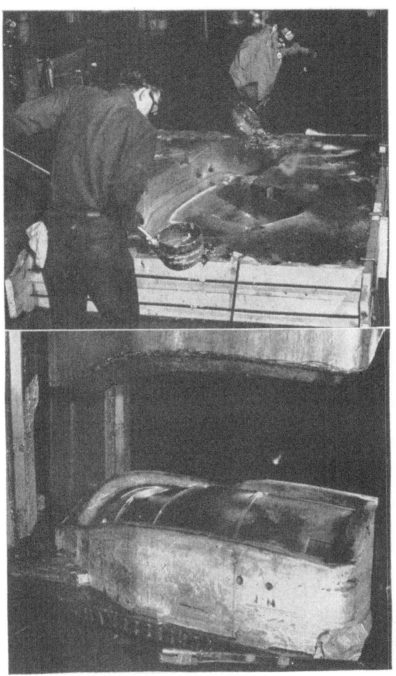

Above. Workmen casting a stamping die for cabin of Army P-39 Airacobra.
Below. Die of Airacobra cabin in drop-hammer press ready to be stamped out.
*Courtesy of Bell Aircraft Corp., Buffalo, N. Y.*

*Chapter X*

# Use of the Drop Hammer

Because of the importance of the drop hammer in the mass production of aircraft, it deserves a chapter by itself, apart from other tools. The drop hammer is definitely a sheet metal tool, one which makes possible the forming of aluminum, aluminum alloys, magnesium, stainless steels, and other steel alloys. Its major use is in power bumping, and it plays an important part in the making of airplane engine and nacelle cowling, exhaust pipes, collector rings, tail stacks, fairing, fittings, ribs, frames, and in fact all parts made from sheet metal that require forming into contours or drawn shapes. The drop hammer is doing much of the work formerly done in airplane factories by the hydraulic press. Some of the reasons for the increasing use of the drop hammer are: it can accomplish more work over a given period of time; the dies are very simple, and their cost is in proportion.

The head of the drop hammer, which is raised by means of a friction drum, is manipulated with ease and flexibility. Dogs which may be released by a foot treadle hold the hammer head up when the mechanic works with the dies. A rope coiled a few turns around the constantly revolving drum, Fig. 1, needs only to be pulled slightly to obtain enough friction to lift the head. This gives the operator the feeling that he is lifting the hammer head by himself, facilitating the co-ordination between mind and movement which is helpful in bumping pieces extremely difficult to form. This co-ordination is an important factor in operating the drop hammer, because each time a setup is made with a change of dies, the operator must follow the technique best suited to that particular job, taking on, for the duration of the job, a new set of habits which may be quite different from those used previously.

In airplane factories, drop hammers are installed in batteries in departments or buildings set aside for their use only. To explain the basic principles of drop-hammer forming, pattern making, and die casting, we will work with a single unit and with all the accessories that are required for it. In setting up a drop-hammer department, a separate room for making dies and patterns is recommended. This would be called the *pattern and foundry department*. However in small factories or repair bases this department may be a part of the drop-hammer department, housed in the same room.

**TOOLS AND EQUIPMENT.** Equipment for the pattern and foundry department consists of:

Large sink with running water
Compressed air outlets
Electric outlets
Band saw (metal cutting)
Grinding wheel
Drill press
Electric hand drill
Grinder, flexible shaft with angle disc head
Overhead track, with trolley and 2-ton hoist
Portable hoist with winch, straddling the drop-hammer base and low enough to clear the hammer head, for changing dies
Steel shrink rules (one 24", 1/8" shrink to foot; and one 24", 5/32" shrink to foot)
Molders' tools, double-end slick and spoon, finishing trowel, etc.
Heavy work bench with vise
Welding outfit complete with large size torch
Facing bellows
Venturi style suction hose attached to air outlet
Gas-fired lead pot
Gas-fired zinc pot
Ladles
500 pounds (minimum) of good molding clay
Clay tub on castors
Molding plaster
Molding sand, 1 cu. yd.
Lead, alloyed with antimony, 2,000 lb.
Electrolytic zinc with a purity of 99%, 2,000 lb.
Molding flasks, assorted sizes
Facing powder, lycopodium recommended
Garden-type sprinkling can, large
Scoop shovel
Sand sifter, centrifugal, oscillating, or vibratory
Mechanic's sheet metal tools

In the drop-hammer room, in addition to the hammer, we will need a low bench with: an assortment of stakes, such as round nigger-head, hatchet, beakhorn, blowhorn, and square; a heavy anvil with an assortment of mallets and smoothing hammers; and heat-treating and quenching tanks near the hammer, for annealing and heat-treating aluminum alloys.

**WORKING TOLERANCES FOR FINISHED PARTS.** Since the drop hammer is not classed as a precision machine, it is impractical for the engineering department to demand close tolerances on formed parts. With each operation the dies have a tendency to expand or spread, resulting in a progressive growth of the finished parts.

# DROP HAMMER

*Note:* As the female die spreads with use, the male die conforms to its growth. This is the advantage of lead and zinc dies over dies of other materials, because when metal clearance has grown

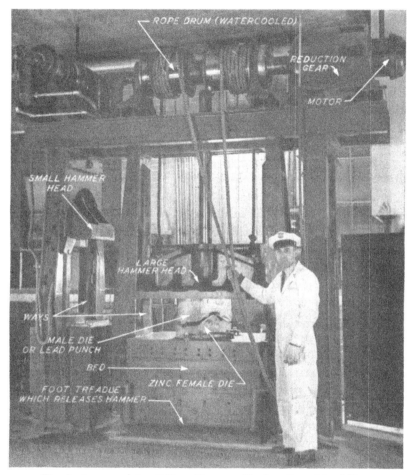

Fig. 1. Large and Small Drop Hammers Used in Sheet Metal Shop
*Courtesy of United Air Lines, Chicago, Ill.*

too great, a few blows with the hammer, given with no formed sheet between the dies, will lessen the clearance. In other words, if the lead die is dropped into the female zinc die, the lead die will expand or form to the zinc die. This practice is not usually necessary, however, except in the development of deep draws with heavy gage materials, or in certain difficult formations of stainless steels.

**PROCEDURE OF DROP-HAMMER FORMING.** The preparation for producing a finished part by drop hammer is outlined as follows:

1. Make a clay model of the part that is to be stamped out or formed by drop hammer.

2. Cast a plaster pattern against the clay model.

3. Place the plaster pattern within a drag* where molding sand is rammed solidly around it.

4. Next pull the plaster pattern out of the sand mold and pour the molten zinc into the sand mold, thus casting a zinc die like the plaster pattern.

5. Cast a lead punch (male die) directly in the zinc die.

6. Install the zinc and lead dies in the drop hammer, ready for the operation of forming the sheet metal.

**Use of Shrinkage Rule.** In laying out or developing patterns, all measurements must be made with the use of a shrinkage rule, because the mold made by the plaster pattern in the sand is filled with molten metal at very high temperature, and as it cools and solidifies it contracts. To compensate for this, the patternmaker must add to the size of the pattern. In order that exact relations may be maintained for all dimensions, a shrinkage rule is used. Such a rule is longer than the standard rule and is graduated accordingly; it can be purchased in ranges from $\frac{1}{8}''$ to $\frac{7}{16}''$ longer to the foot, graduated in 8ths, 16ths, 32nds, and 64ths. The shrinkage of lead and zinc castings is identical, $\frac{5}{16}''$ to the foot. Shrinkage allowance ordinarily made on plaster patterns is from $\frac{1}{8}''$ to $\frac{5}{32}''$ per foot. The shrink rule with $\frac{1}{8}''$ per foot allowance is the one most commonly used while one with $\frac{5}{32}''$ per foot is used for the bulkier or more evenly dimensioned patterns. This sounds confusing since the shrinkage of zinc is $\frac{5}{16}''$ per foot, however the zinc in dies is alloyed with other metals which tend to reduce the shrinkage, such as Kirksite. The amount of shrinkage, in any case, depends somewhat upon the shape and size of the casting. A casting that is long and wide in comparison to its thickness, especially one with a concave surface, from which most airplane parts are stamped, will contract differently from one that is more compact—even though they may have the

---

* A *drag* is the bottom part of a flask or mold; the *cope* or upper portion of the flask is not generally used for casting drop-hammer dies, as these dies are not cast in closed molds.

same bulk or weight and have been cast from the same material. It is hard to give any fixed rule governing shrinkage allowances as it is dependent upon localized conditions, trial and error, and practice.

When marking lines in layout work on a new damp or wet plaster pattern, use an indelible pencil which penetrates deeply into the plaster and makes a permanent line or mark.

**Developing a Clay Model.** A part to be formed in the drop hammer is first modeled in soft clay. When not in use, this clay must be kept soft by adding water, to maintain the consistency of glazier putty. Care must be taken to keep the clay free from dirt, metal shavings, plaster, etc., as the presence of foreign hard particles will result in poor surface smoothness of the model.

The following is the procedure for producing a clay model from which the plaster pattern will be made for the cowling shown in Fig. 2:

1. Prepare a baseboard similar to the one shown at (C) in Fig. 3; this board should be somewhat larger than the piece of cowling. The drawing (Fig. 2) shows the piece of cowling to be 12" square, and since the baseboard is to be larger, we will fasten two boards together, say two that are each 1x8x16, with 1x2 or 1x4 cleats nailed to the bottom; see (C), Fig. 3. These baseboards should be saved for future use, as any of larger size than the model being made can be used. The baseboard is used as a platform upon which to mold the clay and fasten the contour or guide boards; the guide boards support the modeling clay and are sometimes used, as in this instance, as a guide for shaping and smoothing the clay.

2. Prepare the contour or guide boards 1 and 2, (C), Fig. 3. If 12" boards are sufficient for profiles of a model, then wood may be used for template material. If models are so large that boards over 12" wide would be required, thick plywood usually is chosen, to save having to piece boards together or to stock those over 12" in width.

Refer again to the drawing shown at (A), Fig. 2, where the cowling is shown to be 12" square. Dies must be made about 1" larger so that wrinkles may be trimmed off around the edges of the formed piece of cowling after it has been stamped out. Therefore, our model would be 13" square, instead of 12". Then we must make our model longer still so that there will be enough material for fastening at the corners (see how contour boards 1 and 2 at (C) in Fig. 3 are extended at the ends for fastening to 3 and 4). Boards of 1" are actually only 3/4" thick, so we must add 1½" to 13", or 14½" over-all. The contour at 5 and 6 in (C) of Fig. 3 should be the same as that at 7 in Fig. 2, or that edge of the cowling *a* to *b* in Fig. 2 at (B). The width of these boards should be cut down, making them as narrow as possible, allowing only enough to fasten the ends, with clearance for the tab of the scraping template to pass throughout the length of the model without contacting the baseboard.

246　　　　　　　　AIRCRAFT SHEET METAL WORK

3. Prepare contour guide boards or templates *3* and *4*, as shown at *(C)*, Fig. 3, from dimensions shown at *(A)*, Fig. 2. These boards need be only

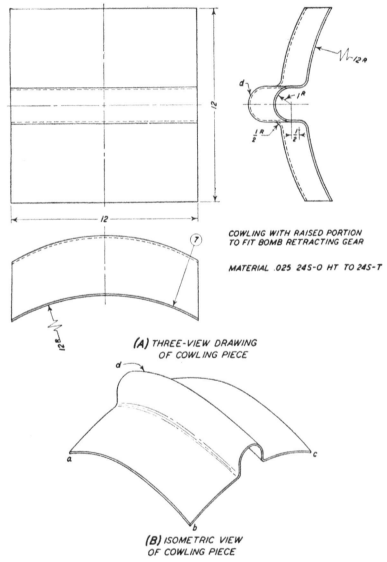

Fig. 2. Typical Cowling Piece for Bomb Retracting Gear

13" long, no material being necessary for the splice; however, the layout is more complicated, since it includes the cross section of the 2" round channel portion.

## DROP HAMMER

As you can see from Fig. 3, the scraping template at (A) and the contour boards at 3 and 4 at (C) could be marked out from a single template, thus we would need to make the layout only once. Choose a metal for the template, about 28-gage galvanized iron or terneplate, which may be cut out

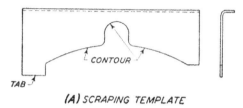

(A) SCRAPING TEMPLATE

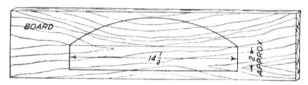

(B) CONTOUR GUIDE BOARD

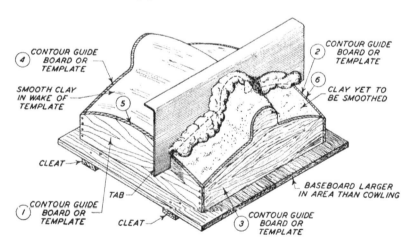

(C) DEVELOPING A CLAY MODEL

Fig. 3. Scraping Template (A), and Contour Guide Board (B), Used for Developing a Clay Model (C)

easily with the tin snips. Next, lay out the two contour boards and the scraping template from the template you have just made.

Now! If you have agreed with the text in the preceding paragraph you have been misled, because even though the method explained seems to be a step toward work simplification, there is a still better way. Simply lay out the scraping template first, and after it is cut out on the band saw and filed smooth along and down to the scribe marks, use it as a template for laying out the end pieces, thus saving material, time, and labor over the first method. You will run across many such instances in a sheet metal shop where intelligent planning will bring out time- and material-saving procedures.

4. Nail the contour boards together at the corners and tack or toe-nail the whole to the baseboard. This toe-nailing must not be too firmly done, because these boards must be disassembled before making the plaster form and pouring the plaster.

The next step is to shape off the top edges of the contour boards, forming a bevel to the contour of the 12" radius, by use of a woodworking plane or vixon file. Sandpaper the finished edges so that the scraping template will ride smoothly on them. Any irregularities on these surfaces will transfer to the surface of the clay; since the finished piece of cowling will be no smoother than the clay model unless much hand work is done later on either the plaster pattern or the zinc die, it pays to practice good workmanship at this stage of the work.

5. The scraping template, (A) in Fig. 3, is laid out on heavy gage (.040 to .064) black or galvanized iron, from the drawing at (A), Fig. 2. After the piece is cut out, it will fit over and match the shape of the contour boards 3 and 4. A tab is left on one side which would be 7¼" from the center, since the outside dimension across the box is 14½". As at (A), Fig. 3, this scraping template is cut out with an angle broken along the top edge, for stiffening. The tab is used as a gage, sliding along the side of the contour guide board as the template is drawn through the clay and controlling not only the forming of the raised portion which is the shape of the round channel portion of the cowling piece but also the shape of the whole piece.

The preceding illustrates a typical procedure; each job must be altered to suit the work at hand. This work is very interesting and will keep you on your toes planning ways to simplify and perfect it.

To sum up, contour guides or templates like those shown at (C), Fig. 3, are used to assist in forming the contours of a model. Contours or profiles of the desired shape are cut from plywood, board, or in some cases, sheet metal, and mounted at their proper locations on a wood baseboard. These profile guide pieces or templates, the shape of the model as viewed from the side or in cross section, may be left imbedded in the clay as an integral part of the model where a guide is needed on which to rest a scraping template.

The clay is kneaded, worked, and firmly thrown by hand into its approximate location, building it to sufficient mass so that it will be smoothed out when the scraping template is pulled across its crown or surface, (C) in Fig. 3, guided by the contour boards at the edge of, or in, the clay model. If the clay does not show a smooth surface in the wake of the scraping template, brush water on it, and pass over the course again.

You need considerable practice to become an expert builder of clay models since much free-hand work, similar to the work of a sculptor, is involved.

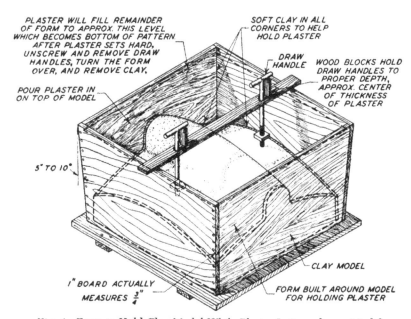

Fig. 4. Form to Hold Clay Model While Plaster Is Poured over Model

**Casting the Plaster Pattern.** When the clay model is completed, the next step is to build a wood form, like that shown in Fig. 4, around the clay model and pour the molding plaster over the clay model. The height of this form must be sufficient to give the required strength and thickness to the plaster pattern, indicated by the dotted line near the top of the wood form in Fig. 4; the height of this form likewise determines the strength and thickness of the zinc female die. In other words, *the female die is an exact replica of the plaster pattern.*

Boards of 1″ thickness, Fig. 4, are usually strong enough for the average job and should be set at an angle of from 5° to 10°, as shown, to facilitate pulling the pattern from the sand mold, later. Set the form box with the smaller opening down, and seal all joints and cracks with moist clay so as to hold the liquid plaster. (For making patterns, molding plaster is more economical than wood, though usually wood is used in foundries where many parts are cast. Wood stands up better under long usage or rough treatment, whereas plaster chips or breaks easily.) Weight the form down, or nail it to the base.

To determine the proportions of water and molding plaster, simply sprinkle the plaster over the surface of a pail full of water by handfuls until the consistency of the mixture is such as to prevent the last handful from sinking. Let set a few seconds but do not stir. If the mound on the surface sinks into the mixture and the mixture can apparently absorb more plaster, add more. Let set again, then run your hand to the bottom and work up with fingers spread. Work all particles or lumps until the mixture is creamy, then pour into the form, on top of the clay model. Be sure the form stands level.

As soon as the form is full of plaster, screw nuts on the draw handles, Fig. 4, and insert in a vertical position in the plaster before it sets. Prop draw handles in place, as shown, with wood blocks. As soon as the plaster sets, remove the draw handles by unscrewing (draw handles to be used later on to remove the pattern from the sand, since there is no way to get hold of it when its bottom is up and flush with the sand); turn the form over and remove the form boards. Remove any particles of plaster or foreign material from the clay, then tamp the clay into the clay tub where it is stored.

The next step is to smooth up the plaster pattern. Plane off the sides and bevel the corners $3/8''$ to $1/2''$ deep with a carpenter's sharp jack plane. Sandpaper, and apply two coats of orange shellac diluted with denatured alcohol. It is recommended, after the shellac has dried, to coat with clear nitrate dope, especially where the patterns are large. This use of dope prevents the sand from sticking to the pattern when it is being pulled. Dope applied directly to a green pattern would peel off, which is the reason for first using shellac and alcohol. The alcohol mixes with and absorbs water from the damp pattern, and the shellac acts as a bond for the coating.

**Casting the Zinc Die.** Now that we have the plaster pattern made, we are ready to cast a zinc die from it. Figs. 5 through 9

## DROP HAMMER

show progressive steps from casting zinc from the plaster pattern to the casting of the male die. Place the molding board level on the floor, and set the pattern face up with the drag on the board as shown in Fig. 5. Dust the surface of the plaster pattern with lycopodium, a fine yellowish powder which should be used sparingly as it retails at approximately $9 a pound. Riddle\* molding sand over the pattern to a depth of about 1″. Press the sand firmly around the edges of the pattern, against the molding board, with the fingers.

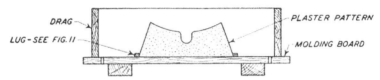

Fig. 5. Plaster Pattern on Molding Board, Drag in Place, Ready for Sand

Then fill in the molding sand, heaping full, as shown in Fig. 6. Ram around the sides of the drag first, and then around and on top of the pattern.

Only by practice will you learn how hard to ram a mold. The sand must be rammed solidly within the drag and around the

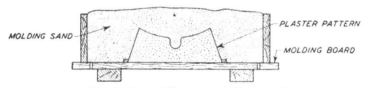

Fig. 6. Drag Filled with Molding Sand

pattern, as well, but in the vicinity of the plaster pattern where breakage might occur, care must be taken not to ram too hard. At the same time, it must be packed firmly enough to withstand the flow and pressure of the molten metal and to set the sand so that it will not upset when the drag is turned over. With soft ramming, the casting may be larger than desired, or softer spots may cause bulges or lumps on it. On the other hand, if the sand is rammed harder than necessary, blowholes † are apt to result in the casting, because the hard ramming of the sand has made it so dense that

---

\* Riddle here means *sift*, a term used by foundrymen. A riddle is a sieve used for sifting sand.
† Holes in the casting caused by trapped air or gas.

there is little chance for steam and other gases to escape through it. You will have to expect some trouble, at first; this is one of those cases where experience is the best teacher. However, considerable skill may be acquired in a surprisingly short time.

With a straightedge, strike off the sand above the edges of the drag. Place the bottom boards\* over the sand and bolt or clamp to the molding board as shown in Fig. 7. Then turn the drag over and remove the molding board from the top, as shown in Fig. 8. Brush off loose sand around edge of pattern. Swab the sand next to the pattern with water, sparingly. Blowholes may result in the casting

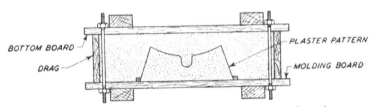

Fig. 7. Sand Rammed and Leveled Off, Bottom Board in Place, and Mold Ready to Be Turned Over

if the sand is too wet. The swabbing is done by wetting a sponge or cloth, then holding it close over the parting line between pattern and sand, squeezing it slightly so that the water will drip along the edge in the sand. This practice helps to keep the sand from breaking away around the pattern when the pattern is being drawn from the sand.

Screw the draw handles, which up to this point have been removed, back into the plaster pattern, as shown in Fig. 8; rap on all sides to loosen; and draw the pattern carefully from the sand, so as not to break the sand. If the sand breaks or chips off, these places must be repaired, and all loose sand in the mold must be cleaned out, using molders' tools called lifters or slicks, or the venturi suction hose.

*Pouring the Die.* When your sand mold is ready for pouring the zinc, be sure the molten metal is hot enough to prevent solidification around the edges between each ladle full, causing *cold shut*, the junction where two pourings of metal run together but do not

---

\* The molding board is the bottom board (but not called so) until the drag is turned over; then the board which was placed on top after leveling off the sand (Fig. 7) becomes the bottom board.

fuse. On the other hand, heavy castings should not be poured with the metal too hot, because cooler metal will *cut* the sand much better; that is, will form closer to the sand, resulting in a smoother casting. The melting point of pure zinc is about 787° F., but since we use an alloy approximately conforming with the S.A.E. Specification No. 925, the melting point is under 725° F. The following are the temperature ranges for pouring certain of the metals against sand or metal:

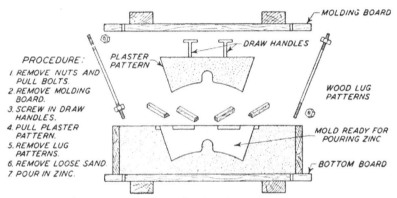

Fig. 8. Plaster Pattern Removed and Mold Ready for Female Zinc Die to Be Poured

Zinc poured in sand—795° F. to 825° F.

Zinc poured on zinc or aluminum—850° F. to 960° F.

Lead poured on metal—800° F. to 975° F.

Aluminum poured in sand—1350° F. to 1375° F.

However, pouring temperatures are determined mostly by experience. One thing to remember, the pouring operation should be done as quickly as practical. Temperatures above 900° F. should be avoided in the zinc pot, unless the die is important enough to justify contaminating the zinc with iron, a harmful impurity which affects the surface smoothness of the die, increases shrinkage, "draws" at corners, and causes brittleness. A clay wash applied to the inside of the pot and baked dry helps to prevent the amalgamation of the iron into the zinc at high temperatures.

In casting aluminum, pour the metal as cold (the 1350° to 1375° temperature range is considered cold) and as quickly as possible. Ram the molds softly; use the swab sparingly; and keep the sand as dry as practicable. Aluminum alloy is not as likely as other metals to wash away portions of the mold, because of its lightness.

Vent all molds well by piercing the sand with $1/16''$ wire through to the pattern.

*Leveling the Die.* After a die has cooled at the edges, the center begins to shrink and fall away, causing a concave surface, as shown at Fig. 10. This may be leveled up by the *capping* method; that is, by pouring a 1″ layer of zinc on top of the casting after solidification has taken place, as shown. This operation should be performed while the die is still hot, to insure a firm bond between the die proper and the cap. A better and recommended method is the use of an oxyacetylene torch for leveling the bottom of the die. A large size tip is used in the torch, and the die is leveled simply by melting the higher areas until the whole surface flows level. This method reduces the distortion due to stresses set up by the solidification of the cap in the capping method.

Warping of the dies can be reduced to a minimum by retarding the solidification of the exposed surface of the die while in the mold, by covering it with a thick asbestos blanket or sheet iron covered with a layer of sand; also by keeping doors and windows closed and thereby preventing cold drafts from flowing over the casting throughout the cooling period.

*Finishing the Die.* Before the lead punch, or male die, can be cast against the zinc die, all irregularities of the zinc surface must be removed, and the die finished smooth and polished bright. A flexible shaft grinder may be used for the accessible places, but elsewhere the die must be scraped, filed, and sanded by hand.

**Casting the Punch or Male Die.** A false belief, held by many who are just becoming acquainted with the lead and zinc die process, is: since zinc melts at 787° F. and lead at 621° F., thus it is possible to pour molten lead into zinc without melting the zinc only when the lead is less than 787° F. Actually, lead is poured into the female die at a temperature higher than the melting point of zinc, and is used where a soft punch is permissible. For stamping out stainless steel, where wrinkles form to an extent which would deform the surface of the lead punch, zinc is used for the male die also. In this case zinc is poured in the zinc female die just as molten

# DROP HAMMER

lead would be poured. The fact is that the contact with the mass of the cold die solidifies the molten metal enough to prevent the melting of the metal in the female die.

In preparing the mold for the punch, the female die is laid on the leveled molding board face up, (A) in Fig. 9. Then sand is

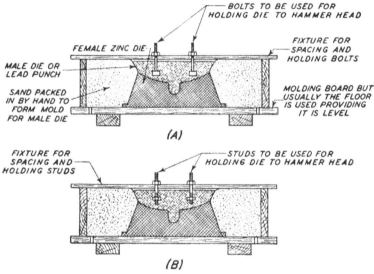

Fig. 9. (A), Lead Punch Poured into Zinc Die with Fixture Holding Bolts to Be Used for Fastening Die to Hammer Head; (B), Fixture Holding Studs Which May Be Used Instead of Bolts

packed in, to form the desired shape and size around the face. An oxyacetylene torch with a carbonizing flame (with the oxygen turned off) is played upon the zinc surface, forming a black film of carbon which acts to separate the two dies and prevent sticking.

When pouring the punch or male die, care should be taken not to pour the hot metal in one place. Each ladleful must be poured in a different place and the ladle moved while pouring to prevent the hot metal from melting out a spot in the bottom die.

As soon as the punch is poured and before solidification, a bolt spacing fixture with properly spaced holes containing bolts for later use in fastening the male die to the hammer head is placed at the proper height; see (A) in Fig. 9. Another very desirable method of fastening dies to hammer heads is using studs instead of bolts; see (B) in Fig. 9. (The same fixture is used for locating either bolts or studs.) Using studs makes storage easier because the studs can be removed and stacked one on another. Also, the studs may be used in other punches; thus, fewer studs would be used than bolts since bolts would be stored with each die. The studs should be made about $1/16''$ oversize and coated with graphite which makes the extraction (unscrewing) of the studs easier. The nuts are attached

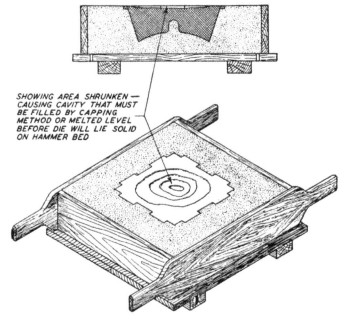

Fig. 10. Concave Shrinkage on Bottom Surface of Zinc Die, Caused by Cooling

to both ends of the studs and the studs are inserted into the molten mass where they are left until the metal has solidified; then the studs are removed by unscrewing, leaving the nuts embedded in the punch. It is now easy to insert standard size studs in the oversize holes for attaching to the hammer head. This method is similar to that used in the plaster pattern, Fig. 4, in that it leaves the nuts embedded in the lead punch like the nuts on the ends of the draw handles were embedded in the plaster. However, the stud is threaded on both ends to help in adjusting for depth location.

**Fastening Dies to Head and Base.** When the dies have cooled and been cleaned up, they are set on the base or anvil of the drop

# DROP HAMMER

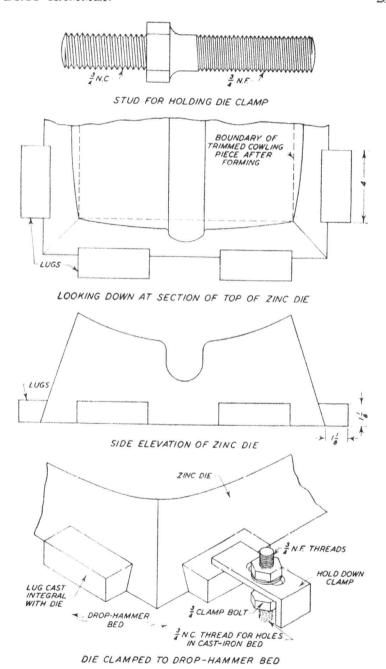

Fig. 11. Lugs and Clamps Which Hold Die to Drop-Hammer Bed

## 258 AIRCRAFT SHEET METAL WORK

hammer, one die in the other; the bolts (or studs) protruding from the male die are lined up with corresponding holes in the hammer

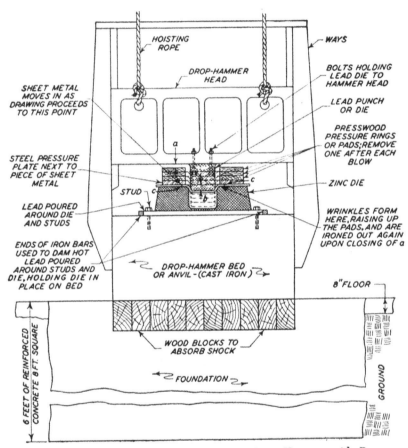

Fig. 12. Typical Job Setup of Deep Draw in Drop Hammer with Pressure Plate and Pressure Pads Employed for Controlling Wrinkles. Note Space *a* When Punch Is Drawing Sheet Metal Through *b* Distance. This Causes Wrinkles to Form at *c*; They Will Be Ironed Out Again Upon Closing of Hammer Head at *a* on Pressure Pads

head, and washers and nuts are applied and tightened securely. The hammer head, of course, has been lowered for this operation.

After the male die is fastened, the female die must be secured to the drop-hammer bed. Details of the clamping process are illustrated in Fig. 11. The die may also be secured by the leading-in process; see Fig. 12. One reason why clamping the die is preferred

# DROP HAMMER

is because the die can be installed at any time without firing up the lead pot, as is necessary for the leading-in process. For the use of clamps, lugs must be cast to the die when it is poured. This is accomplished by laying blocks of wood, about $1\frac{1}{8}'' \times 1\frac{1}{8}'' \times 4''$, around the edge of the plaster patterns (see Fig. 8). These blocks are removed from the sand after the pattern is pulled from the sand. When the molten zinc is poured into the mold, Fig. 10, the lugs are integral with the die; the die is fastened to the hammer bed as shown in Fig. 11. This system is preferable to leading-in, also, because the die when installed will be solid on the bed, whereas lead shrinks away from the die when cool, allowing slight movement. Leading-in is not permitted in some plants because of the likelihood of contamination in the remelting process. However, if the mechanic will exercise care in separating the lead from the zinc when doing the "tear down," there should be little danger of contamination.

In the leading-in process, an assortment of different lengths of inch-square iron bars are used as a dam, placed at a distance surrounding the female die; see Fig. 12. Inside this dammed-up area a few studs are screwed into tapped holes in the bed before the lead is poured, thus securing the die to the bed.

Except for checking for metal clearance, the hammer is now ready for operation.

**REVERSE DIES.** In large dies, especially those including vertical side walls and where a zinc punch is desired, we are forced to reverse the build-up procedure, because the solidification shrinkage causes too much metal clearance between punch and die. With a lead punch, clearance can be reduced by a few blows of the hammer head, causing expansion into the die, but zinc is too hard to respond to this treatment. Therefore it is necessary to cast the punch first and the die over it. This method does away with any metal clearance, because the shrinkage is toward the punch instead of away from the die, as would be the case if the procedure were not reversed. To visualize this, look first at (A) of Fig. 9 and consider how the hot lead for the male punch contracts on cooling and thus allows clearance between the lead die and the zinc die. Now, turn (A) of Fig. 9 upside down and think of the female die as the hot metal. You can see that the contraction would be toward the punch and would make for a closer fit.

To allow sufficient clearance so that the male and female dies can be separated, after the contraction, spray or brush a clay wash on the hot punch, which bakes on; then spray on a solution of water and iron oxide and sodium silicate (water glass), with the punch at about 500° F.; then brush over a coating of graphite paint, or smoke, with the oxyacetylene torch. Another way is to spray the following solution on the punch before pouring the zinc: one part of whiting powder; one part of silica flour; add water to make thin enough to use in spray gun; add one-half cup of molasses per gallon of solution.

To give additional clearance, heat the punch to about 450° F. before pouring the die. To follow this technique, of course, you must start with the plaster pattern, modeling it as the punch instead of the die.

**METALS USED IN DIES.** No doubt considerable improvement is yet to be made in the design and making of dies for the drop hammer. The most popular of the dies are made of lead for the male and zinc for the female dies. However, both male and female dies may be cast of zinc; or of aluminum alloy for female and either zinc or lead for the male die. *Kirksite* is a zinc-aluminum alloy already prepared for those plants and repair bases which do not alloy the zinc themselves. Dies of aluminum are sometimes employed for stamping stainless steel parts, especially when it is desirable to store the dies for future use. Aluminum alloy die castings are not only easier to handle for storage but are also cheaper, considering storage expenditures. For example, cast aluminum weighs 160 lb. per cubic foot; zinc weighs 448 lb. If aluminum is worth 20¢ a pound, and zinc sells for approximately 8¢ a pound, this means $32.00 for a cubic foot of aluminum compared to $35.84 for zinc. In periods of national shortage, however, neither metal can be stored for extensive periods.

Aluminum dies should not be used for deep draws or difficult forming of aluminum alloys where there is much movement of the metal over radii. Soft metals are likely to gall (abrade through friction) when passed tightly over cast aluminum, whereas stainless steel performs very satisfactorily. For stainless steel stampings in considerable volume, such as exhaust pipes, etc., steel dies are used.

**TYPES OF DIES. Wedge.** Wedge dies tend to split under

# DROP HAMMER

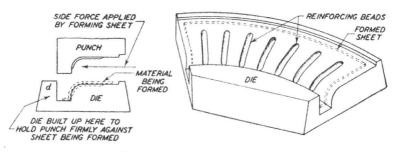

(A) DIAPHRAGM PUNCH AND DIE
(THE DIAPHRAGM IS THE STAINLESS STEEL SHROUD BEHIND THE EXHAUST COLLECTOR RING)

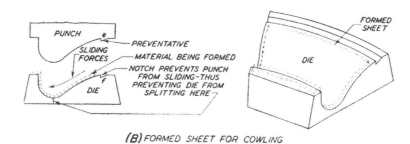

(B) FORMED SHEET FOR COWLING

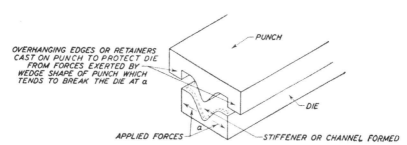

(C) PUNCH AND DIE FOR FLANGED V SECTION CHANNEL

Fig. 13. (A), Die Built Up at $d$ to Hold Punch Against Sheet Being Formed; (B), Notches at $e$ and $f$ Hold Punch in Place Where Sloping Surfaces Are Involved, (C), How to Prevent Die from Spreading or Breaking

the impact of the wedge-shape punch unless provision is made in design. Retainers, tongue and groove, or other devices built into the dies may prevent the zinc die from spreading or splitting. One such device is shown at (C) in Fig. 13. Where one side of a die is used for forming the sheet, as at (A) in Fig. 13 and where the punch tends to be pushed sidewise by the resistance in forming the sheet, a heavy portion is built into the die to counteract these lateral forces and to hold the punch tight against the sheet which is being formed. The die shown at (B) in Fig. 13 has a retainer notch or stop groove in a portion of its face where it will not interfere with forming the sheet. The punch is notched to fit into this groove, thus halting the sliding action which if permitted would split or spread the zinc die.

**Large Dies.** In working with large dies that extend over the edge of the base of the drop hammer, extensions are provided to support the overhang. The extensions are installed by bolting to the side of the base, the shear load supported by keyways, ledge, or tongue and groove; note bolt holes on bed in Fig. 1. Curved U, Vee, Zee, or hat channel sections with large radii can be formed by progressively feeding the work through the die. This type of die is made so as to gradually form the metal to the desired shape as it is fed through; it is sometimes called a *progressive die*.

**Blanking and Forming Dies.** When metal is to be formed by blanking and forming dies, the lower die is grooved, leaving a sharp edge. The metal is inserted between the dies, and the hammer head drops, blanking the metal to a general outline. Then, the hammer head is lifted, and rubber blocks are inserted on top of the metal and above the grooves. Again the hammer head drops, forcing the blocks and the metal into the grooves and thereby cutting the metal.

Sometimes, where simple shallow draws are to be made, it is unnecessary to cast a male die or punch. Instead, a heavy rubber blanket of sufficient thickness is used to form the sheet to the die.

**TECHNIQUES IN DROP-HAMMER FORMING.** When your setup is complete, the next thing is to start stamping out the pieces from the sheet metal. First, you must go to stock and charge out the number of sheets of the thickness called for on the drawing. These sheets are cut on the squaring shears, somewhat larger than the size of the part to be stamped. If many parts are to be stamped from flat stock, patterns should be made and arranged on the sheet so that the largest possible number of blanks can be cut from each sheet, thus reducing waste to the minimum.

**Making Simple Pieces.** Having cut out the blanks, we center one of them on top of the zinc die. To facilitate the forming of sheet metal, it is common practice to employ some lubricant such as

# DROP HAMMER

vegetable oil or drawing compound. Take hold of both ropes after switching on the motor (see Fig. 1); pull the hammer head off the safety dogs; step on the treadle which disengages the dogs; and let the hammer and punch fall. Again pull on the ropes, taking your foot off the treadle, and restore the hammer head and punch to rest position. Remove the stamped piece from the die and insert the next blank. This is all there is to it when making simple pieces.

**Progressive Forming.** If the required form is complex and there is a possibility of the metal wrinkling into folds or tearing, progressive forming is used, the metal being gradually worked into shape by successive blows of the punch.

Progressive dies are used only rarely in drop-hammer forming. They are used in a series, each die being nearer the shape of the finished part than its predecessor. The piece is thus formed in gradual steps.

Progressive steps may be taken to control forming in connection with deep or severe draws with the aid of half-hard rubber sheets of different thicknesses, cut to conform with the inside of the female die and placed in the bottom of it. The effect is similar to that of progressive dies. After each blow of the hammer, one thickness of rubber is removed, leaving a deeper die to receive the next blow.

Where possible in connection with severe draws, it is advisable to employ the progressive drawing method in which pressure pads are used. This method of progressive forming with one set of dies, Fig. 12, limits the depth of travel of the male die and at the same time shrinks and smooths out developing wrinkles in the flange, by using a pressure plate or collar of $\frac{1}{4}''$ steel plate, with additional pieces of pressed wood cut to the same shape as the steel plate. These pads are piled to the desired height on top of the pressure plate, which in turn is centered on the sheet to be formed, resting at the parting ledge of the dies and clearing the line of motion of the male die. Removing one thickness of pressed wood after each blow of the hammer brings the sheet metal a step nearer the final shape, while avoiding undue wrinkling. These pads or plates do not hold pressure on the flanges of the work until the male die starts drawing the metal and creating the wrinkles which are flattened out by the progression of the hammer head. When one or more of the shims are removed from between the dies after each blow or two of the hammer, any wrinkles which have formed in the sheet are simultaneously ironed out, so that they do not become folds in the walls of the stamping. This preforming process of shrinking the metal before it starts into the draw is the secret of successful deep draws with the drop hammer.

**Other Forming Techniques.** The drop-hammer operator quickly acquires techniques for a variety of forming operations. Sometimes he may let the hammer fall with its full weight. Or he may allow the male die to press the work very lightly into the female die, using a series of short punches, until he has observed the action of the metal. Then if the work is forming to his satisfaction, he gives it a final blow to drive the die home, resulting in as perfect a stamping as is required.

*Controlling Wrinkles.* The drop hammer has an advantage over the hydraulic press, in that the metal does not as readily stretch to the tearing point in deep drawing as it would in regular hydropress practice. There is a tendency for the sheet metal to wrinkle around the area of the unused stock, or at the top of the die. The maximum depth of drawing depends upon how successfully you can control the wrinkles developing in the sheet metal around the area at the top of the die, converting them into a thicker gage by shrinkage. In other words, drawing may continue as long as the wrinkles are kept under control. When they become too great to be taken up by shrinkage as they are drawn into the die, they must be shrunk by the use of a mallet for soft metal, or a lead or steel hammer for steel.

For heavy wrinkles, like those encountered when working with stainless steel, you will need the assistance of a steel crowfoot, Fig. 14, which straddles the wrinkle, held tight by the weight of the hammer head lowered upon it, and which holds the metal from crawling elsewhere or springing back. Try to visualize a flat sheet being drawn into a deep draw such as a stew pan. You can see that the flat sheet is not thick enough to stretch into the area necessary to make the pan. Then where is this metal to come from? It is drawn in from the outer surplus metal in the flange, Fig. 14. As the necessary material is drawn in, its circumference is decreased; the natural result is that it forms into wrinkles. These wrinkles diminish, if controlled in size, as they approach the top radius of

# DROP HAMMER

the die where the metal is shrunk as to width but thickened by the force of the operation. These wrinkles are called *out of control* if they grow so great that folding over results; see Fig. 14. Note how

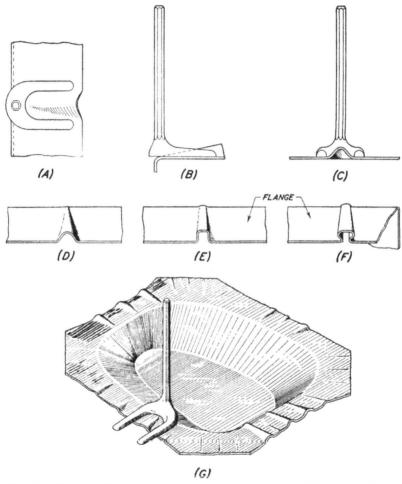

Fig. 14. Showing Use of Crowfoot in Straightening Out Wrinkle; It Keeps Metal Which Surrounds a Wrinkle from Forming Other Wrinkles

the crowfoot is used to help save the stamping. In Fig. 14, (A) is the top view of a crowfoot astride a wrinkle on the flange of sheet; (B) is the side view; and (C) is the front view. The wrinkle at (D) can be shrunk so that the metal will lie flat, if the crowfoot is placed as at (A) and if it holds the metal firm. The undesirable wrinkle at (E) is somewhat out of control, but it can be saved by first shaping the wrinkle with the mallet so it can be shrunk with the aid of a crowfoot. (F) shows a fold developing; it is too late to do much with this unless, toward the end of the draw, part of the wrinkle may be shrunk by removing metal from the die and straightening out over a round-headed stake. (G) is a picture view of a crowfoot over a wrinkle on a flange.

*When a crowfoot is used, the shaft extending up to the head must be sufficiently long to raise the hammer head off the safety dogs.* Never let the head come below the safety dogs to the shaft of a crowfoot when pounding out wrinkles between crowfoot prongs, because a misaimed blow might result in knocking out the prop and allowing the head to fall. For average work, wrinkles may be pounded out by holding the part being worked on in place by hand. In some cases, pieces of half-hard rubber placed in or on wrinkled areas help to preform or shrink the metal to a degree where a finishing blow will result in a smooth and finished product.

*Forming Nacelle Cowlings.* In forming nacelle cowlings of large convex surfaces, the use of clamping bars spaced about 3" or 4" apart on two sides of the female die, opposite each other, is recommended; thus the sheet to be formed is clamped to a bar which is secured rigidly to the die by means of brackets screwed into previously drilled and tapped holes for relatively large bolts. Clamp one edge of the sheet to the die and force the sheet down until it touches the high portions of the die, before clamping the remaining edge. If light strokes are used, the metal stretches and conforms to the male die before coming in contact with the female die, where the final setting of the metal is given. In forming of this kind, it is wise to use some lubricant such as vegetable oil, cylinder oil, or special drawing compounds.

*Forming Iron Templates.* When making cowlings or subassemblies, it is sometimes necessary to stamp out a template or component of a steel jig which fits the contour of the stamping or drawn part of the job in process. These parts usually are made from galvanized iron or black iron, and are used as drill jigs, cutting templates, or jigs with cutouts, where fittings are attached for locating stiffeners, angles, brackets, etc. Such jigs are especially used where interchangeable parts are desired in a fleet of airplanes. It is best to form the iron just before removal of the dies at the finish of a particular job, because the forming of the heavier gage iron which may be required is rather hard on the dies. In production work where the dies must be replaced with a new set because of excessive wearing in spots, the iron forming may be done by using both sets, progressively. Restamping of these pieces may be done with the new dies as soon as installed, provided the metal thickness is the same, or no thicker than, the original job unit, and that there are no violent wrinkles present.

**PARTS FORMED BY DROP HAMMER.** Parts formed by the drop hammer are made from annealed stock, and if of a metal which has a tendency to workharden quickly through drawing operations, the annealing process may have to be repeated two or three times during the forming operations. This is particularly true when forming stainless steel. Parts made from annealed alloys having heat-treatable characteristics are heat-treated after the forming operation to the desired strength. After heat treament, if the parts or pieces are large and there is danger of their warping from the heat, it is the practice to place them in the die and restrike them immediately, before the material develops its full hardness.

**Tubular Parts.** In making exhaust collector rings and stacks, or other tubular parts, the dies are made for punching right- and left-hand half-round pieces, which must be trimmed and assembled in jigs for welding or riveting together. Fig. 15 shows a workman assembling in a jig an exhaust collector ring for a Pratt and Whitney twin wasp, 14-cylinder engine; parts of the ring were formed by drop hammer and then welded. On the floor are clamps and accessories for the jig to hold stampings in place and for fitting, trimming, and tack welding. Trimming steel half-sections made on the drop hammer is by a band saw or a special deep-throat power shear.

**Large Parts.** In production work in factories, the drop hammer is limited to the smaller group of sheet metal parts, but in the repair bases large areas of metal sheets, three feet wide and four feet long or over, often are stamped out by attaching extensions to the drop-hammer bed. In the factory such work would be done on the hydraulic press. Lead-zinc dies may be used with the hydraulic press, also.

268                AIRCRAFT SHEET METAL WORK

**REMOVING DIES FROM BASE.** A large cold chisel is used to break up the dies for easy handling and remelting when the dies are held in place by the leading-in process. The chisel is held on the dies with long-handled tongs, while the drop-hammer head is

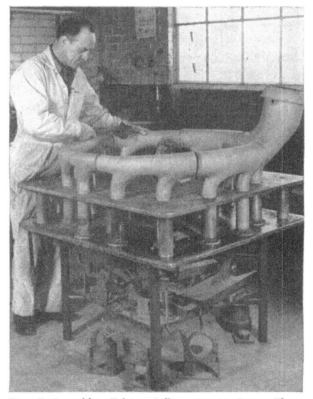

Fig. 15. Assembling Exhaust Collector Ring in Jig; on Floor, Clamps and Accessories for Jig
*Courtesy of United Air Lines, Chicago, Ill.*

utilized for striking power. The chisel should be made of $1\frac{1}{4}''$ hex stock, with a 2" blade, and should be 12" to 18" long. All particles of one die should be removed before breaking up another, unless both are made of the same material.

**CECOSTAMP.** The cecostamp, Fig. 16, is a machine that uses the same dies as the drop hammer. The chief difference between the two is that the cecostamp is air operated, and is much faster than the friction lift drop hammer. The cecostamp is sold

# DROP HAMMER

in 30" x 24" to 10' x 10' sizes. The machine features a built-in safety latch, and a cylinder built in above the operating cylinder which contains a piston that is under air pressure while the machine is in operation. Should the working piston for some reason go beyond its upper limit of travel, it will strike the safety piston, forcing

Fig. 16. Cecostamp, a Machine Which Uses Same Dies as Drop Hammer
*Courtesy of Chambersburg Engineering Company, Chambersburg, Pa.*

it against the air cushion and arresting further movement. Movement of the ram is controlled by a balanced piston valve which admits air to or exhausts air from the cylinder. This valve is operated either by a control level or by cams. The ram goes up or down as this lever is moved up or down. Moving the lever rapidly or slowly moves the ram accordingly.

### QUESTIONS

1. Why does the zinc punch not melt the zinc die when poured into it?

2. Why is the drop hammer used instead of the hydraulic press?

3. What would be your reason for routing a job to the hydraulic press instead of to the drop hammer?

4. At what temperature does zinc melt?

5. What causes the hammer head to raise when the rope is pulled?

6. What happens when lead and zinc are both melted in the zinc pot?

7. Where must the hammer head be, in relation to the drop hammer, when using a crowfoot or when handling work in the dies? Why?

8. What method is used to prevent dies from sticking together when poured one against the other?

9. When is rubber used in drop-hammer technique?

10. Which is an exact model of a lead punch, the plaster pattern, or the clay model?

11. Explain the shrinkage rule. Why is it used?

12. Tell as briefly as possible how to mix a batch of molding plaster.

13. How is a plaster pattern treated before using?

14. What prevents the sand from sticking to the pattern when it is pulled from the mold?

15. By what means is a plaster pattern pulled from the sand mold?

16. What is the highest temperature to which zinc should be heated in an iron pot?

17. What metal would you choose for the punch that is to be used in forming stainless steel exhaust pipes?

18. What is meant by leading-in?

19. How would you level up the bottom of a die of large size which has shrunk in solidifying?

20. Name the two methods of fastening the lower die to the drop-hammer bed. Which is the better way, and why?

Workers finishing installation of control fittings, inspection holes, and miscellaneous parts on outer wing sections of attack bombers.
*Courtesy of Douglas Aircraft Co., Inc., Santa Monica, Cal.*

*Chapter XI*

# Assembly, Repairs, Miscellaneous Techniques and Projects*

The objective of this chapter is to give you practice, through several varied projects, in using your hands in co-ordination with your mind, thus increasing your skill and overcoming any awkwardness; also to teach you to apply some of the things you have learned in preceding chapters. New things will be taken up, of course, in connection with the projects themselves.

You are urged to carry out each project to the very best of your ability. If any are not clear to you, go over them again at various times, until you have a thorough understanding of them; the additional practice will be well worth your while. Save your finished projects, as a hobby. They will prove interesting to your friends, also, as examples of the workmanship that goes into aircraft manufacture and maintenance.

**ASSEMBLY.** In the aircraft industry the assembly mechanic is one of the most important links in the chain of craftsmen who are responsible for the dependability and the commercial value of the airplane. Broadly speaking, the assembly mechanic is the last man to contribute his intelligence and skill to the finished product. For that reason, his is the work by which the quality of workmanship of the entire product may be judged. Thus, it is apparent that the assembly mechanic not only must work swiftly and efficiently but also must direct his skill toward keeping the work unscathed; that is, to prevent toolmarks, scratches, dents, and like blemishes which would make the work look like a used job.

When working on assembly, then, be careful in handling the formed parts and subassemblies. Use heavy paper on the bench top, and keep it free from drilling chips or tools that might scratch the surface of the metal.

---

* A portion of the material in this chapter and Figs. 2–9 are from the Civil Aeronautics Manual 18.

## AIRCRAFT SHEET METAL WORK

**EXTRUDED SECTIONS.** Extruded sections, such as those shown in Fig. 1, have greatly simplified aircraft design. They make

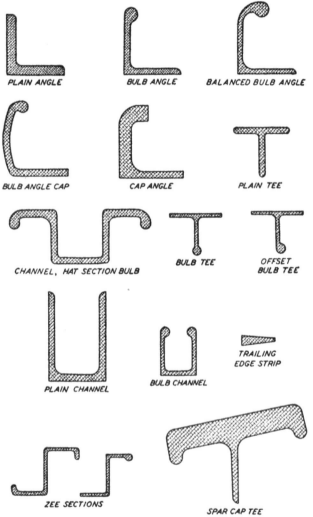

Fig. 1. Extruded Sections of Aluminum Alloys Used in Aircraft Fabrication

it possible to produce a better appearing job than did previous methods of fabricating or forming the various angles, channels, and like units. The extrusion process is an operation which forces the material, while in plastic condition, through dies. Almost any shape

# ASSEMBLY, REPAIRS, PROJECTS

needed by the industry is available; should a new design call for a new shape, the cost of making the necessary dies for its production is relatively small.

The bulb angle is the most popular of the extruded sections. Fuselage stringers and wing stringers usually are made of bulb-angle and bulb-zee sections. The bulb angle is formed in jigs around radii and offsets or joggles. Joggles are made by employing special dies, which unfortunately are too complicated to be made as a project at this stage. However, if joggle dies are available, the student should by all means seize the opportunity to use them.

**Bending of Tubes and Sections.** A tube is usually bent on a bending machine where an internal mandrel is used to prevent the tubing from flattening or collapsing at the point of bend. Among the materials used for filling thin wall tubing prior to bending are sand, resin, spiral springs, and Cerrobend or Bend Alloy (alloys of bismuth), besides the mandrel used with the tube bending machine. Lead, though sometimes used, is not considered satisfactory, as it shrinks in solidifying, thereby lessening its value as a support. For small tubing with relatively heavy wall thickness, the bending operation may sometimes be done without any filler, by use of the hand tube benders; however for very sharp bends, small tubing should be filled also.

The bismuth alloy Cerrobend, among others, has the properties that make it the ideal filler for tube bending, as follows: (a) low melting point, 160° F. (b) expands in solidifying, approximately .006" per inch (c) elongation factor, 140 to 200%. Cerrobend, or other similar bismuth alloys, must be melted in a double boiler, or placed in a clean ladle and held in boiling water. First, the tube should be swabbed with a rag soaked in thin machine oil, to prevent cold sets when the bend metal is poured into the tube; then the tube should be plugged with wood or cork, next, some of the hot water should be poured in, to preheat the tube. The water will run out as the alloy displaces it. When the tube is filled with alloy, it should be plunged immediately into a tank of cold water. This will rapidly solidify the alloy inside the tube, and the alloy will have the desired fine-grained ductile structure; the fine-grained structure is important because, in a severe bend, the bend alloy (alloy inside the tube) must stretch but not break. If the alloy is not doused in cold water and attains a coarse-grained structure, it will break. The filled tube must be left in the cold water long enough to cool thoroughly; this will take some time because of the low thermal conductivity of the bend alloy.

After the bending, the tube is set in boiling water so that the alloy will melt and run out. The tube is then swabbed out to remove any remaining drops of alloy.

Sections of angle or channel, either in the extruded or preformed shapes, may be embedded in a bar of Cerrobend, bent to the desired shape and melted out through the same procedure as used for tubing, except the alloy surrounds the section; instead of filling it.

Forming of this kind often is practical in repair or experimental work, where forming dies would be too expensive for the amount of work at hand. It is used to a limited extent in factories, also.

**ALUMINUM ALLOY STRUCTURES. Proof of Strength.** Extensive repairs to damaged stressed skin of monocoque* types of aluminum alloy structures ought to be made at the factory of origin, or by a repair station rated for this type of work. In any event, such work should be undertaken even by a certificated mechanic, only if he is thoroughly experienced in this type of work, and the repairs should be made in accordance with the specific recommendations of the manufacturer of the airplane.

In many cases, repair parts, joints, or reinforcements can be designed and proof of adequate strength shown, without calculation of actual loads and stresses, merely by proper consideration of the material and the dimensions of the original parts and the riveted attachments. An important point to bear in mind in making repairs on monocoque structures is that a repaired part must be as strong as the original with respect to all types of loads and to rigidity.

**Unconventional Attachments.** *Rivnuts, Lok-skrus,* drive screws, self-tapping screws, and other unconventional or untried attachment devices should not be used in primary structure, unless approved by either the Civil Aeronautics Administration for civil aircraft or by the Army and Navy for military aircraft.

**Aluminum Alloy Bolts.** Aluminum alloy bolts of less than $\frac{1}{4}''$ diameter should not be used in primary structure. Furthermore, no

---

* A type of construction of a fuselage in which the loads in the structure are taken principally by the skin, usually reinforced by bulkheads and stringers.

aluminum alloy bolts and nuts will be permitted where they will be removed repeatedly for purposes of maintenance and inspection.

**Aluminum Alloy Nuts.** Aluminum Alloy nuts may be used on steel bolts in shear in land planes, provided the bolts are cadmium plated; they must not be used on seaplanes.

**Drilling Oversize Holes.** Great care should be exercised to avoid drilling oversize holes, or otherwise decreasing the effective tensile area of wing spar cap strips; of longitudinal stringers of wing, fuselage, or fin; or of other members with high tensile stress. All repairs or reinforcements to such members should be done in accordance with factory recommendations and with the approval of a representative of the Civil Aeronautics Administration.

**Disassembly Prior to Repairing.** If the parts to be removed are essential to the rigidity of the whole, the remaining structure should be adequately supported prior to disassembly, in such a way as to prevent distortion and permanent damage. Rivets may be removed by special tools developed for that purpose or by center punching the heads, drilling not quite through with a drill the same size as the rivets, and shearing the heads off by a sharp blow from a small cold chisel. Riveted joints adjacent to the damaged parts should be inspected after repair has been made for partial failure (slippage) by removing one or more rivets to see whether the holes are elongated or the rivets have started to shear.

**Simple Test for Identifying Heat-Treated Alloys Which Contain Copper.** If for any reason the identification of the alloy is not on the material, it is possible to distinguish between heat-treatable alloys which contain copper and those which are not heat-treatable alloys, by immersing a sample of the material in a 10% solution of caustic soda (sodium hydroxide). The heat-treated alloys will turn black due to their copper content, whereas the others will remain bright. In the case of Alclad, the surface will remain bright but a dark area will appear in the center, when viewed from the edge.

**SELECTION OF ALUMINUM ALLOY FOR REPLACEMENT PARTS.** In selecting the alloy, ordinarily it is satisfactory to use 24S–T in place of 17S–T, since the former is stronger. On the other hand, it is not permissible to replace 24S–T by 17S–T, unless the deficiency in strength of the latter material is compensated for by an increase in material thickness or unless the required structural strength has been substantiated by tests or analyses. The choice of temper depends, of course, upon the severity of the subsequent forming operations. Parts which are to have single curvature (one curve in one direction and therefore straight bend lines) with large bend radii may be formed advantageously from S–T material; whereas a part such as a fuselage frame would have to be formed from S–O (soft annealed sheet) and heat-treated to the S–T condition after forming. Sheet metal parts which are to be left unpainted should be made of Alclad (aluminum coated) material.

All repairs involving 24S–RT alloy members should be made using the same material.

All sheet metal and finished parts should be free from cracks, scratches, kinks, tool marks, corrosion pits, or other defects which might lead to cracking.

**FORMING SHEET METAL PARTS.** Bend lines ought to lie at an angle to the grain of the metal (preferably 90°). Before bending, all rough edges should be smoothed, burrs removed, and relief holes drilled at the ends of bend lines and at corners to prevent cracks. For material in the S–T condition, the bend radius should be large.

**HEAT TREATMENT FOR ALUMINUM ALLOY PARTS.** All structural aluminum alloy parts should be heat-treated in accordance with heat-treatment instructions issued by the material manufacturers. If the heat treatment produces warping, the parts should be straightened immediately after quenching. Parts to be riveted together should be heat-treated before riveting, as heat-treating after riveting causes warping, and also because if riveted assemblies are heated in a salt bath, the salt cannot be entirely washed out of the crevices and will cause corrosion.

**REPAIR METHODS FOR ALUMINUM ALLOYS. Wing and Tail Surface Ribs.** Damaged aluminum alloy ribs, either of the stamped sheet metal type or the built-up type employing special sections of square or round tubing, may be repaired by the addition of suitable reinforcements.

Acceptable methods of repair are shown in Figs. 2 and 3. These examples deal with types of ribs commonly found in small and medium-sized aircraft. Any other method of reinforcement should

ASSEMBLY, REPAIRS, PROJECTS 277

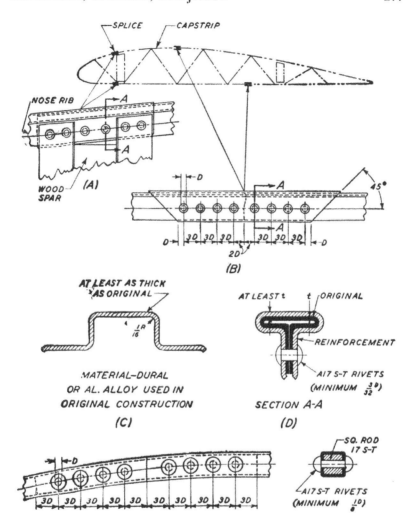

NOTE: COMMONLY FOUND ON SMALL AND MEDIUM SIZE AIRCRAFT (SEE FIG. 3)
Fig. 2. Typical Repair for Buckled or Cracked Formed Metal Wing Rib Capstrips

be approved specifically by a representative of the Civil Aeronautics Administration.

Fig. 2 at (A) shows how to splice to a rib a replacement for a nose section which has been damaged. The detail at (A) slants at a degree which indicates it is that portion at the top of the spar.

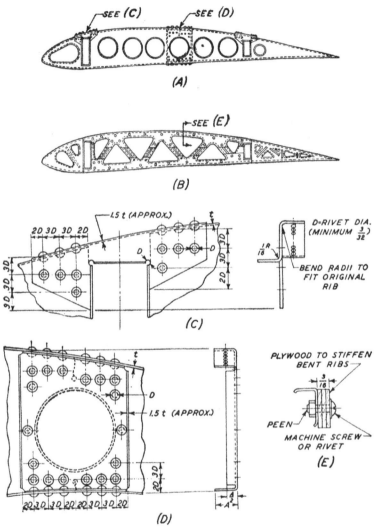

NOTE: COMMONLY FOUND ON SMALL AND MEDIUM SIZE AIRCRAFT
Fig. 3. Typical Metal Rib Repairs

The same procedure is applied at the lower part of the spar, also indexed with an arrow. The splice or repair of a broken capstrip between two attaching points of web stiffeners is shown at (B), Fig. 2. The hat section channel shown at (C) is the way a capstrip splice is formed before its final arrangement over a damaged portion of a capstrip shown at (D).

## ASSEMBLY, REPAIRS, PROJECTS

In Fig. 3, which illustrates typical metal rib repairs, (A) and (B) show metal ribs of two designs. (C) is a typical repair over the cutout for the spar. (D) shows a typical reinforcing patch for a cracked portion of rib (A). (E) illustrates how plywood is used as a reinforcement for a rib which has become bent or buckled.

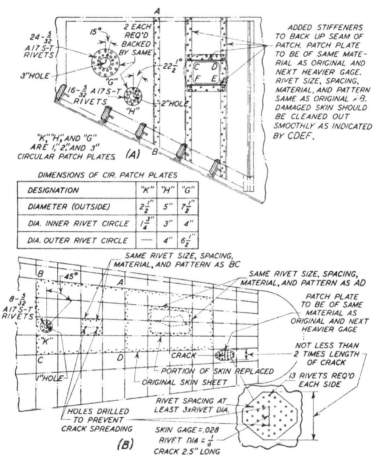

Fig. 4. Typical Repairs of Stressed Sheet Covering

**Patching of Small Holes and Replacement of Portions of Skin Panels.** Small holes in skin panels which do not involve damage to the stiffening members may be patched by covering the hole with a patch plate in the manner shown in Fig. 4. (A) illustrates typical patch plates, with added stiffeners for strength and circular patches with rivet size and spacings; (B) shows methods for making typical patches and repairs on the skin of a wing panel.

In case metal skin is damaged extensively, repair should be made by replacing an entire sheet panel which extends from one structural member to the next. The repair seams should be made to lie along stiffening members, bulkheads, etc., and each seam should be made exactly the same as the parallel manufactured seams at the edges of the original sheet, with regard to rivet size, spacing, and rivet pattern. If the two manufactured seams are different, the stronger one should be copied. Fig. 4 illustrates acceptable repairs.

**Splicing of Sheets.** There are cases in which the method of copying the seams at the edges of a sheet may not be satisfactory; for example, when the sheet has cutouts or doubler plates at an edge seam, or when other members transmit loads into the sheet. In these instances, the splice should be designed to carry the full allowable tension load for the sheet.

**Straightening of Stringers or Intermediate Frames.** *Members Slightly Bent.* Members which are slightly bent may be straightened cold and examined, with a magnifying glass, for injury to the material. The straightened parts should then be reinforced to an extent, depending upon the condition of the material and the magnitude of any remaining kinks or buckles. If strain cracks are apparent, complete reinforcements should be added, following recommendations of the manufacturers or the Administration of Civil Aeronautics. Attachment of such reinforcements should be made in sound metal, beyond the damaged portion.

*Local Heating.* Local heating should never be applied to facilitate bending, swaging, flattening, or expanding operations on heat-treated aluminum alloy members. It is unnecessary, in general practice, and it is impossible to control the temperature closely enough to prevent possible damage to the metal or impairment of its corrosion resistance. However, a torch with a large, soft flame is sometimes played over the surface of the cold-worked aluminum of the alloys which are not heat-treatable, to anneal for bending or forming. This practice is permissible for these types of alloys only when it is impracticable to anneal in a furnace or bath. The metal should not be heated to a temperature higher than that indicated by the charring of a resinous pine stick.

ASSEMBLY, REPAIRS, PROJECTS 281

**Repairing Cracked Members.** Figs. 5 through 9 show acceptable methods of repairing typical cracks which occur in structural parts in service, from various causes. Certain general procedures should be followed in repairing such defects.

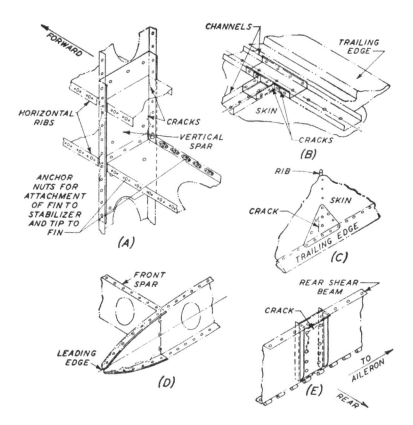

NOTE: ALL REINFORCING PLATES TO BE OF SAME ALLOY AND APPROX. 1.5 THICKNESS OF ORIGINAL

Fig. 5. Typical Methods of Repairing Cracked Leading and Trailing Edges and Rib Intersections

Fig. 5 illustrates typical methods of repairing cracked leading and trailing edges and rib intersections. (A) shows where to look for fatigue cracks in a multicellular structure; (B) shows how channel structures in trailing edges are repaired by the insertion of channel sections of heavier material; (C) is a skin patch over a trailing edge fatigue crack; (D) illustrates typical failure in leading edge nose rib; and (E) shows channel type splice over fatigue crack on false spar at aileron attachment.

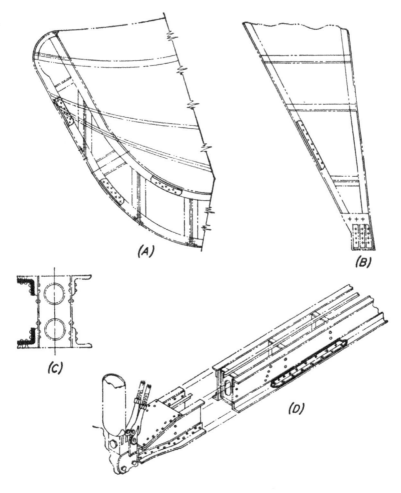

NOTE: STRENGTH INVESTIGATION USUALLY REQUIRED FOR THIS TYPE OF REPAIR
Fig. 6. Application of Typical Flange Splices and Reinforcement

On Fig. 6, (A) and (B) show application of a typical flange splice and reinforcement for angles made from sheet stock; (C) and (D) show application of a flange splice and reinforcement for extruded angles.

Typical methods of repairing cracked members at fittings are

# ASSEMBLY, REPAIRS, PROJECTS

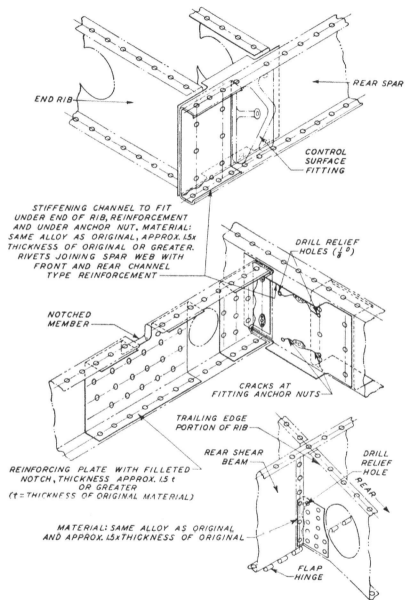

Fig. 7. Typical Methods of Repairing Cracked Members at Fittings

illustrated on Fig. 7. Fig. 8 shows methods of repairing cracked frame and stiffener combination. At (A), failures in a transverse stiffener are shown; (B) illustrates repair of cracked stiffener; and (C) is installation of gusset plate under skin next to stiffener inter-

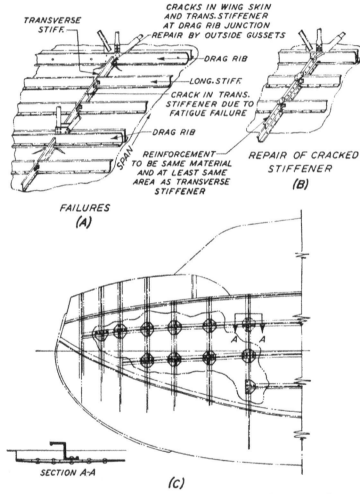

Fig. 8. Typical Methods of Repairing Cracked Frame and Stiffener Combination

sections which have failed through fatigue. Fig. 9 shows typical repairs to a rudder and to the fuselage at the tail post.

Small holes $3/32''$ or $1/8''$ should be drilled at the extreme ends of the cracks to prevent their spreading farther.

Reinforcements, as shown in the figures mentioned, should be

## ASSEMBLY, REPAIRS, PROJECTS

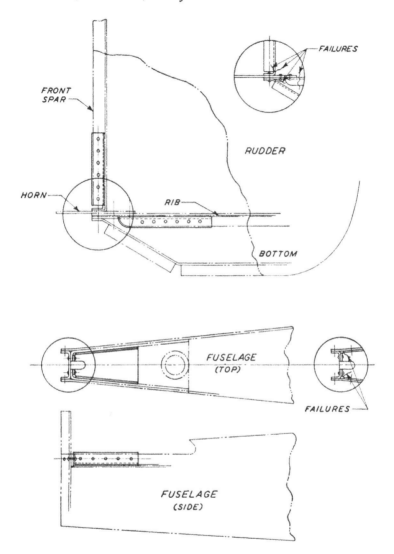

NOTE: USE SAME MATERIAL, NEXT HEAVIER GAGE FOR REINFORCEMENT.
Fig. 9. Typical Repairs to Rudder and to Fuselage at Tail Post

added to carry the stresses across the damaged portion and to stiffen the joints.

Cracks develop at a particular point because of stress concentration at that point, together with repetition of stress (such as produced by vibration of the structure). The stress concentration may be due to the design or to defects such as nicks, scratches, tool marks, and initial stresses or cracks occurring in forming or heat-treating operations. It should be noted that an increase in sheet thickness alone usually is beneficial, but it does not necessarily remedy the conditions which lead to the cracking.

**Splicing of Stringers and Flanges.** Splices should be made in accordance with the manufacturer's recommendations, usually contained in a Civil Aeronautics Administration approved repair manual. Stringers should be designed to carry both tension and compression. Fig. 7 shows some typical repair methods.

*Size of Splicing Members.* When material used for the splicing member is the same as was used in the original part, the net cross-sectional area of the splicing member ordinarily should be greater than the cross-sectional area of the element that it splices; see Fig. 7.

*Number of Rivets in Splice.* The number of rivets required on each side of a cut which is to be spliced in a stringer or flange may be determined as explained in the chapter on rivets. In any case, the rivets should be arranged in the splice so that the design tensile load for the member and splice plate can be carried into the splice without failing the member at the outermost rivet holes.

**CORROSION PROTECTION.** When unpainted alloys are used, except aluminum coated materials (Alclad), protective coatings should normally be applied. An important point is that *dry aluminum cannot corrode;* therefore, if effective means are provided to prevent moisture from coming in contact with aluminum alloys, corrosion will be prevented.

**Anodic Treatment and Priming.** Results of tests and of service experience have demonstrated that, in general, corrosion is most effectively prevented by anodic coating and by using a paint brush in the shop to apply primer on each individual part of the airplane structure *prior to assembly.* This is not a hard and fast rule, for there are certain subassemblies, relatively far removed from immersion and salt spray conditions (for instance, the wing area on a seaplane as compared to the tail section and hull), which have been found, in service, to perform satisfactorily when anodically coated and primed after assembly. For the most severe conditions, however, protection in every part before assembly should be accepted as general procedure.

**Aluminum Alloys in Contact with Wood.** Aluminum alloy elements in close contact with wood must be thoroughly moisture-proofed to prevent corrosion. This may be accomplished effectively by priming the aluminum alloy part with one coat of zinc-chromate primer, then thoroughly painting the wood with one coat of zinc-chromate primer followed by two or more coats of moisture-proof paint (bakelite varnish, for example).

**Joints Between Dissimilar Metals.** Joints between dissimilar metals should receive careful consideration. The best way to prevent corrosion is to insulate properly to keep moisture away. Insulation by placing a thin gasket or fabric, aluminum foil, cellophane, or impregnated fabric between the two parts is not sufficient for airplanes that operate from or over salt water. Water can bridge the insulation at the edges, resulting in electrolytic action, and consequently in corrosion. Therefore, it is of prime importance, in the case of seaplanes, that the edges of the whole joint be sealed with a compound (bakelite varnish, zinc chromate, or bituminous paste) to prevent the entrance of moisture. The following method has been used with success in the case of important structural fittings already riveted in place: sandblast the fitting and the adjacent metal lightly; then spray the whole area with aluminum primer, building up fillets along all seams and around fastenings; finish the assembly with a zinc-chromate primer and an external coating of standard aircraft paint.

*Bolts and Rivets in Joints of Dissimilar Metals or Wood.* Precautions should be taken with steel bolts and rivets which pass through aluminum alloys or wood. In such cases, particularly along the shank, the insulation discussed in the preceding paragraph is not practicable. This makes it necessary, at least in the case of seaplanes, to keep water away by using compound in the joint, dipping the fastenings in primer before inserting, and where possible, using insulating washers on the heads and nuts, and finally filleting all joints and around heads and nuts with compound before the final paint coats are applied.

**PRACTICES NOT CONSIDERED ACCEPTABLE. 1. Quenching in Hot Water or Air.** The quenching of 17S or 24S alloys in water above 100° F. or in air of any temperature, after heat treatment, is not satisfactory. If material is in Alclad form, air quenching is satisfactory when the air temperature is below 100° F.

**2. Transferring Too Slowly from Heat-Treating Medium to Quench Tank.** Insufficiently rapid transfer of 17S or 24S alloys from the heat-treatment medium to the quench tank is unsound practice. An elapsed time of 10 to 15 seconds will, in many cases, result in noticeably impaired corrosion resistance.

**3. Reheating at Temperatures Above Boiling Water.** To reheat 17S or 24S alloys after heat treatment to temperatures above that of boiling water, and to bake primers at temperatures above that of boiling water are not considered acceptable without subsequent complete and correct heat treatment, as these practices tend to impair the effects of the original heat treatment.

**4. Use of Annealed Alloys for Structural Parts.** The use of annealed 17S or 24S alloys for any structural repair of aircraft where corrosion is likely to occur is not considered satisfactory, because such materials are poor resistants to corrosion.

**5. Hygroscopic Materials Improperly Moisture-Proofed.** The use of hygroscopic materials which are improperly moisture-proofed, such as impregnated fabrics, leather, and the like, to effect watertightness of joints and seams is not acceptable.

**6. Leaving Traces of Welding Flux After Welding.** The leaving of any trace of welding flux after welding is not acceptable.

**7. Use of Paint Removers.** The use of paint removers which contain strong caustic compounds and of so-called thin paint removers which may have a tendency to run into the joints, instead of those which have a jelly-like consistency, is not considered satisfactory. Polishes and cleaners which have excessive abrasive action are not considered satisfactory. A suggested testing procedure for determining *safe* cleaners for aluminum, acceptable to the Civil Aeronautics Administration, is given in the following paragraphs:

*Tests for Cleaners Which Are to Be Used in Aqueous Solutions.* Specimens about 3" x .75" x .064" of aluminum alloy of the type under consideration should be exposed at 80° C. for five hours, to each of the following concentrations of the cleaner: .25, .50, .75, 1, 1.5, 2, 3, 5, 10, and 20%. For each specimen, 50 cubic centimeters of solution should be used.

ASSEMBLY, REPAIRS, PROJECTS 289

Some of the undiluted cleaner should be placed on other specimens which are stored in an atmosphere saturated with water vapor at 25° C. for 24 hours.

In addition to the two tests described, specimens should be cleaned following precisely the instructions furnished by the manufacturer of the cleaner; such cleaning operations should be repeated twenty times.

If none of the specimens submitted to these three tests become discolored, etched, or pitted, the cleaner can be considered *safe*. In the case of anodically coated material, the same procedure should be followed as outlined above.

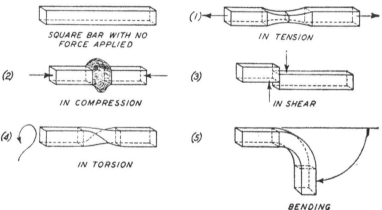

Fig. 10. Examples of Stresses

*Polishes, Abrasive Cleaners, Oily Cleaners, Etc.* Specimens of the aluminum alloy for which the cleaner is intended should be cleaned twenty times, following in detail the instructions furnished by the manufacturer.

Some of the cleaner should be placed on additional specimens which are stored in an atmosphere saturated with water vapor at 25° C. for 24 hours. If the specimens are not discolored, etched, or pitted, the polishing material can be considered *safe*. To be classified as completely safe, however, it must not abrade the aluminum alloy to an extent greater than does No. 0 steel wool.

**BASIC STRESSES.** The stresses to which an airplane or any structure is subject, and which it must be built to withstand, are illustrated in Fig. 10. Tension, compression, and shear are basic stresses, while torsion and bending are combinations of the three basic stresses.

**(1) Tension.** A stress tending to stretch or pull apart a member, making it longer. The lower portion of a spar is in tension while in flight.

**(2) Compression.** The stress that tends to compress or push together a member, making it shorter. The top portion of a spar is in compression in flight.

**(3) Shear.** The stress in shear is that which tends to slide one layer of fibers over the adjacent layer. An example would be that of rivets, which are always in single or double shear, as shown in Fig. 11.

RIVETS IN SINGLE SHEAR        RIVETS IN DOUBLE SHEAR

Fig. 11. Plates Riveted Together with Rivets in Single and Double Shear

**(4) Torque or Torsion.** The act of twisting, or the state of being twisted; the stress of a shaft, axle, etc., in a turning motion under load. Torsion is a combination of the three basic stresses.

**(5) Bending.** Bending also is a combination of the three basic stresses—tension, compression, and shear. When load is applied in such a manner as to cause a shear stress, tension results on one side and compression on the other. All three result in bending.

In airplane construction, we deal with all types of stress applied in different ways. Much improvement is possible; the mechanic will play an important part in obtaining the best possible strength-weight ratio through methods of fabrication. In thin aluminum alloy structure, little difficulty is experienced in tensile stresses. Compression and torsion, however, present a field for much improvement, and considerable research and experimentation are directed toward that improvement.

Engineers are required to design aircraft which can be built within a reasonable price range and without too radical experimentation. In the past, aircraft factories have not always been able to gamble with any but slight departures from accepted designs. Departures that look too radical on paper, even though they might result in advancement in design if they could be tried, are not advisable; however, experimental work is constantly in process, with the Army, the Navy, the airlines, and the factories each doing much toward advancement in design and procedure.

**LOCATING INACCESSIBLE HOLES FOR BOLTS OR RIVETS.** When installing a replacement fitting, bracket, angle, or any part in which holes are to be drilled to match holes already drilled in the structure, or in an installation where close fit is required but where marking through the original holes is difficult, proceed as follows:

Spread a layer of soft modeling clay, about $\frac{1}{8}''$ thick, over the blank part in the area where the holes must be marked, and press it in place, pushing the clay against the rivet holes or bolt holes. When the clay is removed, the impression of the rivet or bolt holes will appear in the clay; center punching and drilling accordingly will assure a satisfactory fit.

**ROLL STRAIGHTENING AND WORKHARDENING.** Either straightening (where a sheet of soft metal has become warped or bent) or workhardening (where it is desirable to stiffen a sheet of soft metal) may be accomplished through rolling. First, roll the sheet into a cylinder; then reverse and roll into a cylinder again, turning the first one inside out. Rolling for straightening or for workhardening is based on the principle of bending the metal beyond its elastic limit in one direction and then taking that bend out by reversing the process, by adjusting the rolls, and repeating until the sheet comes out straight.

**SHEET THICKNESS.** Aluminum alloys are rolled to the Brown & Sharpe system of sheet metal gages, but the thickness is referred to, both in engineering and in the shop, in one-thousandths of an inch. Metal gage standards are shown in the chapter dealing with *Steel in Aircraft Construction*.

**COLD CHISEL CUTTING ANGLE.** For general purposes, the chisel is ground to a cutting angle of 65° for relatively soft metals, increasing the angle for harder materials.

### PROJECT 1. MAKING A CLIP

**Objective**

How to make a pattern for a simple part when a dimension is missing from a drawing, as shown in Fig. 12

## 292  AIRCRAFT SHEET METAL WORK

**Tools and Equipment**

1. Layout tools
2. Aircraft snips—left and right
3. 8" mill file
4. Bending brake
5. Punch press (hand operated)

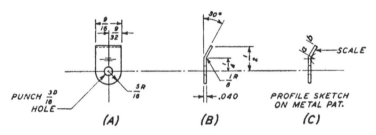

NOTES: 1-MATERIAL 24 ST ALCLAD

Fig. 12. At (A) and (B) Typical Drawings, for Mechanic's Use, of Small Part (Dump Chute Clip) with Dimension Missing; At (C) Mechanic's Drawing

**Material**

Small pieces of .040" 24S–T Alclad

**Information**

This project calls for making the clip shown in Fig. 12, where the dimension is not given for the length of the angle section, bent to 30°. In a part as simple as this, one could have the part laid out by trial and error within the time it would take to figure out the exact length by geometry. Therefore we recommend the following procedure:

**Procedure**

1. Let *(A)* and *(B)* represent the drawing, and *(C)* the mechanic's sketch. Lay out sketch *(C)* on metal, and take the dimension from *a* to *b* by actual measurements on the 6" scale.

2. Cut a 9/16" strip of .040" metal. Lay out that portion below the radius and add bend allowance and that portion found by scaling the laid-out profile.

3. Finish part and present for inspection.

### PROJECT 2. MAKING ANGLE, AND ASSEMBLING NUT PLATE WITH RIVET

**Objectives**

1. To make a simple assembly which includes layout, bending, and riveting, as shown in Fig. 13
2. Practice in making bend allowance

**Tools**

1. Layout tools
2. Hammer (riveting or ball peen)
3. Drill motor
4. Drills 25/64 (sheet metal), No. 40, No. 17
5. Countersink (72°)
6. Drill press
7. Bending brake
8. Aircraft snips

## ASSEMBLY, REPAIRS, PROJECTS

**Material**

1. 3"x3"x.064, 24S-T
2. One 8-32 nut plate (fiber) as called for on drawing
3. One 8-32x1/2" machine screw
4. Two 3/32x1/4 rivets (countersink head 78°) A17S-T

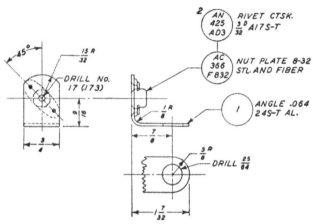

Fig. 13. Typical Drawing of Angle Bracket, Including Fiber Nut Plate

**Procedure**

1. Lay out angle bracket with nut plate from drawing as shown in Fig. 13, using the bend allowance method described in the chapter on bends.
2. Cut part out with aircraft snips.
3. Drill No. 17 hole in location for center of nut plate.
4. Drill 25/64" hole with sheet metal drill.
5. Bend angle in bending brake, making sure you use the proper procedure.
6. Fasten nut plate on the angle at an angle of 45° as shown in Fig. 13, using a 1/2" 8-32 machine screw.
7. Insert No. 40 drill through one of the small holes in the nut plate and drill a hole through the angle. Do the same using the other small hole as a guide.
8. Countersink outside of angle for 3/32 rivets and assemble.
9. Present for inspection.

### PROJECT 3. MAKING AN INSIDE AND OUTSIDE FLANGE
### (See Fig. 14A)

**Objectives**

1. To teach how to shrink metal when forming a flange
2. To teach how to stretch metal when forming a flange

## Tools

1. Layout tools
2. Hermaphrodite dividers
3. Mallet (with ball end)
4. Planishing hammer
5. Aircraft snips (left and right)
6. File, 12", second cut
7. Band saw
8. Sandpaper

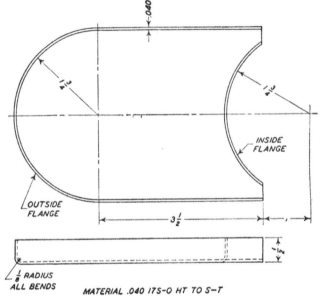

Fig. 14A. Typical Drawing of Part (Cover Plate) Which Requires Stretching and Shrinking in Forming

## Material

1. Piece of template material
2. Piece of hardwood (maple) 3/4" thick, twice the area of template
3. Piece of hardwood or fiber 1"x1"x5" (Cut one end off at a 45° angle, and round all edges to approximately 1/8" radii.)
4. Sheet of .040, 17S-O, 6"x8"

## Procedure

1. Lay out template on the template material to inside dimensions (disregarding flanges) of cover plate as shown in drawing, Fig. 14B.
2. Cut out the template and place it on the 3/4" piece of hardwood; mark around outline with a finely sharpened pencil, Fig. 14C. Saw out, leaving the pencil line. Repeat, because two blocks are needed.
3. Clamp blocks together in a vise and file their edges smooth. See that pencil outline is filed out so that blocks will be same size as template. The template may be clamped in with the blocks of wood and used as a guide in filing. If a disc or belt wood sander is available, it may be used here to good advantage, but the inside radius will have to be done by hand

# ASSEMBLY, REPAIRS, PROJECTS

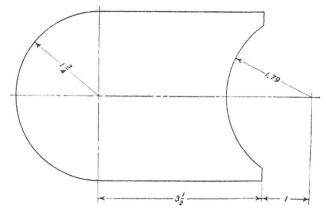

Fig. 14B. Template Laid Out to Inside Dimensions of Cover Plate

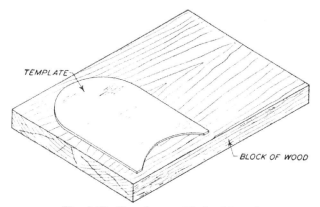

Fig. 14C. Template on Block of Wood

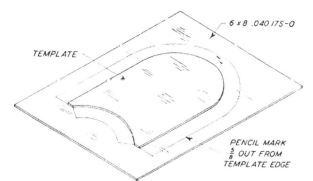

Fig. 14D. Pencil Mark Made on 17S–O Material, Using Template as Guide

filing and sandpaper. The block edge over which the metal is to be flanged should be shaped using a 1/8" radius so the flange will have a 1/8" bend.

4. Place template on the 6"x8" piece of .040 17S-O and mark approximately 5/8" (plus or minus 1/16") out from the edge with a soft pencil, which gives allowance for the flange, Fig. 14D.

5. Cut out along pencil line and smooth the edge with a file.

6. Center the piece of Dural between the two blocks and clamp both with vise and C clamps, Fig. 14E. The C clamps are used so the pieces will not fall apart each time the blocks are turned in the vise throughout the flanging operation.

7. Start the flanging operation with the inside radius (see Fig. 14A) because it is easier to stretch metal, as required for the inside flange, than to shrink it, as required for the outside flange. First smooth the edge of

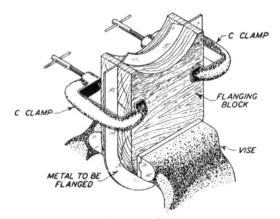

Fig. 14E. Metal Centered Between Forming Blocks of Wood

the metal with fine grade emery cloth so as to avoid cracks which might form if the edge were rough. Flange the metal over, using the ball end of your mallet. Distribute your blows evenly so that the flange will develop smoothly and without unnecessary dents or blemishes which would have to be ironed out.

8. When the metal has reached the form of the block for the inside radius, turn the project in your vise and partially turn down the two sides as shown in Fig. 14F.

9. To flange the outside radius, Fig. 14A, use the 1"x1"x5" block of wood or fiber and the mallet, as shown in Fig. 14F. As you progress around the outside flanged radius, you will notice the metal has a tendency to wrinkle. Such wrinkles can be avoided by distributing your blows carefully and not hurrying the process. Experience will improve your technique. The edge of the flange must be kept turned up, as shown in Fig. 14F, as the metal is drawn in toward the turned up edge, thus causing it to shrink; this also prevents wrinkles.

## ASSEMBLY, REPAIRS, PROJECTS

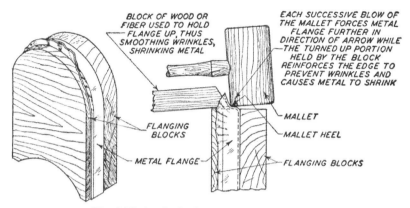

Fig. 14F. Method of Forming Outside Flange

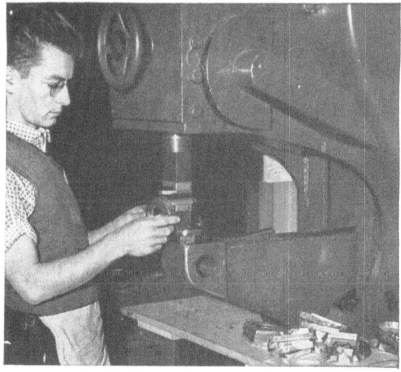

Fig. 14G. Workman Forming Part on Punch Press
*Courtesy of Douglas Aircraft Company*

298  AIRCRAFT SHEET METAL WORK

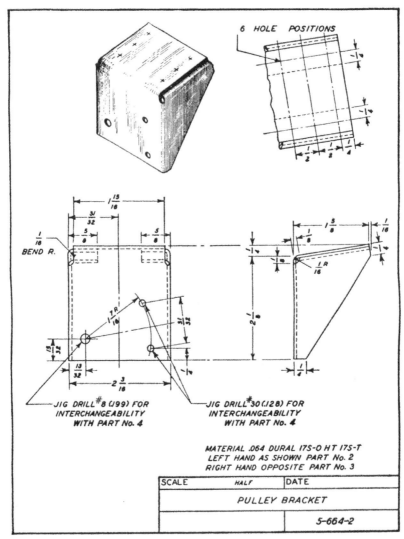

Fig. 15. Typical Drawing of Airplane Pulley Bracket

In forming flanges of this kind around sharper curves or with a larger flange, the block of wood may be relieved, i.e., tapered or beveled back from the radius of the bend to allow for spring back. In other words, when a flange is being formed to 90° and will not stay down as it should, the wood block is tapered or beveled; the flange is then formed to 90°.

10. Hammer smooth with a planishing (smoothing) hammer, and remove from flanging blocks.

# ASSEMBLY, REPAIRS, PROJECTS 299

11. Mark the flange off at 1/2" with hermaphrodite dividers or lay flat on surface plate with flange up, and scribe the cutting line with a universal surface gage where the point of the scribe has been set, at a height of 1/2" above the surface of the plate.

12. Trim off surplus metal with aircraft snips and file smooth.

13. Heat-treat, and present for inspection.

Fig. 14G shows this operation being done by machine.

### PROJECT 4.  MAKING BRACKET FOR PULLEY ASSEMBLY

**Objective**

1. To lay out and assemble a group of parts, as shown in Figs. 15 and 16, without the usual detailed instruction

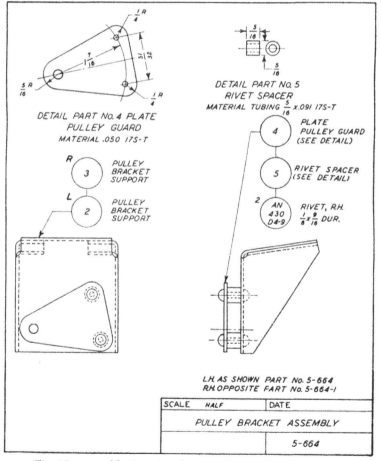

Fig. 16. Assembly Drawing Showing Parts Made According to Specifications on Fig. 15

**Tools**

Choose those you think you will need.

**Material**

Using information on the face of the drawing, choose all the material you will need.

**Procedure**

Lay out all parts and make them ready for bending and fabrication.

**Note**

Flanging blocks would seem necessary because the folded flanges are neither in a straight line nor on the same plane. Do not hurry; this project will take some time, possibly extending through several periods. If you spoil any of the parts, that is to be expected at this stage. Simply start over, determined to do a perfect job.

### PROJECT 5. BRACKET FOR HOLDING LORD SHOCK
### (Make Bracket—Omit Lord Shock)

**Objective**

To practice layout and blueprint reading by working a project without instructions.

**Tools**

Choose the tools you need.

**Material**

Find the material needed from the drawing, Fig. 17. If there is anything you do not know, either ask your instructor, or read that part of the text where it is explained.

Present for inspection.

### PROJECT 6. MAKING 90° BRACKET WITH LAPPED TABS

**Objective**

To teach how to use bending brake in making offset bends by folding up a fitting such as shown in Fig. 18

**Tools**

Choose those needed.

**Material**

Choose from drawing.

**Procedure**

1. Lay out the pattern and cut out the blank; see Fig. 18.

2. Brake up fitting by using filler pieces of proper thickness (spacers) to form offset in the tabs so they may be lapped together when all bends are made.

The spacers should be cut to approximately conform with tab size, and fastened in place with orange shellac. (Note that in (B) in Fig. 18 the spacers were made larger than the tabs so they could be seen.) Brake jaws

# ASSEMBLY, REPAIRS, PROJECTS

must be set to twice metal thickness for clearance. Spacers must have one edge filed to desired radius (1/16 for this particular bracket) and must be stuck in place with rounded edge flush with brake reference line, Fig. 18, for outside tabs. The center tab will be extended too far in the brake; for

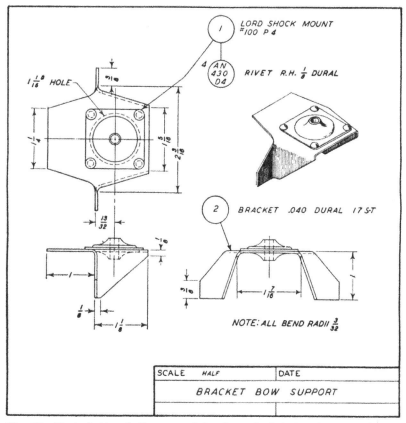

Fig. 17. Typical Aircraft Drawing of Bracket Which Is Hard to Visualize in Flat Pattern; Note Isometric at Upper Right

this reason, oil top of center tab and bottom of center spacer and brake up before the shellac dries, permitting it to draw out through the bending operation.

*Caution: Do not let upper brake jaw clamp metal too tightly.*

3. Remove spacers and clean off the shellac, with alcohol if necessary. Brake up triangular sides by inserting each side in that end of the brake which gives it clearance for angle already bent up.

4. Present for inspection.

## PROJECT 7. MAKING 6" COLLAR WITH WIRED EDGE 2" DEEP

**Note**

In early models of airplanes, a great portion of the cowling and metal parts were rolled wire edged for reinforcement, mainly because of the softer alloys used. Today there is occasional need for the sheet metal mechanic to

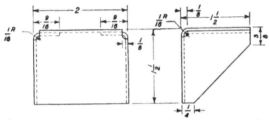

*(A)* 90° BRACKET-MATERIAL .064 17S-O HT TO S-T

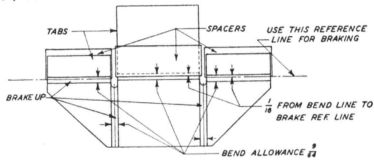

*(B)* STRETCH-OUT OF BRACKET

*(C)* SHOWING HOW SPACERS ARE INSERTED WITH
TABS TO FORM JOGGLED BRAKE LINE

Fig. 18. At (A), Typical Drawing of Bracket; At (B), Stretch-Out; At (C), Bracket Inserted in Brake Using Spacers from Same Material to Form Offset (View from in Front of Brake Jaws)

perform a wiring job; the process will be learned by working this project. The wiring of 17S-T and 24S-T over 1/8 wire is limited to .032 and .025 gage respectively. Heavier gages may be wired in the soft condition. The use of the wiring machine is not considered practical in wiring aluminum alloys; the work must be done by hand, using wiring tongs. The allowance for wiring is found by the following formula: *Add twice the diameter of the wire to four times the thickness of the metal.*

# ASSEMBLY, REPAIRS, PROJECTS

As an example, suppose we construct a tube with 6" diameter and rolled wire edge. The wire is enclosed in the edge of the metal in the flat sheet before it is rolled into shape, the rolls having a groove at one end of the rollers for clearance for the wired edges.

**Objectives**

1. To teach hand wiring procedure
2. To learn more uses for the bending brake

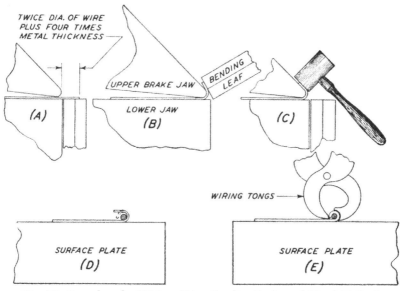

Fig. 19. Procedure for Wiring, Using Bending Brake and Hand Method

**Tools**

1. Rawhide or pyralin mallet
2. Wiring tongs
3. Bending brake
4. Snips
5. Mill file
6. Scale, 6"

**Material**

1. Dural .025 17S–T, 2"x20"
2. Wire, 1/8 Dural, 20" long

**Procedure**

1. Mark with soft pencil the location of wire allowance for bending. Cut to length, allowing 1/2" for seam (6" dia.).

2. Clamp and bend up as far as bending leaf will permit, as shown at (B), Fig. 19.

3. Use mallet and form angle against jaw, as shown at (C).

4. Remove from brake and place on surface plate; insert wire and form edge of metal around wire, as shown by dotted line at (D), with mallet.

One end of the wire should be placed about 1" from the seam, or in other words 1½" in from the end, including the 1/2" lap. (The wire must extend this amount on the other end, because, after the cylinder is rolled, this wire is inserted in the 1½" portion previously left open.)

*Note:* A setting hammer is used for wiring heavy black or galvanized iron. Aluminum alloys mar too easily for this procedure.

5. After the edge is turned down as far as can be done with the mallet, finish the setting operation with the wiring tongs, as at *(E)*, Fig. 19. Let the end where the wire is set back remain open, tapering off the wiring operation there.

6. Run the work through the rolls so that the wire will be on the outside of the cylinder. Notch out portion for seam in open rolled end, slip protruding wire in place, and form the end which was left open around it.

7. Rivet seam and present for inspection.

### PROJECT 8. MAKING CORRUGATION—SQUARE TYPE

**Objective**
1. To teach another use of the bending brake

**Tools**
1. 6" scale
2. Large adjustable wrench (12" crescent)
3. Large screw driver
4. Protractor head and scale
5. Bending brake

**Material**
1. Dural 17S-T, 6"x8", .016 or .018

**Note**
Brake the entire piece of metal (6"x8"x.016) into a piece of square corrugation similar to that shown in Fig. 1, Chapter 6. Make it so the finished corrugations will be 5/16" thick, with 100° web.

**Procedure**

1. Remove the 1/4" steel bar from the edge of the bending leaf with the screw driver to allow clearance for making the narrow reverse bend necessary for this size corrugation.

2. Set bending leaf stop with wrench, so that by bending a small piece of metal it will check to 100° with the protractor.

3. If upper jaw of brake has a 1/32 radius, it may be set back for excessive metal clearance, so that each time a brake is made, the edge of the bending leaf may be used as a gage, showing how far to insert metal for making each succeeding bend. Place the sheet lengthwise in brake, so the first brake will be 6" long, and brake it up 5/16" to an angle of 100°. Remove sheet, turn it over, and make two more brakes, and so on. Occasionally measure the over-all of the corrugations to be sure you are not gaining on one side, that is, going fan-wise; allow for correction in future brakes if necessary.

4. Present for inspection.

# ASSEMBLY, REPAIRS, PROJECTS

## PROJECT 9. TIGHTENING NUTS

### Objective

To learn the feeling of the exact pressure needed to tighten the nut on various sized aircraft bolts.

### Tools

1. End wrenches: 3/8 open end, 7/16 open end, 1/2 open end, 9/16 open end
2. Micrometer, 1"
3. Vise

### Material

1. Two each of 3/16, 1/4, 5/16, and 3/8 nuts and bolts (Bolts should be the shortest available.)
2. Oil can with light oil
3. Enough washers of each size to reach well above bearing surface into threaded portion

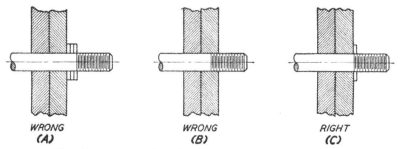

WRONG (A)   WRONG (B)   RIGHT (C)

Fig. 20. Wrong and Right Choices of Lengths in AN Bolts

### Note

One of the most important and also most difficult procedures for a new mechanic to learn is to tighten nuts properly. They are tightened either too loose or too tight, usually the latter which causes the metal in the bolt to reach its elastic limit. (Elastic limit is the maximum intensity of stress to which any material may be subjected and still return to its original shape upon the removal of the stress.) When a bolt has been reinstalled several times, it stretches beyond its usefulness. This can be seen by the naked eye, in that it acquires a bottleneck shape at the point between the nut and the bearing surface.

Aircraft bolts are made from SAE 2530 steel, heat-treated to 125,000 lb. per sq. in. tensile strength. Although this steel possesses great strength, a mechanic can easily stretch a bolt as he applies pressure, because of the great leverage exerted through screw pitch and length of wrench handle.

AN bolts should always be selected by the grip length which should reach through as in (C) in Fig. 20. No thread should be allowed in bearing surface as at (B). A bolt with too long a grip length, causing the excessive use of washers, should likewise be avoided; see (A) of Fig. 20.

Wrenches used in aircraft work are the socket, box, or end types. They must be kept polished so as not to scratch the metal. A torque wrench is used for checking tightness of nuts and in assembling engines, hydraulic units, etc. If a torque wrench is obtainable, use one in practice to help acquire the feeling of the correct pressure to be applied. The maximum torque loads for AN bolts are given in the following table:

| Size of AN Bolts | Maximum Torque Load |
|---|---|
| 3/16 | 40 inch-pounds |
| 1/4 | 90 inch-pounds |
| 5/16 | 150 inch-pounds |
| 3/8 | 350 inch-pounds |
| 7/16 | 46 foot-pounds |
| 1/2 | 75 foot-pounds |
| 9/16 | 125 foot-pounds |
| 5/8 | 150 foot-pounds |
| 3/4 | 250 foot-pounds |

**Procedure**

1. Place a 3/16 aircraft bolt less than 1" long in the micrometer. Measure its length. Put your findings down on paper.

2. Place the head of the bolt in the vise, threads up.

3. Place enough washers on the bolt to cover about two threads and screw on nut.

4. Tighten nut to where you judge it should be taut enough.

5. Remove from vise and mike° the over-all length. If properly tightened it should measure from one to two 1/1000ths inch longer. Remove the nut and see if the bolt has returned to its original length. If it does not, you have tightened the nut too much. If it does, you have tightened the nut properly, providing it showed a gain in length when miked° with the nut tightened.

6. Repeat the procedure with each of the other bolts.

7. Repeat the procedure using oil on threads. Note the results.

### PROJECT 10. MAKING CYLINDER
### 3" Diameter by 6" Long, with Grooved Seam

**Objective**

To learn how to make a seam similar to that used in stovepipe

**Tools**

1. Layout tools
2. Mallet
3. Hammer, 1 lb. ball peen
4. Folder
5. Sheet metal rolls
6. Hand groover
7. Shears (hand)

*Note:* The hand groover is a stock tool, but may be made from a piece of fiber or hard wood.

**Material**

28 Ga. (gage) galvanized iron, 6"x12"

---

° *Mike*, or *miked*, is a term used by mechanics and machinists when they mean the act of measuring by a micrometer.

# ASSEMBLY, REPAIRS, PROJECTS

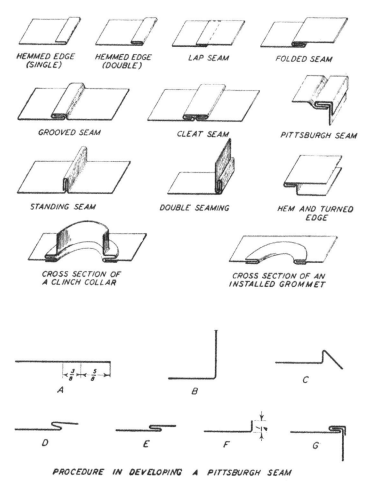

Fig. 21. Above, Common Seams; Below, Progressive Steps in Making Pittsburgh Seam When Unobstructed Inside Corner Is Desired (Compare with Pittsburgh Seam Above)

**Note**

The amount of material to be added to the pattern for making a grooved seam depends upon the width of the single edges turned on the folder, and upon bend allowance, and like factors. Three times the width of the single edge may be added for fairly close work. If very close diameter of the pipe is necessary, or if heavier material is used, the actual amount of material taken up by the bends must be added. In this case, take a 1" strip of metal 4" long and cut it in two parts. Groove the seam and close it down with a mallet and then measure accurately the length of the strip. The difference between this dimension and the length of the piece before cutting and seaming will be the amount of material which should be added for the seam.

**Procedure**

1. Lay out the length for a 3" pipe and add three times the width of the hem (1/4"); cut off surplus metal and save piece; see Fig. 21 for illustration of grooved seam.

2. Mark a line 1/4" from the edge that is to be folded, and place in folder for adjustment. Adjust folder so that line will show about the thickness of the metal out from the jaw; fold metal over all the way. Remove metal from folder and place the strip of metal you trimmed off the end of the sheet in the fold of the metal to act as a spacer. Place the folded edge on top of the jaw; pull the lever until the bar stands vertical; push the hem against the bar; and pull lever, pressing the hem flat against the spacer. Repeat the operation on the other end and opposite side of the sheet.

3. Set the rolls so that the hems will not be closed together; then roll the sheet into a tube. Hook the hems together, then slip the pipe over a conductor stake or piece of shafting for grooving. Set the metal, as shown in the grooved seam in Fig. 21, with hand groover and hammer. Pound the seam with the mallet to close it down, leaving it tight and smooth.

4. Present for inspection.

### PROJECT 11. MAKING AND INSTALLING SHEET METAL GROMMET

**Objective**

1. To teach the process of forming metal
2. To improve the quality of workmanship

**Tools**

1. Layout tools
2. Mallet
3. Hammer, ball peen
4. Hermaphrodite dividers
5. Punch, hand lever
6. Band saw
7. File, 12" half round
8. Aircraft snips, left and right

**Material**

1. Dural, .032x4"x6", 17S–T
2. Aluminum, 2S .050 4"x4"
3. Emery cloth, 120 grit or finer
4. Maple board, 4"x8"x1/2"

**Procedure**

1. Find exact center in .032x4"x6" sheet of Dural and punch a 2" hole.

2. Find exact center in aluminum .050x4"x4" and punch a 1" hole there. Smooth edge of hole with fine emery cloth.

3. Cut maple board in two pieces, each 4"x4", and find exact center of one of them. Clamp boards together, and drill a 1/8" hole in two opposite corners at a point 1/2" from each edge (see (A) in Fig. 22), and insert a short length of 1/8" welding rod for dowel pins (see (B) in Fig. 22). Next, make a 2" hole through both pieces, which can be done in several ways. Mark the circumference of the hole with dividers and saw out with jig saw;

## ASSEMBLY, REPAIRS, PROJECTS

or, employ a brace and expansive bit; or, use the drill press and a 2" Forstner (bit) hole cutter. Be sure to place a board to back up the work on the drill table, using clamps to hold down the blocks while drilling. Choose one board on which to form the flange, and form a small radius (1/32") on one edge of the hole with emery cloth.

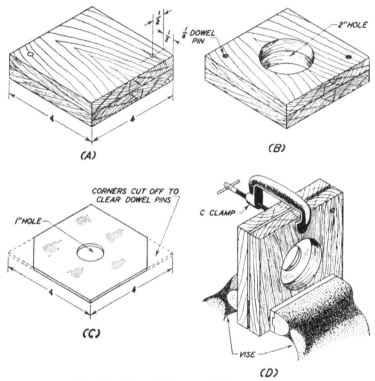

Fig. 22. Some of Steps in Making a Grommet

4. Notch out the piece of aluminum on two opposite corners to clear dowel pins, see (C) in Fig. 22. Center hole in sheet metal with hole between forming blocks. Clamp the forming blocks, with the sheet of aluminum in place, in the vise as in (D), Fig. 22; also clamp the upper portion of the forming blocks together with a C clamp.

5. With a ball end mallet, or the polished peen of a 1 lb. ball peen hammer, flange the extruding metal to the contour of the hole in the wood forming blocks. Stretch the metal by striking first just outside the hole, and working carefully around it so as to set the metal down gradually. If the metal starts to crack at the edge of the hole, file out and smooth with fine emery cloth.

6. After the flange has been set down to conform to shape of the 2" hole in the flanging blocks, remove and mark the flange to a width of

3/8", trim off, and smooth with file and emery cloth. Mark the outside diameter of the grommet to 2¾", cut out, and smooth edge.

7. Insert the grommet in the hole made in the piece of 17S–T, filing clearance if necessary. Lay flat with the flange up on a plate; then turn flange down, completing the grommet. Try to make a smooth neat job.

8. Present for inspection and comment.

### PROJECT 12. MAKING A PIANO HINGE

**Objective**

To practice making a part from a drawing, using knowledge gained from completed projects

**Tools**

Select those needed for the job.

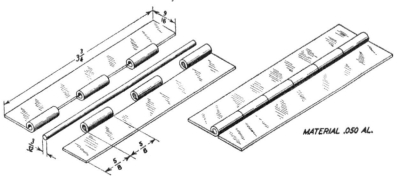

Fig. 23. Piano Hinge Used in Aircraft Work

**Material**

Select from the drawing.

**Procedure**

Lay out and make the section of piano hinge shown in Fig. 23.
Present for inspection and comment.

**BOLTS AND SCREW THREADS.** For years the industries have been urging the adoption and use of the United States Standard Thread for all screw threads, wherever possible, and the SAE (Society of Automotive Engineers) Standard where finer pitches are required. Those two standards are in general use throughout the aviation industry, as is also the ASME (American Society of Mechanical Engineers) Standard for machine screws.

The National Screw-Thread Commission has recommended the name National Standard Thread in place of United States Standard Thread, using the letters *N.C.* for National Coarse, and *N.F.* for Na-

## ASSEMBLY, REPAIRS, PROJECTS

tional Fine, in place of the former method of marking—U.S.S. and SAE; National Special (N.S.) is used for other pitches of this same shape thread. Therefore, in referring to threads, dies, or taps, bear in mind that:

N.C. is the same as the former U.S.S.;

N.F. is the same as the former SAE;

N.S. is the same as the former U.S.S.S. (United States Standard Special).

**Taps.** Taps are furnished in tapering, plug, or bottoming styles, as shown in Fig. 24. The sheet metal mechanic uses the taper

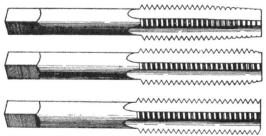

Fig. 24. From Top to Bottom, Taper, Plug, and Bottoming Taps; Also Frequently Used in This Order

more than the other taps, because of the necessity for chasing (cleaning out) old holes, or tapping through relatively thin sections. In jig or assembly work, however, it is often necessary to use all three types in making threads to the bottom of a hole.

*Caution:* A tap should never be run into the cotton cellulose fiber of an elastic stop nut.

In tapping holes, use plenty of suitable oil (lard oil or cutting oil). Do not try to turn the tap when it becomes bound, especially when tapping threads in hard material; when binding takes place, do not force the tap, but back it out and clean the chips out of the hole and out of the tap flutes before proceeding. (This cleaning applies also to the use of the thread cutting dies, except that they do not have to be removed; instead they are backed up one turn and cleaned out with air. Cutting threads with a die is a much more difficult operation than tapping. The die is adjustable and requires manipulation in order that the finished job will not present a too tight or too sloppy fit.)

# 312 AIRCRAFT SHEET METAL WORK

## PROJECT 13. TAPPING OR CUTTING INTERNALLY THREADED PARTS

**Objective**
1. To teach procedure of drilling proper sized holes for tapped threads
2. To teach proper use of a complete set of taps in any given size

**Tools and equipment**
1. Ball peen hammer
2. Center punch
3. Twist drill (refer to thread cutting table for proper size)
4. 1/4"x28 taps (taper, plug, bottoming)
5. Tap wrench
6. Drill press
7. Drill vise
8. Drill, to be selected

**Material**
1. Mild steel block, approximately 1/2"x1"x1"
2. Cutting oil

**Procedure**
1. Center-punch hole location in center of a 1/2"x1" face of the steel block.
2. Select proper size drill for tapping 1/4"x28 N.F. thread, and chuck drill in press.
3. Place steel block in drill vise, making sure that top surface is level.
4. Set drill for 3/4" depth and drill hole in steel block, using cutting oil.
5. Remove work and place in bench vise.
6. Mount taper tap in tap wrench. Lubricate hole and tap; then tap hole to its full depth.
7. Replace taper with plug tap, and repeat operation 6.
8. Replace plug with bottoming tap, and repeat operation 6.
9. Present work for inspection.

## PROJECT 14. THREADING A BOLT OR STUD

**Objective**
1. To teach how to cut external threads to fit a tapped hole

**Tools and equipment**
1. 1/4"x28 N.F. die (adjustable round split)
2. Die stock with guides
3. 6" steel rule
4. Small screw driver
5. Hack saw
6. Smooth file, 8"

**Material**
1. 1/4" used aircraft bolt with at least 1" unthreaded portion between threads and head. (AN4-14 to AN4-16, or longer)
2. Cutting oil

**Procedure**
1. Cut off threaded portion of aircraft bolt and chamfer the end slightly with a file.

# ASSEMBLY, REPAIRS, PROJECTS

2. Secure bolt upright in smooth-jawed vise (or, copper or aluminum protectors may be used in the vise jaws) allowing bolt to protrude above the vise from the bolt head.

3. Adjust die to make a medium cut, and place it in the die stock.

4. Start die on bolt, using an even pressure and keeping the die square with the work. Use the proper guide in the stock.

5. After starting cutting operation, reverse the rotation of the die at frequent intervals, to clear the chips. Use plenty of cutting oil.

6. After threading a short portion of the bolt, remove the die and try it in the threaded hole made in the steel block of Project 13. If fit is too loose or too tight, adjust the die with the screw, decreasing or increasing its cutting depth accordingly, and repeating steps 4, 5, and 6 if necessary.

7. If the guide strikes the bolt head before 3/4" of threads are cut, remove the die from stock, turn it over, and finish cutting the remainder of the threads without the use of the guide.

8. Present work for inspection and comment.

## PROJECT 15. EXTRACTING A BROKEN STUD

**Objective**

To teach procedure of removing broken stud without injuring surrounding area (Actually, in this procedure, you start with a bolt. However, when it is cut off, the part left is like a broken stud.)

**Tools and equipment**

1. Ball peen hammer
2. Center punch
3. Assortment of twist drills
4. Hand drill
5. Screw extractor (easy out)
6. Tap wrench or adjustable wrench
7. Hack saw

**Material**

1. Tapped steel block made in Project 13
2. Threaded steel bolt made in Project 14

**Procedure**

1. Tighten bolt in steel block and cut off bolt flush with surface of block.

2. Center punch exact center of stud, as nearly as eye can measure.

3. Drill stud to full depth, using a drill size 3/32", or a number drill of approximate size. Be careful not to drill into steel block.

4. Drill stud with largest drill possible, leaving a thin shell between drill and threads.

5. Remove cutting chips from hole. *If using compressed air, turn face away during operation.*

6. Insert extractor in drilled hole; tap lightly with hammer.

7. Place wrench on extractor and ease out stud.

*Note:* Since the extractor tightens as it turns to the left, it should not slip; if it does, however, tap with hammer and hold down with left hand while turning with wrench.

**Reamers.** Because airplanes are made with high strength-weight ratio, every advantage is taken mechanically to obtain greatest possible strength through workmanship. In this effort, the reamer plays an important part.

In the assembly of spars, attaching points of stabilizers, struts, and superstructure, standard bolts of Army and Navy specifications often are measured with the micrometer (to take advantage of tolerances allowed), and holes are reamed accordingly.

Where extreme accuracy is required, special reamers are used which ordinarily are marked on the shank in thousandths of an inch. An undersize $21/64$ is marked .327, or 3265; a $5/16$ is marked .312 ($1/2$-thousandth under), etc.

When a reamer of proper size is not available, the mechanic must hone a standard reamer to the required size. However, since reamers are so expensive, this is not recommended for the student until he has been shown how by a machinist or master mechanic who is familiar with the procedure. (Reamers should not be stored without some means of protection for the cutting edges.)

When dulled through wear, reamers should be hand stoned, first on the face of the flutes, then on the top of the flutes. The stone should be held perfectly flat with face and clearance, so that the original shape of the flutes will be maintained. A stone especially designed for this purpose, which gives quicker results than an ordinary oil stone, is shown in Fig. 25.

Fig. 25. India Stone for Hand Stoning of Reamers
*Courtesy of Behr-Manning Corp., Troy, N. Y.*

## ASSEMBLY, REPAIRS, PROJECTS

A micrometer is used in determining cutting edge depth. The first cutting edge to be honed is marked with a red pencil to check your starting point. Hone each cutting edge the amount determined by checking with the micrometer against the opposite cutting edge through its diameter. When half around, and you are checking against the edge you honed first, hone to double the allowed figure on the micrometer. When finished, mark with electric pencil or etch the size on the shank in thousandths of an inch.

Very little difficulty is experienced in reaming holes for taper pins. However, large taper bolt holes are more difficult to ream and require more skill. If a reamer chatters in the hole, change from a straight to a spiral flute, or vice versa. Roughing reamers are used to hog out the greater portion of the metal. Use a steady pressure in turning the taper reamer in the hole. Bluing in (as described in procedure 5 of the following project) should be adopted when reaming large tapered holes.

### PROJECT 16. REAM HOLE AND FIT TAPER PIN

**Objective**
1. To teach proper method of reaming a tapered hole
2. To teach procedure of making sure a taper pin fits the hole properly

**Tools and equipment**
1. °Reamer, taper pin (taper 1/4" per foot)
2. °Twist drill
3. Tap wrench
4. Cutting oil
5. Ball peen hammer
6. Center punch
7. Drill press and vise

**Material**
°Taper pin; mild steel block, 1"x1"x1"; small tube of Persian blue

**Procedure**
1. Center punch steel block in exact center of any side.
2. Select proper drill for taper being used. Drill through block with aid of drill press and vise.
3. Place block in bench vise; ream hole; use plenty of cutting oil.
4. Clean out hole, and try taper pin. Consider whether it will be in approximate center by the time it is driven tight.
5. When hole seems about the right size, coat the pin with a very thin film of Persian blue, to a dry consistency; the blue will adhere to the surface and show high spots. If too heavy a coating is applied, however, it will be forced into the areas of poor fit also, and no story will be told.
6. Tap pin into hole gently, then remove. Present to instructor.

---

° Sizes of taper reamer, drill, and taper pin depend on sizes available. Use as large a size as possible, no less than No. 8, 9, or 10, to simulate actually installing taper bolts.

Above. Sheet metal and fabricated items on hand for overhauling and maintaining Mainliners. Note reflections of tubular parts on sheet at lower left.
Below. Replacement parts in another section of stockroom.
*Courtesy of United Air Lines, Cheyenne, Wyo.*

*Chapter XII*

# Aluminum and Related Metals

**ALUMINUM AND ITS ALLOYS.** Aluminum stands fifth among commercial metals in tonnage consumed. If the comparison is made by volume instead of by weight, aluminum holds fourth place in metal production, despite the fact that the electrolytic process for producing it was discovered scarcely more than fifty years ago. Its development in so short a period can be explained only by its wide range of uses, due to the diversity of its properties.

**Qualities.** The outstanding quality of aluminum is its light weight; by volume, it weighs only about one-third as much as most of the commonly used metals. Moreover, it has many other useful properties, and in many applications a fortunate combination of several desirable characteristics makes aluminum the most practical and economical material to use. Other qualities possessed in marked degree by aluminum are: resistance to the corrosive action of the atmosphere and of a great variety of compounds; thermal and electrical conductivity; reflectivity of radiant energy of all wave lengths; and ease of fabrication. Aluminum may be welded by all commercial methods, and may be finished economically in a variety of textures and colors. Finishes are painting, etching, dyeing, anodizing, polishing, wire buffing, sandblasting, hammering, and burnishing.

**Comparison with Steel.** Comparison of costs should be based on the finished article as made from various possible materials, rather than on the relative price of these materials per pound. Since the volume of metal is much the same ordinarily, the cost per pound of aluminum should be divided by the ratio of specific gravities (approximately 3 for most of the common metals) when comparing material costs. In addition, labor costs may be saved by the greater ease with which aluminum is fabricated and finished, and savings may result from the lower unit cost of distribution due to the lighter weight of the metal. Such economies may more than overcome an otherwise unfavorable cost comparison with, for example, the common grades of steel. In such comparisons, also, the higher scrap value of aluminum, when the article finally is discarded, is a factor to be considered.

Although commercially pure aluminum in the annealed or cast condition has a tensile strength only about one-fourth that of structural steel, this strength is markedly increased upon cold working.

**Aluminum Alloys.** The word *aluminum* is used here in its popular sense, that is, including not only commercially pure aluminum but also the light-weight alloys of which aluminum is the principal constituent. The addition of other metals in the alloys is another means of increasing the strength and hardness of aluminum. Even the small percentage of impurities in commercial aluminum is sufficient to increase its strength, as compared with that of the pure metal, by about 50%.

The metals most commonly used in the production of commercial aluminum alloys are copper, silicon, manganese, magnesium, chromium, iron, zinc, and nickel. These elements may be added singly, or combinations of them may be used to produce certain desired characteristics in the resulting alloy. If the alloy is to be manufactured in wrought form, the total percentage of alloying elements is seldom more than 6 or 7%; in casting alloys, appreciably higher percentages often are used. (*Cast* means the condition of a metal after it cools from a molten state; *wrought* means the condition of a cast metal after it has been rolled, so that its structure is fine grained enough to be malleable.)

*Alloy Tensile Strength.* The tensile strength of aluminum alloys in the cast or the annealed condition ranges, depending on their composition, up to about double that of commercial aluminum. The wrought alloys may have their strength further increased by cold working. Although the gain in strength which results from alloying and strain-hardening is accompanied by a decrease in ductility, or ease of fabrication, the results of strain-hardening still remain more than adequate for a great variety of commercial uses.

*Other Qualities.* Some time ago it was discovered that certain of the aluminum alloys, when subjected to appropriate heat-treatment processes, showed remarkable increases in tensile and yield strengths, and in hardness. Elongation in many instances is also increased over that of the annealed alloy. Combinations of alloying and heat-treating processes have produced a series of aluminum

# ALUMINUM AND RELATED METALS

alloys having strengths comparable to that of structural steel, yet retaining in large measure the properties of aluminum. Both wrought and cast alloys which respond to heat treatment have been developed; however, the improvement is more pronounced in the alloys in which the cast structure has been broken up by working the metal. It is development of this class of alloys that has made possible the addition of "excellent mechanical properties" to the list of desirable characteristics of aluminum.

In addition to the change in strength brought about by alloying other metals with aluminum, there are accompanying changes in other properties of the metal. These changes vary with the different alloys, with the result that several alloys may have substantially the same tensile strength yet differ widely in yield strength, ductility, thermal and electrical conductivity, and the ease with which they can be cast or fabricated. For many applications, considerations other than strength are the deciding factors in the choice of the material. There is little increase in weight in the commercial alloys of aluminum. Mostly, the increase in specific gravity does not exceed 3%, and several alloys are slightly lighter than pure aluminum.

The properties required in a material, as well as those in which some sacrifice might be made without serious handicap, vary widely with the use that is intended. Many commercial alloys have been developed, each designed to meet the requirements of a certain type of application; thus the compromise usually involved in the choice of material is considerably reduced.

Table 1A gives the nominal composition of the wrought aluminum alloys; Table 1B gives the nominal composition of aluminum sand-casting alloys.

**Alloy Manufacturing Problems.** The ease with which the various alloys can be used in manufacture varies with the nature of the manufacturing process. Some alloys, for example, may be rolled readily into plate and sheet, but present difficulties in the fabrication of tubing or forgings which would make their cost prohibitive. Additional alloys have been developed, primarily to overcome these manufacturing problems. These usually are fabricated only in the forms for which they were developed. Aluminum alloys with a wide range of mechanical properties are available in practically all the forms in which metal is manufactured, though not all of the

### Table 1A—Nominal Composition of Wrought Aluminum Alloys*

| Alloy | Per Cent of Alloying Elements—Aluminum and Normal Impurities Constitute Remainder | | | | | | | | |
|---|---|---|---|---|---|---|---|---|---|
| | Copper | Silicon | Manganese | Magnesium | Zinc | Nickel | Chromium | Lead | Bismuth |
| 2S   | ... | ... | ... | ... | .... | ... | .... | ... | ... |
| 3S   | ... | ... | 1.2 | ... | .... | ... | .... | ... | ... |
| 11S  | 5.5 | ... | ... | ... | .... | ... | .... | 0.5 | 0.5 |
| 14S  | 4.4 | 0.8 | 0.8 | 0.4 | .... | ... | .... | ... | ... |
| 17S  | 4.0 | ... | 0.5 | 0.5 | .... | ... | .... | ... | ... |
| A17S | 2.5 | ... | ... | 0.3 | .... | ... | .... | ... | ... |
| 18S  | 4.0 | ... | ... | 0.5 | .... | 2.0 | .... | ... | ... |
| 24S  | 4.5 | ... | 0.6 | 1.5 | .... | ... | .... | ... | ... |
| 25S  | 4.5 | 0.8 | 0.8 | ... | .... | ... | .... | ... | ... |
| 32S  | 0.9 | 12.5 | ... | 1.0 | .... | 0.9 | .... | ... | ... |
| A51S | ... | 1.0 | ... | 0.6 | .... | ... | 0.25 | ... | ... |
| 52S  | ... | ... | ... | 2.5 | .... | ... | 0.25 | ... | ... |
| 53S  | ... | 0.7 | ... | 1.3 | .... | ... | 0.25 | ... | ... |
| 56S  | ... | ... | 0.1 | 5.2 | .... | ... | 0.1 | ... | ... |
| 61S  | 0.25 | 0.6 | ... | 1.0 | .... | ... | 0.25 | ... | ... |
| 70S  | 1.0 | ... | 0.7 | 0.4 | 10.0 | ... | .... | ... | ... |

* Heat-treatment symbols have been omitted since composition does not vary for different heat-treatment practices.
*Courtesy of Aluminum Co. of America, Pittsburgh, Pa.*

### Table 1B—Nominal Composition of Aluminum Sand-Casting Alloys*

| Alloy | Per Cent of Alloying Elements—Aluminum and Normal Impurities Constitute Remainder | | | | | | |
|---|---|---|---|---|---|---|---|
| | Copper | Iron | Silicon | Zinc | Magnesium | Nickel | Manganese |
| 43   | .... | ... | 5.0  | .... | .... | ... | ... |
| 47   | .... | ... | 12.5 | .... | .... | ... | ... |
| 108  | 4.0  | ... | 3.0  | .... | .... | ... | ... |
| 112  | 7.5  | 1.2 | .... | 2.0  | .... | ... | ... |
| 122  | 10.0 | 1.2 | .... | .... | 0.2  | ... | ... |
| 142  | 4.0  | ... | .... | .... | 1.5  | 2.0 | ... |
| 195  | 4.0  | ... | .... | .... | .... | ... | ... |
| 212  | 8.0  | 1.0 | 1.2  | .... | .... | ... | ... |
| 214  | .... | ... | .... | .... | 3.8  | ... | ... |
| 220  | .... | ... | .... | .... | 10.0 | ... | ... |
| A334 | 3.0  | ... | 4.0  | .... | 0.3  | ... | ... |
| 355  | 1.3  | ... | 5.0  | .... | 0.5  | ... | ... |
| A355 | 1.4  | ... | 5.0  | .... | 0.5  | 0.8 | 0.8 |
| 356  | .... | ... | 7.0  | .... | 0.3  | ... | ... |
| 645  | 2.5  | 1.2 | ...  | 11.0 | .... | ... | ... |

* Heat-treatment symbols have been omitted since composition does not vary for different heat-treatment practices.
*Courtesy of Aluminum Co. of America, Pittsburgh, Pa.*

alloys are available in all of these forms. Standard commodities of wrought alloys are shown in Table 2. These merely represent the alloys regularly manufactured in these forms.

# ALUMINUM AND RELATED METALS

**Table 2—Alcoa Aluminum and Its Alloys**
(Commodities marked ° are standard.)

| Alloy | Sheet | Plate | Wire | Rod and Bar | Rolled Shapes | Extruded Shapes | Tubing and Pipe | Rivets | Forgings |
|---|---|---|---|---|---|---|---|---|---|
| 2S | ° | ° | ° | ° | .... | ° | ° | ° | .... |
| 3S | ° | ° | ° | ° | .... | ° | ° | ° | .... |
| 11S | .... | .... | ° | ° | .... | .... | .... | .... | ° |
| 14S | .... | .... | .... | .... | .... | .... | .... | .... | ° |
| 17S | ° | ° | ° | ° | ° | ° | ° | ° | ° |
| Alclad 17S | ° | ° | .... | .... | .... | .... | .... | .... | .... |
| A17S | .... | .... | (1) | .... | .... | .... | .... | ° | .... |
| 18S | .... | .... | .... | .... | .... | .... | .... | .... | ° |
| 24S | ° | ° | ° | ° | .... | ° | ° | ° | .... |
| Alclad 24S | ° | ° | .... | .... | .... | .... | .... | .... | .... |
| 25S | .... | .... | .... | .... | .... | .... | .... | .... | ° |
| 32S | .... | .... | .... | .... | .... | .... | .... | .... | ° |
| A51S | .... | .... | .... | .... | .... | .... | .... | .... | ° |
| 52S | ° | ° | ° | ° | .... | .... | (2) | .... | .... |
| 53S | ° | ° | ° | ° | ° | ° | ° | ° | ° |
| 56S | .... | .... | ° | .... | .... | .... | .... | .... | .... |
| 61S | ° | ° | .... | .... | .... | ° | ° | .... | .... |
| 70S | .... | .... | .... | .... | .... | .... | .... | .... | ° |

[1] Rivet wire only is a standard product; other sizes and tempers are not regularly produced.

[2] Only a limited number of standard sizes which are in common use are regularly produced.

*Courtesy of Aluminum Co. of America, Pittsburgh, Pa.*

**PHYSICAL PROPERTIES OF ALUMINUM. Specific Gravity.** Aluminum of commercial purity weighs .098 lb. per cu. in., or has a specific gravity of 2.71. (Data for wrought alloys are shown in Table 3, for cast alloys in Table 4.) For practical purposes it is easy to remember that aluminum and its alloys weigh about a tenth of a pound per cubic inch.

**Thermal Conductivity.** Aluminum of 99.6% purity has a thermal conductivity of .52 in C.G.S. units (calories per second, per sq. centimeter, per centimeter of thickness, per degree Centigrade) which is equivalent to 1,509 B.t.u. per hour, per sq. ft., per inch of thickness, per degree Fahrenheit. Thermal conductivity of various aluminum alloys is shown in Tables 3 and 4, also.

**Thermal Expansion.** Thermal expansion is the increase of length and volume upon heating. The coefficient of thermal expansion of aluminum is slightly more than twice that of steel and

## Table 3—Typical Properties of Wrought Alloys

| Alloy | Specific Gravity | Weight, Lb. per Cu. In. | Electrical Conductivity, Per Cent of International Annealed Copper Standard | Thermal Conductivity at 100° C., C.G.S. Units |
|---|---|---|---|---|
| 2S–O | 2.71 | 0.098 | 59 | 0.54 |
| 2S–H | 2.71 | 0.098 | 57 | 0.52 |
| 3S–O | 2.73 | 0.099 | 50 | 0.45 |
| 3S–¼ H | 2.73 | 0.099 | 42 | 0.39 |
| 3S–½ H | 2.73 | 0.099 | 41 | 0.38 |
| 3S–H | 2.73 | 0.099 | 40 | 0.37 |
| 11S–T3 | 2.82 | 0.102 | 40 | 0.37 |
| 14S–O | 2.80 | 0.101 | 50 | 0.45 |
| 14S–T | 2.80 | 0.101 | 40 | 0.37 |
| 17S–O | 2.79 | 0.101 | 45 | 0.41 |
| 17S–T | 2.79 | 0.101 | 30 | 0.28 |
| A17S–T | 2.74 | 0.099 | .. | .... |
| 18S–O | 2.80 | 0.101 | 50 | 0.45 |
| 18S–T | 2.80 | 0.101 | 40 | 0.37 |
| 24S–O | 2.77 | 0.100 | 50 | 0.45 |
| 24S–T | 2.77 | 0.100 | 30 | 0.28 |
| 25S–T | 2.79 | 0.101 | 40 | 0.37 |
| 32S–O | 2.69 | 0.097 | 40 | 0.37 |
| 32S–T | 2.69 | 0.097 | 35 | 0.32 |
| A51S–O | 2.69 | 0.097 | 55 | 0.50 |
| A51S–W or T | 2.69 | 0.097 | 45 | 0.41 |
| 52S–O | 2.67 | 0.096 | 40 | 0.37 |
| 52S–H | 2.67 | 0.096 | 40 | 0.37 |
| 53S–O | 2.69 | 0.097 | 45 | 0.41 |
| 53S–W or T | 2.69 | 0.097 | 40 | 0.37 |
| 56S–O | 2.64 | 0.095 | 29 | 0.28 |
| 56S–H | 2.64 | 0.095 | 27 | 0.26 |
| 61S–O | 2.70 | 0.098 | 45 | 0.41 |
| 61S–W or T | 2.70 | 0.098 | 40 | 0.37 |
| 70S–O | 2.91 | 0.105 | 40 | 0.37 |
| 70S–T | 2.91 | 0.105 | 35 | 0.32 |
| Brass | 8.4–8.8 | 0.304–0.319 | 26–43 | 0.29–0.44 |
| Copper | 8.94 | 0.322 | 100 | .... |
| Magnesium | 1.74 | 0.063 | 38 | 0.37 |
| Monel | 8.8 | 0.318 | 4 | 0.06 |
| Nickel | 8.84 | 0.319 | 16 | 0.14 |
| Steel | 7.6–7.8 | 0.276–0.282 | 3–15 | .... |
| Tin | 7.3 | 0.265 | 15 | 0.15 |
| Zinc | 7.1 | 0.258 | 30 | 0.27 |

*Courtesy of Aluminum Co. of America, Pittsburgh, Pa.*

### Table 4—Typical Properties of Sand-Casting Alloys

| Alloy | Specific Gravity | Weight, Lb. per Cu. In. | Electrical Conductivity, Per Cent of International Annealed Copper Standard | Thermal Conductivity at 100° C., C.G.S. Units |
|---|---|---|---|---|
| 43 | 2.66 | 0.096 | 37 | 0.34 |
| 43 annealed[°] | 2.66 | 0.096 | 42 | 0.39 |
| 47 | 2.65 | 0.096 | 40 | 0.37 |
| 47 annealed[°] | 2.65 | 0.096 | 32 | 0.39 |
| 108 | 2.75 | 0.099 | 31 | 0.29 |
| 108 annealed[°] | 2.75 | 0.099 | 38 | 0.35 |
| 112 | 2.85 | 0.103 | 30 | 0.28 |
| 112 annealed[°] | 2.85 | 0.103 | 38 | 0.35 |
| 122 | 2.85 | 0.103 | 34 | 0.32 |
| 122–T2 | 2.85 | 0.103 | 41 | 0.38 |
| 122–T61 | 2.85 | 0.103 | 33 | 0.31 |
| 142–T2 | 2.73 | 0.099 | 44 | 0.40 |
| 142–T61 | 2.73 | 0.099 | 37 | 0.35 |
| 195–T4 | 2.77 | 0.100 | 35 | 0.33 |
| 195–T62 | 2.77 | 0.100 | 37 | 0.34 |
| 214 | 2.63 | 0.095 | 35 | 0.32 |
| 214 annealed[°] | 2.63 | 0.095 | 35 | 0.32 |
| 220–T4 | 2.56 | 0.092 | 21 | 0.20 |
| A334 | 2.73 | 0.099 | 31 | 0.29 |
| 355–T4 | 2.68 | 0.097 | 35 | 0.33 |
| 355–T6 | 2.68 | 0.097 | 36 | 0.33 |
| 355–T51 | 2.68 | 0.097 | 43 | 0.40 |
| A355–T51 | 2.71 | 0.098 | 32 | 0.31 |
| 356–T4 | 2.65 | 0.096 | 39 | 0.36 |
| 356–T6 | 2.65 | 0.096 | 39 | 0.36 |
| 356–T51 | 2.65 | 0.096 | 43 | 0.39 |
| 645 | 2.94 | 0.106 | 33 | 0.31 |
| 645 annealed[°] | 2.94 | 0.106 | 35 | 0.33 |

[°] While castings are not commonly annealed, similar effects on conductivities may result from the slower rate of cooling of thick sections as compared with thin ones and other variables in foundry practices. Comparison of the values for as-cast and annealed specimens will show the extent to which variations may be expected, depending upon differences in thermal conditions in the production of different types of castings.

Courtesy of Aluminum Co. of America, Pittsburgh, Pa.

cast iron, as shown in Table 5. For the aluminum alloys, it is the same or only slightly less than that of pure aluminum, except for

# AIRCRAFT SHEET METAL WORK

**Table 5—Average Coefficient of Thermal Expansion per Degree Fahrenheit**

| Alloy | TEMPERATURE RANGE | | |
|---|---|---|---|
| | 68–212° F. | 68–392° F. | 68–572° F. |
| 2S, 3S, 4S | 0.0000133 | 0.0000138 | 0.0000144 |
| 14S, 17S, 18S, 24S, 25S | 0.0000122 | 0.0000130 | 0.0000138 |
| 32S | 0.0000108 | 0.0000114 | 0.0000119 |
| 51S | 0.0000130 | 0.0000136 | 0.0000141 |
| 53S | 0.0000130 | 0.0000136 | 0.0000141 |
| 70S | 0.0000136 | 0.0000144 | 0.0000150 |
| 12 | 0.0000125 | 0.0000130 | 0.0000136 |
| 43 | 0.0000122 | 0.0000127 | 0.0000133 |
| 47 | 0.0000111 | 0.0000113 | 0.0000119 |
| 108, 109, 112, A113 | 0.0000122 | 0.0000127 | 0.0000133 |
| 122 | 0.0000122 | 0.0000127 | 0.0000130 |
| A132 | 0.0000105 | 0.0000111 | 0.0000116 |
| 142 | 0.0000125 | 0.0000130 | 0.0000136 |
| 144 | 0.0000119 | 0.0000125 | 0.0000127 |
| 195 | 0.0000127 | 0.0000133 | 0.0000138 |
| B195 | 0.0000122 | 0.0000127 | 0.0000133 |
| D195 | 0.0000127 | 0.0000133 | 0.0000138 |
| 212 | 0.0000122 | 0.0000127 | 0.0000133 |
| 214, 216 | 0.0000133 | 0.0000138 | 0.0000144 |
| 220 | 0.0000136 | 0.0000141 | 0.0000147 |
| A334, 355 | 0.0000122 | 0.0000127 | 0.0000133 |
| A355 | 0.0000119 | 0.0000125 | 0.0000130 |
| 356 | 0.0000119 | 0.0000127 | 0.0000130 |
| Brass | 0.0000107 | | |
| Cast Iron | 0.0000059 | | |
| Copper | 0.0000093 | | |
| Lead | 0.0000150 | | |
| Monel | 0.0000078 | | |
| Nickel | 0.0000057 | | |
| Steel | 0.0000061 | | |
| Zinc | 0.0000165 | | |

*Courtesy of Aluminum Co. of America, Pittsburgh, Pa.*

those alloys which contain relatively high percentages of silicon. In these, the thermal coefficient is appreciably lowered. In spite of the difference in expansion under thermal changes, composite structures of steel and aluminum alloys give satisfactory performance.

**Modulus of Elasticity.** Young's modulus, or the ratio of internal stress to strain in the elastic range, is approximately the same in aluminum and in aluminum alloys. The average value for the modulus of elasticity is about 10,300,000 lbs. per sq. in., which value is increased by relatively larger additions of alloying elements.

Because of the lower value of this constant in aluminum as compared with steel, it is necessary to use deeper sections in the aluminum alloys, in order to prevent deflection under load. However, these deeper sections in aluminum can produce a structure having the same deflection as steel under load, and with an even higher ultimate strength than would be obtained with structural steel; at the same time they would accomplish a saving in weight of more than a pound for each pound of aluminum alloy used.

The lower modulus of elasticity is an asset when impact loads are to be resisted, since other things being equal, the lower the modulus the greater the ability to absorb energy without permanent deformation. The lower modulus is also advantageous in supports, or other fixed deflections, accidental or intentional. The modulus of rigidity is 3,850,000 lbs. per sq. in. for aluminum and also for its commercial alloys.

**MECHANICAL PROPERTIES OF ALUMINUM.** Typical mechanical properties of aluminum (2S) and of the various wrought aluminum alloys are shown in Table 6. These values may be used in comparing the alloys with each other, or with other materials.

The common practice of straightening or flattening material by stretching causes a substantial increase in the tensile yield strength in the direction of the applied stress, with comparatively little change in the properties measured at right angles to the direction of stretching. There is a much smaller increase in the compressive yield strength of the material. The properties guaranteed for sheet are determined at 90° to the direction of stretching (except in intermediate tempers of the softer alloys). The minimum yield strengths of tubing, shapes, and bar are specified conservatively on the basis

### Table 6—Typical Mechanical Properties of Wrought Aluminum Alloys

| Alloy and Temper | Tension | | | | Hardness | Shear | Fatigue |
|---|---|---|---|---|---|---|---|
| | Yield Strength (Set = 0.2%), Lb./Sq. In. | Ultimate Strength, Lb./Sq. In. | Elongation, % in 2" Sheet Specimen ($\frac{1}{16}"$ Thick) | Elongation, % in 2" Round Specimen ($\frac{1}{2}"$ Diameter) | Brinell, 500-kg. Load 10-mm. Ball | Shearing Strength, Lb./Sq. In. | Endurance Limit, Lb./Sq. In. |
| 2S–O   | 5,000  | 13,000 | 35 | 45 | 23 | 9,500  | 5,000 |
| 2S–¼H  | 13,000 | 15,000 | 12 | 25 | 28 | 10,000 | 6,000 |
| 2S–½H  | 14,000 | 17,000 | 9  | 20 | 32 | 11,000 | 7,000 |
| 2S–¾H  | 17,000 | 20,000 | 6  | 17 | 38 | 12,000 | 8,500 |
| 2S–H   | 21,000 | 24,000 | 5  | 15 | 44 | 13,000 | 8,500 |
| 3S–O   | 6,000  | 16,000 | 30 | 40 | 28 | 11,000 | 7,000 |
| 3S–¼H  | 15,000 | 18,000 | 10 | 20 | 35 | 12,000 | 8,000 |
| 3S–½H  | 18,000 | 21,000 | 8  | 16 | 40 | 14,000 | 9,000 |
| 3S–¾H  | 21,000 | 25,000 | 5  | 14 | 47 | 15,000 | 9,500 |
| 3S–H   | 25,000 | 29,000 | 4  | 10 | 55 | 16,000 | 10,000 |
| 11S–T3 | 42,000 | 49,000 | .. | 14 | 95  | 30,000 | 12,500 |
| 11S–T8 | 44,000 | 57,000 | .. | 14 | 100 | 33,000 | ...... |
| 17S–O      | 10,000 | 26,000 | 20 | 22 | 45  | 18,000 | 11,000 |
| 17S–T      | 40,000 | 62,000 | 20 | 22 | 100 | 36,000 | 15,000 |
| Alclad 17S–T | 33,000 | 56,000 | 18 | .. | ..  | 32,000 | ...... |
| A17S–T     | 24,000 | 43,000 | .. | 27 | 70  | 26,000 | 13,500 |
| 24S–O   | 10,000 | 26,000 | 20 | 22 | 42  | 18,000 | 12,000 |
| 24S–T   | 45,000 | 68,000 | 19 | 22 | 105 | 41,000 | 18,000 |
| 24S–RT  | 55,000 | 70,000 | 13 | .. | 116 | 42,000 | ...... |
| Alclad 24S–T  | 41,000 | 62,000 | 18 | .. | .. | 40,000 | ...... |
| Alclad 24S–RT | 50,000 | 66,000 | 11 | .. | .. | 41,000 | ...... |
| 52S–O   | 14,000 | 29,000 | 25 | 30 | 45 | 18,000 | 17,000 |
| 52S–¼H  | 26,000 | 34,000 | 12 | 18 | 62 | 20,000 | 18,000 |
| 52S–½H  | 29,000 | 37,000 | 10 | 14 | 67 | 21,000 | 19,000 |
| 52S–¾H  | 34,000 | 39,000 | 8  | 10 | 74 | 23,000 | 20,000 |
| 52S–H   | 36,000 | 41,000 | 7  | 8  | 85 | 24,000 | 20,500 |
| 53S–O | 7,000  | 16,000 | 25 | 35 | 26 | 11,000 | 7,500 |
| 53S–W | 20,000 | 33,000 | 22 | 30 | 65 | 20,000 | 10,000 |
| 53S–T | 33,000 | 39,000 | 14 | 20 | 80 | 24,000 | 11,000 |
| 61S–O | 8,000  | 18,000 | 22 | .. | 30 | 12,500 | 8,000 |
| 61S–W | 21,000 | 35,000 | 22 | .. | 65 | 24,000 | 12,500 |
| 61S–T | 39,000 | 45,000 | 12 | .. | 95 | 30,000 | 12,500 |

*Courtesy of Aluminum Co. of America, Pittsburgh, Pa.*

of tension test results, taking into account the relation to known compressive properties.

# ALUMINUM AND RELATED METALS

**Effect of Temperature on Mechanical Properties.** In common with other materials, the tensile strength, yield strength, and modulus of elasticity of aluminum alloys are lower at high temperatures than at ordinary temperatures. Elongation usually increases as the temperature is raised, until at a temperature a little below the melting point it drops nearly to zero. This represents the *hot-short* range of the metal.

Tests made at 114° F. show both strengths and elongations higher than those obtained at ordinary temperatures. (Tests are made at this temperature to show that an alloy is all right to be used in planes that will travel in tropical countries.) Table 7 gives the tensile properties of a number of wrought aluminum alloys, determined at elevated temperatures by storing a number of specimens at each testing temperature, with findings noted at increasing intervals of time. When successive tests gave the same results, these were considered representative of the alloy in equilibrium at the test temperature.

**CHEMICAL PROPERTIES OF ALUMINUM.** The resistance of aluminum to the attack of many chemicals is due to its property of forming a thin, firmly adherent coat of oxide over its surface while it is being rolled hot and in contact with air. The oxide film forms to a certain thickness, then stops; it prevents further action.

**Resistance to Corrosion.** The resistance of a material to corrosion is relative, involving comparison with other metals or with other alloys of the same metal. No commercial metal is immune under all conditions to which structural materials are exposed. It is possible to overstress the danger of corrosion; on the other hand, if this factor is ignored, metal failures may result, which could have been avoided by simple protective measures.

*Resistance of Wrought Alloys.* Commercial aluminum contains, at a maximum, 1% of impurities. This metal, designated 2S in the wrought condition, is widely used because of its high resistance to ordinary conditions of exposure. Selected grades of even higher purity are more resistant to most forms of attack. The addition of the alloying elements does not usually improve, and may impair, the resistance of the metal. However, magnesium, manganese, and chromium have no adverse effect; silicon has but little.

All of the commercial aluminum alloys are properly classed as

**Table 7—Typical Tensile Properties at Elevated Temperatures of Wrought Aluminum Alloys (After Prolonged Heating at Testing Temperature)**

| Alloy | Temperature, Deg. F. | Strength, Lb./Sq. In. Yield | Strength, Lb./Sq. In. Tensile | Elongation, % in 2" | Alloy | Temperature, Deg. F. | Strength, Lb./Sq. In. Yield | Strength, Lb./Sq. In. Tensile | Elongation, % in 2" |
|---|---|---|---|---|---|---|---|---|---|
| 2S–O | 75 | 5,000 | 13,000 | 45 | 17S–T | 75 | 40,000 | 62,000 | 20 |
|  | 300 | 3,500 | 7,500 | 65 |  | 300 | 34,000 | 40,000 | 16 |
|  | 400 | 3,000 | 6,000 | 70 |  | 400 | 21,000 | 26,000 | 25 |
|  | 500 | 2,000 | 3,500 | 85 |  | 500 | 9,500 | 13,000 | 35 |
|  | 600 | 1,500 | 2,500 | 90 |  | 600 | 3,500 | 5,500 | 90 |
|  | 700 | 1,000 | 1,500 | 95 |  | 700 | 3,000 | 4,000 | 100 |
| 2S–½H | 75 | 14,000 | 17,000 | 20 | A17S–T | 75 | 24,000 | 43,000 | 27 |
|  | 300 | 10,000 | 13,000 | 22 |  | 300 | 26,000 | 30,000 | 20 |
|  | 400 | 6,500 | 9,500 | 25 |  | 400 | 13,000 | 17,000 | 30 |
|  | 500 | 2,000 | 3,500 | 85 |  | 500 | 5,500 | 8,000 | 45 |
|  | 600 | 1,500 | 2,500 | 90 |  | 600 | 3,000 | 4,500 | 70 |
|  | 700 | 1,000 | 1,500 | 95 |  | 700 | 2,000 | 3,000 | 90 |
| 2S–H | 75 | 21,000 | 24,000 | 15 | 18S–T | 75 | 47,000 | 63,000 | 17 |
|  | 300 | 14,000 | 17,500 | 16 |  | 300 | 44,000 | 49,000 | 10 |
|  | 400 | 3,000 | 6,000 | 70 |  | 400 | ...... | ...... | .. |
|  | 500 | 2,000 | 3,500 | 85 |  | 500 | 7,000 | 11,000 | 32 |
|  | 600 | 1,500 | 2,500 | 90 |  | 600 | ...... | ...... | .. |
|  | 700 | 1,000 | 1,500 | 95 |  | 700 | 2,500 | 4,000 | 85 |
| 3S–O | 75 | 6,000 | 16,000 | 40 | 24S–T | 75 | 45,000 | 68,000 | 22 |
|  | 300 | 5,000 | 11,000 | 47 |  | 300 | 35,000 | 42,000 | 21 |
|  | 400 | 4,500 | 8,000 | 50 |  | 400 | 23,000 | 28,000 | 25 |
|  | 500 | 3,500 | 5,500 | 60 |  | 500 | 10,000 | 14,000 | 40 |
|  | 600 | 2,500 | 4,000 | 60 |  | 600 | 6,000 | 7,500 | 65 |
|  | 700 | 2,000 | 3,000 | 60 |  | 700 | 3,500 | 4,500 | 100 |
| 3S–½H | 75 | 18,000 | 21,000 | 16 | 25S–T | 75 | 35,000 | 57,000 | 18 |
|  | 300 | 15,000 | 18,000 | 17 |  | 300 | 28,000 | 35,000 | 14 |
|  | 400 | 9,000 | 14,000 | 22 |  | 400 | 13,500 | 19,000 | 24 |
|  | 500 | 5,000 | 10,500 | 25 |  | 500 | 4,500 | 6,500 | 45 |
|  | 600 | 3,000 | 6,000 | 40 |  | 600 | 4,000 | 4,500 | 50 |
|  | 700 | 2,000 | 3,000 | 60 |  | 700 | 3,000 | 3,500 | 55 |
| 3S–H | 75 | 25,000 | 29,000 | 10 | 32S–T | 75 | 46,000 | 56,000 | 8 |
|  | 300 | 16,000 | 23,000 | 12 |  | 300 | 33,000 | 39,000 | 9 |
|  | 400 | 8,000 | 17,000 | 15 |  | 400 | 11,000 | 16,000 | 34 |
|  | 500 | 5,000 | 10,500 | 25 |  | 500 | 6,500 | 8,500 | 50 |
|  | 600 | 3,000 | 4,500 | 55 |  | 600 | 3,500 | 6,000 | 60 |
|  | 700 | 2,000 | 3,000 | 60 |  | 700 | 2,000 | 3,500 | 120 |
| 11S–T3 | 75 | 42,000 | 49,000 | 14 | A51S–T | 75 | 40,000 | 47,000 | 20 |
|  | 300 | 17,000 | 25,000 | 24 |  | 300 | 15,000 | 19,000 | 28 |
|  | 400 | 12,000 | 18,000 | 34 |  | 400 | 5,500 | 7,500 | 58 |
|  | 500 | 4,500 | 8,000 | 44 |  | 500 | 4,500 | 5,500 | 59 |
|  | 600 | 1,500 | 4,000 | 90 |  | 600 | 3,500 | 4,500 | 60 |
|  | 700 | 1,000 | 2,500 | 106 |  | 700 | 3,000 | 3,500 | 65 |
| 14S–T | 75 | 55,000 | 70,000 | 14 | 52S–O | 75 | 14,000 | 29,000 | 30 |
|  | 300 | 39,000 | 43,000 | 14 |  | 300 | 13,500 | 23,000 | 55 |
|  | 400 | 13,000 | 17,000 | 28 |  | 400 | 11,000 | 18,000 | 65 |
|  | 500 | 8,500 | 10,500 | 32 |  | 500 | 8,000 | 12,000 | 100 |
|  | 600 | 4,500 | 6,000 | 45 |  | 600 | 4,000 | 7,500 | 105 |
|  | 700 | 3,500 | 4,000 | 55 |  | 700 | 2,500 | 5,000 | 120 |

*Courtesy of Aluminum Co. of America, Pittsburgh, Pa.*

corrosion resistant. Some are more resistant than others and are chosen for applications in which this is of major importance.

Alloy 3S has practically the same resistance as 2S to the atmosphere and to salt water. Alloy 52S appears to be more resistant to salt water than 2S, from the standpoint of retaining both its mechanical properties and its appearance.

Considerable study has been made as to the effect of the temper of these alloys in their resistance to corrosion. In general, it may be said that any differences in this property as a result of strain-hardening are less than the small differences which occur from one lot to another of commercial materials.

Aluminum alloys are used, ordinarily, without any protection other than the usual precaution to avoid electrolytic action from contact with dissimilar metal. Under severe exposure, as on shipboard, or where the metal is in continuous contact with wood or other absorbent material in the presence of moisture, protective paint coatings are used as an added precaution.

Of the heat-treatable alloys, 53S is most resistant, being equal to 2S in its resistance to industrial atmospheres and to salt spray, although less resistant to some chemicals. Under atmospheric exposure, there is little difference between 53S and 61S; in salt-water spray, 61S shows somewhat more attack. Both 53S and 61S are definitely more resistant than 17S or 24S. Also, unlike the other alloys, their resistance to attack is about the same in all tempers (O, W, and T); even when quenched slowly during heat treatment, there is a minimum of harmful effect.

Both 17S and 24S, while more susceptible to losses in mechanical properties when exposed to the weather than are 53S and 61S, may still be classed as resistant material. Tests of standard structural shapes of 17S alloy showed no significant loss in tensile properties after exposure on the seacoast for one year. Specimens taken after four years showed only slightly more loss than those taken at one year, indicating that the attack tends to be self-limiting.

In the design of aircraft, where reduction of weight has made for improved performance, thin sections must be used; however, these sections must still retain the necessary margin of safety. Protective systems, including anodic oxidization, primers, and aluminum paint, have been developed which make possible the maintenance of seaplanes built of 24S alloy. The use of Alclad 24S sheet has made it unnecessary to paint the skin of landplanes. Unpainted seaplane floats of the same material are now being used in trial installations.

It should be noted that the resistance to corrosion of 17S–T and 24S–T depends upon the proper methods of heat treatment. These alloys, for example, are never used in the annealed temper because of their inferior mechanical properties and corrosion resistance in that temper. A rapid quench from the heat treatment temperature is necessary to develop corrosion resistance, even though a slower quench may produce the specified physical properties. Corrosion resistance of these alloys is lowered, also, if they are heated after quenching. *For that reason, hot forming, hot riveting, and welding processes are not recommended on aluminum except where the assembly can be heat-treated afterward.*

*Resistance of Cast Alloys.* Among the cast alloys, also, there are differences in resistance to corrosive environments. Under most exposures the alloys in which magnesium is the alloying agent (214, 220, and die-cast alloy 218) show the least attack. Only slightly less resistant are the alloys in which silicon or silicon and magnesium are the hardening elements (43, 47, 356, B214, and die-cast alloy 13). The alloys in which copper or nickel are present in substantial percentages show less resistance.

It is seen, then, that while all of the aluminum alloys in commercial use are resistant to corrosion they are not all equally resistant. Applying protective treatment, therefore, depends upon the alloy as well as on the type of service it is to perform.

**PRODUCTION OF WROUGHT ALUMINUM ALLOYS.** In the production of the wrought alloys, an ingot is first cast in the size and form best suited for the equipment on which it is to be processed. It is then worked hot, in order to break down the cast structure under conditions of maximum plasticity. It may be reduced to final dimensions without any cooling beyond that which normally occurs during fabrication, or the final working may be done cold.

**Tempers and Temper Designation.** As the metal is cold-worked, it becomes strain-hardened; the increase in strength and hardness depends on the amount of reduction which the metal re-

# ALUMINUM AND RELATED METALS

ceives. If it is subsequently heated to annealing temperature, the effects of cold working are removed and the metal is in its soft temper, designated by the symbol $O$ following the alloy designation. (In the case of Alcoa wrought alloys, this alloy designation is a number followed by the letter S, as 2S.) A letter preceding the alloy symbol indicates a minor change in composition from that of the basic alloy.

In some alloys, the strain-hardening process is the only means of increasing the tensile properties. Alloys 2S, 3S, and 52S are of this type, and their various tempers are produced by subjecting them to definite reductions by cold work after they have been annealed during their fabrication.

The hard temper, designated $H$, is defined as the tensile properties resulting from the maximum amount of cold working which it is practicable to perform with commercial fabricating equipment.

Tempers intermediate between soft and hard are produced by varying the amount of cold work by selection of the thickness at which the metal is given its last annealing. The tempers are designated by the fractional symbols, $\frac{1}{4}H$, $\frac{1}{2}H$, and $\frac{3}{4}H$, indicating an increase of the strength of the annealed alloy by the corresponding fraction of the spread between the soft and the hard tempers.

Alloys 2S, 3S, and 52S are available in definite, controlled tempers other than soft $O$ only in those commodities produced by cold work from the hot mill slab or bloom, such as sheet, tubing, and wire. Both rolled and extruded bar, rod, plate, and shapes are finished directly from hot ingots, and are supplied *as-fabricated*. Bar and rod may be hot-worked slightly oversize to permit a cold finishing operation, but this is done only to improve surface finish and dimensional accuracy, since the amount of cold working during cold finishing is not sufficient to cause change in the mechanical properties of the alloy.

During the hot-working process, the metal cools gradually; the smaller the size of the finished product, the lower the temperature. Consequently, there is a variable amount of strain-hardening of as-fabricated commodities; heavy sections are in practically the soft temper, while thin sections may have properties approaching those of the half-hard temper. Different lots of the same material (that is, the same size, shape, and alloy) show reasonable uniformity of properties because of the standardized manufacturing processes.

In another group of wrought aluminum alloys, improved mechanical properties are produced by heat treatment, or by a combination of heat treatment and strain-hardening.

The symbol $T$ following the alloy number indicates that the alloy is in its fully heat-treated and age-hardened condition. In the heat treatment of aluminum alloys, some of the alloys (17S, A17S and 24S) age-harden fully on standing at room temperature after they have been quenched from the solution heat-treatment temperature. Others (53S, 61S) show some improvement in properties at room temperature, yet to develop their maximum strength, they must be aged artificially by heating to a moderately high temperature. The symbol $W$, used only with the alloys which require artificial aging or precipitation heat treatment, indicates that the alloy has been subjected to the solution heat treatment, but has not been artificially aged. This temper, $W$, is sometimes called *as-quenched*, although the name is not strictly accurate, since the alloys 53S and 61S experience some increase in strength on standing at ordinary temperatures. The practice of forming the heat-treatable alloys immediately after quenching, before age-hardening has progressed appreciably, is discussed later in this chapter.

The temper which results from strain-hardening an alloy after it has been heat-treated is designated by the symbol $RT$. In case an alloy is supplied with mechanical properties produced by a modification of the usual heat-treatment and strain-hardening processes as described, the temper is denoted by the letter $T$, and a number—as 11S–T3 (the 3 designating a modified heat treatment).

**CHOICE OF ALUMINUM ALLOY.** There is no limit to the number of alloys of aluminum that might be produced, but manufacturing efficiency demands that the number be as small as possible while still providing the necessary combinations of properties to meet the needs of industry. Even so the list is long. The choice of alloy for a particular product depends upon which of the qualities is most essential for the use intended. Thus the determining factor might be maximum mechanical properties; resistance to corrosion; ease of machining, welding or forming; minimum cost; or even commercial availability. Commodities such as sheet and plate are available in practically all of the wrought alloys except those few which have been developed primarily for the production of forgings, machining rod, or like specific purposes.

**Choosing Wrought Alloys.** The wrought alloys are of two types: one, those in which the harder tempers are produced by strain-hardening after annealing, such as 2S, 3S, 52S, and 56S; and two, the heat-treatable alloys which, as the name implies, respond to thermal treatments by improvement of mechanical properties. These are 17S, 24S, 53S, 61S, etc.

*Wrought Alloys Which Are Not Heat-Treatable.* For many purposes, 2S and 3S have sufficient strength, and their use involves lower cost and greater ease of fabrication than the harder alloys.

The alloy 52S in the annealed temper has a tensile strength double that of 2S–O; when 52S is cold-worked, it strain-hardens correspondingly more rapidly than S–O. This alloy, 52S, has mechanical properties intermediate between those of 2S and 3S, and the high-strength, heat-treatable alloys. It is also intermediate in cost. The endurance limit of 52S is higher than that of any of the wrought alloys of aluminum.

The alloys 2S, 3S, and 52S cover a range of strength, in their various tempers, from 12,000 to more than 40,000 lbs. per sq. in., produced by strain-hardening the annealed metal. Even in those products such as bar, rod, shapes, and plate which are not cold-worked in the course of manufacture, the use of 52S in the as-fabricated condition gives a tensile strength at least equal to that of 3S in the hard temper and much greater ductility. Where higher strengths are necessary in these forms, however, one of the heat-treated alloys should be chosen.

Alloy 56S is now commercially available only in the form of wire. The various tempers of this alloy are produced by strain-hardening, yet its mechanical properties are quite comparable with those of the heat-treated alloys. In the soft temper, its tensile strength is greater than that of 53S–T; in the hard temper, its strength compares with that of 17S–T.

*Wrought Alloys Which Are Heat-Treatable.* The oldest of these heat-treatable alloys, 17S, is still the most widely used and is regularly produced in all the forms in which structural materials are required. In aircraft, however, it has been replaced largely by 24S, because the higher tensile and yield strengths of 24S–T make possible considerable saving in weight and therefore improved performance. The forming qualities of 17S are better than those of 24S in the same tempers; also its slower rate of age-hardening after solution heat treatment is a distinct advantage in certain types of forming operations. Both 17S and 24S have good mechanical properties and resistance to corrosion when in the heat-treated temper. Cold working after heat treatment causes a marked increase in the yield strength and has little or no effect on corrosion resistance in these alloys.

Since the high mechanical properties of these alloys have been obtained at some sacrifice of resistance to corrosion, they should be protected where their use involves exposure to the weather.

Alloy 53S has the maximum resistance to corrosion of any of the heat-treated alloys, being about equal to commercially pure aluminum in this respect. The corrosion resistance, also, is substantially the same in all tempers. Its mechanical properties are lower than 17S. In the heat-treated and aged condition (53S–T), it can be formed rather more easily than 17S, and in the unaged condition (53S–W), it can be subjected to even more severe fabrication procedures. The formed article can afterward be aged to develop mechanical properties at least equal to 53S–T. A modified heat treatment producing somewhat higher mechanical properties is sometimes specified (53S–T5).

Alloy 61S, like 53S, requires aging at high temperatures to develop its maximum strength. Its resistance to corrosion approaches that of 53S, and, although its ultimate tensile strength is about 20% lower, its yield strength is slightly higher than that of 17S–T. Moreover, 61S–T can be bent around sharper radii and formed more severely than either 17S–T or 53S–T. It is even more workable before artificial aging (61S–W) and, like 53S–W, the formed article can be subjected to the precipitation heat treatment in order to develop the maximum properties of the alloy.

The need for an alloy having mechanical properties comparable to those of 17S–T, but with free-cutting machining properties, has been filled by alloy 11S. Automatic screw machine stock is produced in the modified heat treatment, 11S–T3.

**Commercial Forms of Wrought Aluminum Alloys.** Alumi-

# ALUMINUM AND RELATED METALS

num and its alloys are manufactured in many and diverse forms, from foil so thin that a pound has a covering area of more than 30,000 sq. in., through all ranges of sheet thickness up to plate 3" thick; from wire, one pound of which makes a length of more than 20,000 miles, to rods 8" in diameter; from angles weighing only $4/100$ ths of a pound per lineal foot, to standard 12" structural channels. Seamless tubes are produced, from the size of a hypodermic needle to that used in the mast of an ocean-going ship. These materials, together with castings, forgings, pressings, impact extrusions, rivets, and screw machine products, make aluminum readily accessible for the diversity of uses to which it is put.

**Alclad Products.** *Alclad* is the registered trade-mark used by the Aluminum Company of America to identify alloys of exceptional resistance to corrosion. This property is imparted by means of a surface layer of aluminum of high purity over a high-strength core. The thickness of the surface metal is chosen so as to retain, in the resultant product, the maximum physical properties, consistent with adequate protection of the alloy core. In the commonly used thicknesses of Alclad 17S–T and 24S–T sheet, the tensile and yield strengths are approximately 10% lower than these values for the uncoated alloys.

This surface coating not only protects the alloy which it covers, but by electrolytic action prevents attack on the sheared edges of the sheet or on areas of the base alloy which may have become exposed through scratches or abrasions. Ordinary 17S alloy rivets, used with Alclad sheet, are also considerably protected through this electrolytic process.

Such protection is accomplished at the expense of slight solution (dissolving) of the surface layers; under continued exposure to sea water, for instance, corrosion products may accumulate on the surface of the sheet as a result. However, while the appearance of the metal may be impaired, mechanical test specimens taken from such sheet show that the base metal has not suffered loss of mechanical properties. With proper cleaning the appearance of the surface can be restored without removing the surface metal upon which such protection depends. Test specimens of Alclad sheet, subjected to the standard salt spray test for a period of five years, show no loss in mechanical properties; except for slight solution of the pure metal layer near the machined edges and in a few isolated spots, the sheet appeared bright. A riveted tensile test specimen, in which ordinary 17S–T rivets were used to join the Alclad sheets, had the same strength after three years' exposure to $3\frac{1}{2}\%$ salt spray, as the control specimen which had been carefully stored.

Sheet and plate are available in Alclad 17S and Alclad 24S in all the tempers in which the base alloy is supplied. Alclad 17S wire is manufactured with a corrosion-resistant hard alloy surface. Other Alclad products are in process of development, or in use for specific purposes. Alclad sheet is extensively used in the aircraft industry. The metal clad dirigible airship, ZMC–2, made for the United States Navy, has Alclad 17S–T sheet as outer shell. Although only .0095" thick, and unprotected, test samples from this shell showed no deterioration after eight years.

For most types of service it is not necessary to paint Alclad sheet. However, for seaplane floats and other applications where corrosive conditions are unusually severe, a protective coating of paint may be found desirable. The surface of Alclad sheet may be anodically treated, prior to painting, in order to obtain the maximum adherence of the paint.

The use of Alclad sheet to replace plain sheet of the same alloy sometimes requires using slightly heavier gages, to compensate for its lower tensile properties. This increased metal thickness does not necessarily mean an increase in weight, however, since the weight of the protecting paint film which ordinarily would be applied to the uncoated sheet, might offset the added weight of metal.

**USES OF ALUMINUM IN AIRCRAFT. Forgings.** Forgings make up a relatively small percentage of the total weight of an airframe, but they serve important functions as structural and other fittings, in addition to their use in engine parts and propeller blades. Most forgings are made in dies on hammers, but some, called *press forgings,* are pressed rather than struck. Press forgings require less *draft* (depth) than hammer forgings and permit somewhat closer tolerances. Forgings often are made from alloy A51S–T, while 14S–T ordinarily is used for highly stressed forged fittings.

Some of the alloys used for other wrought products are suitable for forgings. However, there are alloys which have been developed for this specific purpose. The more important of these are 14S and

25S. Alloy 14S is not quite as easily worked as 25S, yet it is suitable for most forging designs. Since forgings of 14S–T have the highest mechanical properties of any of the alloys produced in this form, they find wide application where high strength and good corrosion resistance are required.

Alloy 17S–T is specified for some forgings because of its greater corrosion resistance. The 3S and 2S alloys are occasionally used for forgings because of the ease with which they can be welded and because of their high resistance to corrosion. Another alloy developed for forging work is A51S; this alloy can be forged even more readily than 25S. It is therefore suitable for large and intricate parts which are difficult to produce in the harder alloys, and in which the higher mechanical properties are not required. Alloy 70S, having properties between A51S and 14S, is of use in forgings.

For forged parts such as pistons, in which the retention of strength at high temperatures is essential, alloys 32S and 18S are used. In addition to good mechanical properties at the working temperatures of internal combustion engines, 32S has the advantage of a lower coefficient of thermal expansion than that of other wrought aluminum alloys.

A variety of parts can be produced most efficiently as press forgings. In this process the metal is forced into a die cavity by enormous pressure. It differs from drop forging in that steady pressure is used, instead of repeated impact from the forging hammer. This product can be kept to closer dimensional tolerances and hence requires less machining for finishing.

**Sheet.** A large percentage of the material in the modern airplane is aluminum alloy sheet, used for the outer covering or skin, and for various other structural parts. Since most of this sheet is subject to high stresses under some flight conditions, high-strength alloy 24S–T is almost universally used, although some 17S–T sheet is employed. Nonstructural parts are sometimes made from the alloys that are not heat-treated, or from one of the lower-strength, heat-treated alloys.

Whether the sheet used is 24S or 17S, it is usually in the Alclad form (previously described as a material having a high strength core covered on each side with a coating of relatively pure aluminum which has exceptional resistance to corrosion). These Alclad materials have been in use for over twelve years, and have made a good service record in sea- and landplanes operating in every part of the world. They have been inspected for chemical composition and mechanical properties by the U. S. Navy Department inspector, and their use on both Army and Navy contracts has been approved.

**Coiled Strip.** The aircraft industry is making increased use of 24S and 17S coiled strip. Heat-treated coiled strip is generally used in the fabrication of rolled sections, where longer units are useful. Annealed coiled strip also is used for blanking out small parts which are subsequently heat-treated.

**Extruded Shapes.** Extruded shapes are used as stringers and other important structural parts of the airframe, providing easy assembly as well as maximum structural usefulness. Since high strength is required in most extrusions used in aircraft, 24S-T is the alloy generally used. Extruded shapes not requiring high strength are sometimes made from 53S or 61S in a suitable temper, or from 2S or 3S. These last two alloys are used in the *as-extruded* condition, which is slightly work-hardened and not in specific temper.

**Castings.** Aluminum castings are more extensively used in the airplane engine and for accessories than in the airframe itself. Any shape which can be forged can also be produced as a casting. While forgings are sometimes preferred because of their greater strength and homogeneity (close grain), it may be necessary to produce the more complicated shapes as castings. Many castings are furnished *as-cast*, but in some alloys they are heat-treated for greater strength.

**Wire, Rod, and Bar.** The products in this group find their application in aircraft principally in the form of structural fittings machined from rod or bar, usually of 24S-T or 17S-T. Other uses include screw-machine products. Rivets are made from special rivet wire. The most widely used rivet alloy is A17S-T. Joints requiring higher strength are made with 17S-T or 24S-T rivets.

Nuts, bolts, screws, and various special screw-machine products for aircraft use are made almost exclusively from 24S-T and 17S-T.

**Tubing.** Tubing is used in aircraft construction for fuel lines, oil and other liquid lines, and for instrument lines. It is also used for structural parts either in the airframe itself or in such units as the control system or power plant. Tubing is supplied in 2S, 3S, and 52S; heat-treatable structural tubing, in 17S, 24S, 53S, and 61S.

# ALUMINUM AND RELATED METALS

**THERMAL TREATMENT OF WROUGHT ALUMINUM ALLOYS. Annealing Aluminum Alloys.** If annealed aircraft materials are called for, ordinarily they are bought in that condition from the metal manufacturer. However, the necessity may arise for annealing either heat-treated or cold-worked material.

The strain-hardening which results from cold-working aluminum alloys may be removed by annealing, i.e., by heating to permit recrystallization to take place. The rate at which recrystallization occurs is increased with rises in temperature; also the rate is greater or less great, depending on how severely the metal is worked before the annealing. Complete softening is practically instantaneous for 2S and 52S at temperatures in excess of 650° F.; for 3S, at temperatures of 750° F. or higher. Heating for longer periods at somewhat lower temperatures will accomplish the same results. Provided the metal has reached the instantaneous annealing temperature, the exact temperature is not critical although the recommended temperature should not be greatly exceeded; the metal should not be held at this point for a very long time. The rate of cooling also is not important, except that too rapid cooling may impair the flatness of the material.

In the case of the heat-treatable alloys, greater care is required in annealing. The metal must be raised to a temperature which will permit recrystallization to remove strain-hardening. On the other hand, the temperature must be kept as low as possible in order to avoid heat-treatment effects which would prevent complete softening of the alloy, or the cooling rate must be slow enough to counteract the effect of such heating.

Heating these alloys to 650° F. is sufficient to remove the strain-hardening which results from cold working. This temperature should not be exceeded by more than 10°, yet the metal temperature, in any part of the load, must not be less than 630° F. The rate of cooling from the annealing temperature is not important if the maximum temperature limit has not been exceeded; slow cooling to a temperature of about 450° F. is a desirable precaution, however, in case any part of the load may have been heated above the recommended temperature.

When applied to metal in the heat-treated temper, this annealing practice, in addition to removing the hardening effect of cold working, also removes most of the effects of heat treatment. For more severe forming which requires that the metal be in fully annealed condition, the following process must be used for metal in the heat-treated temper. The alloy is heated at a temperature of 750° F. to 800° F. for about two hours and is then allowed to cool slowly in the furnace to a temperature of 500° F. The cooling rate should not exceed 50° per hour.

**Solution Heat Treatment.** A solution heat treatment of aluminum alloys requires rather close control of temperature and a rapid quench. To accomplish the maximum improvement in mechanical properties, the heat-treatment temperature is chosen as high as possible without danger of exceeding the melting point of any constituent of the alloy, thus impairing its physical properties and possibly resulting in a severely blistered surface. Prompt transfer of the metal from furnace to quench is necessary.

Temperature control is perhaps most readily obtained by means of a bath of fused sodium nitrate. The rapidity with which the metal is brought to temperature is also an important factor when Alclad sheet is being heat-treated, since the shorter the time in the furnace the less tendency there is toward diffusion.

Air furnaces usually are quite satisfactory, provided they are designed to give the required temperature in all parts of the heating zone. Forced air circulation greatly improves temperature uniformity and increases the rate at which the metal is brought to temperature.

However, 24S alloy heat-treated in air shows a tendency to become surface-roughened by minute blisters; prolonged heating of this alloy in air may even result in lowered physical properties. This does not occur if the alloy is heated in a nitrate bath, nor in an air furnace to which a suitable chemical has been added. A small amount of Alorco Protective Compound (Aluminum Co. of America product) placed in an air furnace affords protection to the alloy; the compound decomposes when heated and the gaseous products of decomposition form a protective coating on the alloy. Information on this chemical should be obtained by users of 24S alloy who intend using air furnaces for heat treatment. The use of such a protective compound is not required for Alclad 24S, or for 24S which has been given an anodic oxide finish.

# ALUMINUM AND RELATED METALS 341

Pyrometric control is essential with any type of furnace. Autographic pyrometers should be used to record the temperature throughout the heating cycle. The length of time the metal must be held at temperature will vary with the nature of the material. The total time in the heat-treating furnace will depend, also, on the equipment used and on the size and spacing of the load.

Very short periods are sufficient to develop the required mechanical properties in reheat-treating 17S–T and 24S–T; to be certain that maximum resistance to corrosion is developed also, however, it is a good idea to heat them at the recommended heat-treating temperature and to hold them at that temperature as long as when heat-treating 17S–O or 24S–O.

When heat-treating Alclad sheet, the size and spacing of load should be arranged to permit the minimum time in the furnace which will develop the required physical properties. Tests should be made with the heat-treating equipment to determine the minimum time required for different thicknesses and kinds of product, because prolonged heating causes the alloying elements present in the core to diffuse into the purer surface layers. Tests, made on sheet which was heated until the copper of the base alloy diffused into spots on the surface, showed little loss in corrosion resistance, provided the sheet was quenched rapidly. However, similar sheets which were quenched slowly in an air blast were definitely inferior to sheets which did not show this diffusion. Recommended heat-treating temperatures for the various alloys appear in Table 8.

Equally important to proper control of heat treating is a rapid quench. This is especially true in the case of A17S, 17S, and 24S; even though a slower quench may develop the required physical properties, the resistance to corrosion shown after slow quenching is definitely inferior to that of rapidly quenched alloys.

The quench recommended is a rapid immersion in cold water. The water should be kept as cold as possible by continuous overflow from the mains, and its volume should be sufficient to prevent appreciable rise of temperature during the quench. If heated in a nitrate bath, all traces of the salt must be washed from the metal after quenching.

Rapid quenching is recommended in the case of Alclad sheet also, in order to maintain a maximum corrosion resistance. How-

Table 8—Conditions for Heat Treatment of Aluminum Alloys

| Alloy | Solution Heat Treatment | | | | Precipitation Heat Treatment § | | |
|---|---|---|---|---|---|---|---|
| | Temperature, Deg. F. | Approximate Time of Heating | Quench † | Temper Designation | Temperature, Deg. F. | Time of Aging | Temper Designation |
| 17S  | 930–950 | (°) | Cold water |       | Room               | 4 days(‡)          | 17S–T  |
| A17S | 930–950 | (°) | Cold water |       | Room               | 4 days(‡)          | A17S–T |
| 24S  | 910–930 | (°) | Cold water |       | Room               | 4 days(‡)          | 24S–T  |
| 53S  | 960–980 | (°) | Cold water | 53S–W | 315–325 or 345–355 | 18 hours / 8 hours | 53S–T  |
| 61S  | 960–980 | (°) | Cold water | 61S–W | 315–325 or 345–355 | 18 hours / 8 hours | 61S–T  |

° In a molten nitrate bath, the time varies from 10 to 60 minutes depending upon the size of the load and the thickness of the material. In an air furnace, proper allowance must be made for a slower rate of bringing the load up to temperature. For heavy material a longer time at temperature may be necessary.

† It is essential that the quench be made with a minimum time loss in transfer from the furnace.

‡ More than 90% of the maximum properties are obtained during the first day of aging.

§ Precipitation heat treatment at elevated temperatures is patented.

*Courtesy of Aluminum Co. of America, Pittsburgh, Pa.*

ever, Alclad sheet which has been quenched in air or other relatively slow medium has more resistance than uncoated 24S or 17S, even when these have been heat-treated and quenched under ideal conditions. Consequently, if it is necessary to heat-treat a complicated assembly which cannot be quenched in cold water because of the distortion which would result, the assembly may still be made from Alclad sheet rather than from uncoated alloy. Insofar as possible, however, design should be predicated on the use of material which can be quenched in cold water.

**Precipitation Heat-Treatment Practice.** There is appreciable age-hardening of alloys 53S and 61S at room temperatures. However, precipitation heat treatment at somewhat higher temperatures, producing the fully heat-treated temper 53S–T and 61S–T, develops much higher properties than those of the room-aged alloys 53S–W and 61S–W. Temperatures and length of time for precipitation heat treatment, sometimes called artificial aging, are shown in Table 8. The operation is carried out in air furnaces or ovens heated by electricity, steam coils, or gas. The furnace must be designed to produce uniform temperatures, carefully controlled in all parts of the heating zone. The heating time may vary somewhat more at the lower aging temperature than with more rapid aging at higher temperatures. Material aged for too long a time or at too high temperatures shows a loss in physical properties, compared with properly aged alloys.

**ALUMINUM FABRICATING PRACTICES. Forming.** Commercially pure aluminum, 2S, is outstanding for the ease with which it can be drawn, spun, stamped, or forged. If the metal is in its annealed temper, articles may be subjected to several successive drawing and spinning operations without the necessity of intermediate annealing.

The aluminum alloys are less ductile than the pure metal, require more liberal radii for bends, and are incapable of withstanding such severe forming. The various alloys in their different tempers range, in facility of fabrication, from 3S–O, only slightly less ductile than 2S–O, to 24S–RT, which can be used only where the forming is limited to bending over rather liberal radii.

The alloys 2S, 3S, and 52S cover a wide range of mechanical properties in their various tempers. Since their harder tempers are obtained by cold working during the process of manufacture, the amount of forming which can be done on them is greater in the softer tempers. For many drawing operations, the half-hard temper retains sufficient ductility for good working qualities even in 52S, and some less severe draws are successfully accomplished with this alloy in the hard temper.

Alloys 17S–T and 24S–T can be subjected to a considerable amount of forming, the latter requiring slightly larger radii for bends than the former. More severe operations, requiring sharp bends or substantial depth of drawing, may have to be performed on the annealed alloy, with subsequent heat treatment of the formed article. In many cases, forming is done immediately after the metal has been quenched. Age-hardening then takes place in the finished part, thus avoiding the possibility of distortion or warping during heat treatment, with the necessity for straightening.

The forming qualities of 53S–T lie between those of 17S in the as-quenched condition and 17S–T, while 53S–W can be formed quite readily. Alloy 61S–W has excellent working qualities and, like 53S–W, it can be artificially aged after forming in order to achieve the higher strength of the fully heat-treated temper without danger of distortion or warping.

It should be noted that while 61S–T has a yield strength slightly greater than that of 17S–T, and lower elongation, it is much more readily worked than 17S–T. In fact it can be bent around smaller radii and drawn or stamped more easily than 53S–T, even though it has appreciably higher physical properties than the latter alloy.

In drawing or stamping operations, good results depend somewhat on choice of proper lubricant. In large-scale operations, the light lubricating oils, known to the trade as *metal oil*, are most commonly used. The best lubricant is tallow mixed with a small amount of mineral oil, but, because of its higher cost and the greater difficulty of application and removal, it is used only on the more difficult operations for which metal oil does not prove successful.

The surface finish of the tool is important to results. Tool steel with well-polished surfaces may be needed for more difficult draws of harder alloys. For many jobs, however, cast-steel tools and even cast-iron tools are satisfactory, especially if the number of parts to be made is not large.

In forming aluminum alloys it is necessary to take into consideration their characteristic properties. One requirement for successful working is for the tool to allow a suitable radius for bending and drawing operations. The radius required will vary with the grade of alloy and the thickness of the material; it will also depend somewhat upon the type of bending equipment used. Sometimes a change in tools has been found sufficient, without having to change to a soft temper or a softer alloy. The change might consist of something as simple as the slight rounding of a sharp edge, or a polishing operation to improve the surface sufficiently to prevent the metal from flowing into scratches or flaws in the tools, and thus tearing. In difficult forming operations, it may be necessary to resort to successive draws with intermediate annealing, starting, of course, with annealed material.

Tables 9A and 9B are intended as guides in the choice of material or of proper forming radii, rather than as tabulations of definite operating limits. The final choice of alloy or of working radius should be based on a trial under the conditions to be used in production, since relative ease of forming is also affected by the nature

**Table 9A—Approximate Radii for 90° Cold Bend Aluminum and Aluminum Alloy Sheet**

Minimum permissible radius varies with nature of forming operation, type of forming equipment, and design and condition of tools. Minimum working radius for given material or hardest alloy and temper for a given radius can be ascertained only by actual trial under contemplated conditions of fabrication.

| Alloy and Temper | Bend Classification * | Alloy and Temper | Bend Classification * |
|---|---|---|---|
| 2S–O | A | 24S–O (†) | B |
| 2S–¼H | B | 24S–T († ‡) | J |
| 2S–½H | B | 24S–RT (†) | K |
| 2S–¾H | D | 52S–O | A |
| 2S–H | F | 52S–¼H | C |
| | | 52S–½H | D |
| 3S–O | A | 52S–¾H | F |
| 3S–¼H | B | 52S–H | G |
| 3S–½H | C | | |
| 3S–¾H | E | 53S–O | A |
| 3S–H | G | 53S–W | F |
| | | 53S–T | G |
| 17S–O(†) | B | | |
| 17S–T (†‡) | H | 61S–O | B |
| | | 61S–W | E |
| | | 61S–T | F |

*For corresponding bend radii see Table 9B.

† Alclad 17S and Alclad 24S can be bent over slightly smaller radii than the corresponding tempers of the uncoated alloy.

‡ Immediately after quenching, these alloys can be formed over appreciably smaller radii.

*Courtesy of Aluminum Co. of America, Pittsburgh, Pa.*

**Table 9B—Radii Required for 90° Bend in Terms of Thickness, t**

| | Approximate Thickness | | | | | |
|---|---|---|---|---|---|---|
| B & S Gauge<br>Inch<br>Inch | 26<br>0.016<br>1/64 | 20<br>0.032<br>1/32 | 14<br>0.064<br>1/16 | 8<br>0.128<br>1/8 | 5<br>0.182<br>3/16 | 2<br>0.258<br>1/4 |
| Bend Classification A | 0 | 0 | 0 | 0 | 0 | 0 |
| B | 0 | 0 | 0 | 0 | 0 1t | 0 1t |
| C | 0 | 0 | 0 | 0–1t | 0–1t | ½t–1½t |
| D | 0 | 0 | 0–1t | ½t–1½t | 1t–2t | 1½t–3t |
| E | 0–1t | 0–1t | ½t–1½t | 1t–2t | 1½t–3t | 2t–4t |
| F | 0–1t | ½t–1½t | 1t–2t | 1½t–3t | 2t–4t | 2t–4t |
| G | ½t–1½t | 1t–2t | 1½t–3t | 2t–4t | 3t–5t | 4t–6t |
| H | 1t–2t | 1½t–3t | 2t–4t | 3t–5t | 4t–6t | 4t–6t |
| J | 1½t–3t | 2t–4t | 3t–5t | 4t–6t | 4t–6t | 5t–7t |
| K | 2t–4t | 3t–5t | 3t–5t | 4t–6t | 5t–7t | 6t–10t |

*Courtesy of Aluminum Co. of America, Pittsburgh, Pa.*

of the forming process. Thus the final answer must be obtained by actual trial of different materials on the production tools.

**Hot Forming of Aluminum Alloys.** If the heat-treatable alloys are raised to suitable temperatures, they can be formed around much smaller radii than is possible at ordinary temperatures. Working at a temperature near 400° F. brings about considerable improvement in forming characteristics. If this temperature is not exceeded and if the metal is not held at this temperature more than half an hour, there will be no substantial loss in mechanical properties. This method of forming is especially suitable for 61S–T and 53S–T, since their resistance to corrosion is not impaired by heating, as is that of 17S–T and 24S–T. Even with 17S–T and 24S–T alloys, the effect on resistance to corrosion may not be serious, since hot forming would not be used except on the heavy sections.

Heat-treatable alloy plate can be formed into angles and other shapes by heating the metal and forming it in dies. For some classes of material, the best working temperature is in the heat-treatment range. In these cases, the chilling of the metal in the steel dies may constitute a satisfactory quench, sufficient so that the mechanical properties of the heat-treated temper will be developed in the finished part, after suitable aging. In other instances, the metal must be formed at a temperature lower than that required for heat treatment. Here the effects of quenching in the dies to avoid warping may be obtained by reheating the formed section to heat-treating temperature and replacing it in the dies, instead of quenching in water. Natural aging or precipitation heat treatment, whichever is appropriate for the particular alloy, will then develop the full properties of the metal. Die quenching will give satisfactory results, however, only where the die is in intimate contact with the metal being formed. Dies must also be of sufficient mass to absorb the heat from the alloy and bring it promptly to room temperature.

**Machining Aluminum Alloys.** Aluminum alloys, both cast and wrought, are readily machined. In general, tools for machining aluminum should have appreciably more side and top rake than are required for cutting steel; their shapes are more like those tools used with hard woods. In addition, the edges should be keen and smooth; this can be accomplished by finishing on a fine abrasive wheel, followed by hand stoning or lapping. Tungsten carbide tipped tools offer distinct advantages and are required for best results with aluminum alloys of high silicon content.

Lathe tools are set considerably higher on the work than they usually are for use with steel. Ordinarily, twist drills with large spiral angles are preferred. In general, it is best to use comparatively high speeds and fine to medium feeds—the finer the feed, the higher the speed. A lubricant is desirable and should be used freely. For general use, a mixture of kerosene and lard oil is satisfactory as a cutting compound; for milling, sawing, and drilling, the less expensive soluble cutting oils may be used.

**PROTECTIVE FINISHING.** Government specifications—United States Army, Navy, and Civil Aeronautics Board—cover the general requirements on protective finish for aircraft. Details usually are covered in each individual contract, however, as special procedures may be required by reason of special construction methods or the service for which the planes are intended. The extent of protection needed depends on the corrosion resistance of the materials used and the environment in which the aircraft will be used. Also it depends to some extent on completeness and frequency of cleaning and maintenance practices to be followed.

It should be remembered that slowly quenched Alclad sheet is not so resistant to corrosion as if rapidly quenched in cold water. Therefore Alclad sheet that is not cold-water quenched should receive the same protective finish employed on plain 24S-T or 17S-T parts when it is intended for use under severely corrosive conditions.

As previously explained, the fundamental principle of corrosion protection is that dry aluminum cannot corrode. Aircraft structures are designed to drain freely and as far as possible to make all parts accessible for cleaning, repainting, repair, and inspection. Joints and crevices which would hold moisture are avoided where possible, or are suitably sealed. Dissimilar metals are insulated from adjacent aluminum alloy by painting the faying surfaces before assembly and then inserting nonabsorptive gaskets.

In using many of the protective finishes, particularly under severely corrosive conditions, it is necessary to treat anodically* the aluminum alloy parts. Such anodizing treatments produce an artifi-

---

\* Patents covering the anodizing process are held by the Aluminum Company of America.

cial oxide coating which enhances the inherent corrosion resistance of the alloy and improves paint adherence. Anodic treatments applicable to aircraft are described in United States Army Air Corps and Navy Department specifications. In general, these consist of making the aluminum alloy parts the anode of an electrolytic cell, in which the electrolyte is either chromic or sulphuric acid, chromic acid being more generally used in aircraft.

There are three purposes served by anodizing[*] or forming an aluminum hydroxide surface on aluminum. It gives a decorative effect in addition to the improved corrosion resistance and the excellent bond for painting. It is not a plating process, therefore the coating has nothing added, except where dyes are used to produce red, blue, green, yellow, or black surfaces; these dyes add no measurable thickness. The process has an advantage in weight over Alclad sheet aluminum. For instance, a saving in weight of approximately 1000 lbs. would be gained in a plane the size of a DC–4. At $20 per lb., the figure the airlines use as the cost of carrying excess weight, there would be a saving of $20,000 a year for one plane due to the choice of the metal finish.

If Alclad is to be painted, anodic treatment to provide a bond is recommended. Castings do not require this treatment, if painted, since they have a surface sufficiently rough for good adherence of paint. Anodic treatment of castings, however, gives an excellent appearance and helps prevent corrosion. Aluminum alloys which contain over 5% copper cannot be anodically treated with success.

Parts should be treated before assembly, or forming, otherwise the treatment will not penetrate to the parts most vulnerable to corrosion, around rivet joints or tight places. Subsequent heat treatment, however, will spoil the finish. Care must be taken to avoid abrasion of the anodic film, therefore the surface is sometimes painted with linoil or primer before the assembly is worked on.

Less expensive than anodic treatments are the chemical dip treatments, one of which is identified by the trade-name *Alrok* (put out by the Aluminum Company of America). Another type of chemical treatment involves a 5% chromic acid bath at 140° F. Some of these treatments are only slightly less effective than the anodic treat-

---

[*]Patents covering the anodizing process are held by the Aluminum Company of America.

# ALUMINUM AND RELATED METALS

ments, and the processes are approved by the United States Army Air Corps, under certain specific conditions.

**MAGNESIUM ALLOYS.** Pure magnesium weighs only 65% as much as aluminum. It is the lightest metal commercially available for aircraft construction. Its strength-weight ratio is above that of the aluminum alloys. However, owing to the difficulty of forming, and poor resistance to corrosion, magnesium is little used in the wrought state, as yet. Progress in developing methods for successful use of this metal has been very rapid, however, and aircraft sheet metal workers must expect to work much with magnesium alloy sheet in the future. So instead of grumbling because it is not so easily fabricated, we might as well try to help in the development of better techniques for forming and fabrication, since designers are getting more and more *magnesium-minded*.

The chemical composition of magnesium alloys is probably as complex as that of the aluminum alloys, and space will not be taken to analyze the chemical properties of these alloys, except to say that the alloying elements most commonly used with magnesium are manganese, zinc, silicon, and tin.

Magnesium alloy castings are used for airplane landing wheels, brackets, instrument cases, automatic pilot chassis, and engine parts. Magnesium alloy is excellent for small die casting and machining.

Among the magnesium alloys, the sheet metal worker is most likely to use Dowmetal in sheet form. This alloy is made by the Dow Chemical Company, who designate their alloys as follows: Dowmetal A, Dowmetal G, Dowmetal K, etc.

The American Magnesium Corporation, allied with the Aluminum Company of America, is also a large producer of magnesium alloys. They designate their alloys the same as they do their aluminum alloys, except for the trademark AM preceding the number.

In fabrication of all magnesium alloys, bends as sharp as twice the sheet thickness must be worked, heated to between 500° F. and 750° F., and bent slowly. Use a soft pencil for marking and remove all burrs from the edge before attempting to bend the material.

Dowmetal F, or AM53S, can be successfully welded, and is most often used because of its good mechanical properties. Dowmetal M, or AM3S sheet, has good forming characteristics, and is used where corrosion resistance is desired.

Magnesium alloys are treated for resistance to salt water by the Chrome-Pickle Treatment, or by Dow Treatment No. 7.

The Chrome-Pickle Treatment consists of a dip (from ½ to 3 minutes) for all sheet stock in the following solution:

*Sodium dichromate, 1.5 lb.; nitric acid (concentrated), 1.5 pints; water to make 1 gallon.*

After dipping, sheets should be washed immediately in cold running water, followed by a hot rinse to hasten drying. The treated alloy has an iridescent film which has been formed by the removal of about $1/1000$th″ of material. The solution is used at room temperature.

Dow Treatment No. 7 has two steps: (1) dip in a 15% hydrofluoric acid solution at room temperature, for 5 minutes; (2) boil in a 10% dichromate solution for 45 minutes.

The most common magnesium alloys, with their symbols, are listed in Table 10.

### QUESTIONS

1. What is weight of aluminum, in relation to other commonly used metals?
2. How does commercially pure aluminum compare in tensile strength with structural steel?
3. What metals are used with commercially pure aluminum to produce the aluminum alloys?
4. How does the tensile strength of the aluminum alloys compare with that of commercially pure aluminum?
5. About what percentage of other metals do the aluminum alloys contain?
6. What treatment increases tensile and yield strength of most aluminum alloys?
7. What is the specific gravity of commercially pure aluminum?
8. Describe effects of straightening or flattening aluminum and its alloys by stretching.
9. Describe effects of high temperatures on aluminum alloys.
10. What *corrosive environment* is most encountered by aircraft?
11. What condition must be present for aluminum to corrode?
12. How is hard temper designated in aluminum alloys? How is it brought about?
13. What does the letter *T*, following the number of an alloy, mean?
14. What are two types of wrought aluminum alloys?
15. Name some commercial forms of aluminum alloy materials.
16. What does *Alclad* mean? Describe fully.
17. What does *annealing* mean?
18. Describe solution heat treatment of aluminum alloy.

# ALUMINUM AND RELATED METALS

Table 10—Composition and Uses of Common Magnesium Alloys with Corresponding Symbols of Dow Chemical Company and American Magnesium Company

| Dow Chemical Co. Alloy # | American Magnesium Alloy # | Nominal Composition Per Cent |||||||  Characteristic Uses |
|---|---|---|---|---|---|---|---|---|
| | | Al | Mn | Cd | Zn | Cu | Si | |
| A | AM241 | 8.0 | 0.15 | ... | 0.2 | ... | 0.2 | Permanent Mold and Sand Castings |
| B | AM246 | 12.0 | 0.20 | ... | ... | ... | ... | Sand Castings |
| F | AM53S | 4.0 | 0.20 | ... | ... | ... | ... | |
| G | AM240 | 10.0 | 0.10 | ... | 0.2 | ... | 0.50 | Permanent Mold Castings, High Yield Strength |
| H | AM265 | 6.0 | 0.20 | ... | 3.0 | ... | ... | Castings, Fine Grain |
| J-1 | AM57S | 6.5 | 0.20 | ... | 1.0 | ... | ... | Extrusions, Forgings |
| K | AM230 | 10.0 | 0.20 | ... | ... | ... | 0.50 | Die Castings, Thin Sections |
| M | AM3S | ... | 1.50 | ... | ... | ... | .20 | Sheet, Plate, Extruded Bar and Rod |
| M | AM403 (Cast) | ... | 1.50 | ... | ... | ... | .20 | Sand Castings for Aircraft, Tank Fittings, Easily Welded |
| O-1 | AM58S | 8.5 | 0.20 | ... | 0.5 | ... | ... | Forgings |
| E | ... | 6.0 | 0.30 | ... | ... | ... | ... | Rolled Plate, Sheet |
| L | ... | 2.5 | 0.20 | ... | ... | ... | ... | Forgings |
| P | ... | 10.0 | 0.10 | ... | 2.0 | ... | ... | Sand Castings |
| T | ... | 2.0 | 0.20 | 2.0 | ... | 4.0 | ... | Pistons |
| X | ... | 3.0 | 0.20 | ... | 3.0 | ... | ... | Extrusions, Forgings |
| EX | ... | 6.5 | 0.20 | ... | ... | ... | ... | |
| R | ... | 9.0 | 0.20 | ... | 0.6 | ... | 0.2 | Die Castings |

19. What are characteristics of tools recommended for machining aluminum alloys?
20. What is anodic treatment?
21. What three functions has this anodizing process?
22. How does magnesium compare with aluminum in weight?
23. In ease of forming?
24. In corrosion resistance?
25. Why are aircraft designers increasingly *magnesium minded*, despite some of the defects of the material?

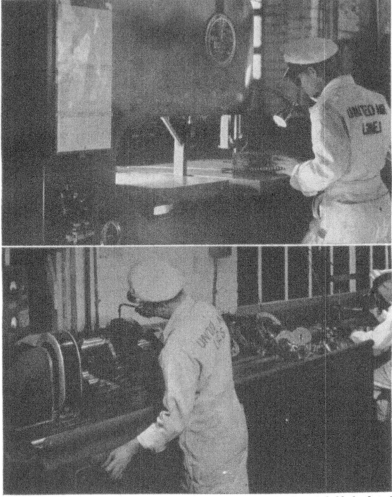

Above. This mechanic will produce a forming die from the three-inch block of steel he is cutting on the Doall machine.

Below. Workman inspecting steel parts with Magnaflux machine which reveals any defects.

*Courtesy of United Air Lines, Cheyenne, Wyo.*

*Chapter XIII*

# Steel in Aircraft Construction

This chapter is designed to give the aircraft mechanic a working knowledge of steels, so that he will understand the engineering reasons for the selection of the materials called for on blueprints, and so that he will have the knowledge required of a master craftsman who is designing or making tools, dies, fixtures, or jigs, when the selection of the material is left to his judgment.

Although aluminum alloys are the predominating structural material in modern airplanes, steel alloys are used for high-strength members or fittings. The ferrous materials most widely used in aircraft for structural purposes are chromium-molybdenum steel (SAE 4130) and nickel steel (SAE 2330). The SAE 4130, called *chrom-molly* in the shops, is used for tubing in the construction of engine mounts, landing struts, fuselage frames, etc. Sheet stock of 4130 is used for making tail, wing, and other structural fittings and parts. The 2330 steel is used for bolts, tie rods, etc.

**SAE NUMBERING OF STEELS.** Each alloy of steel is given an identifying number consisting of several figures, and these figures tell us approximately the composition of that particular material. The following is the system of identifying steels in the Society of Automotive Engineers Specifications. The first figure of each number indicates the type of steel, thus:

| | |
|---|---|
| 1 = carbon steel | 4 = molybdenum steel |
| 2 = nickel steel | 5 = chromium steel |
| 3 = nickel-chromium steel | 6 = chromium-vanadium steel |
| | 7 = tungsten steel |

In the alloys, the second figure indicates the approximate percentage of the predominant alloying element. The last two figures generally indicate the carbon content in points, or hundredths of one per cent. For instance, 1025 is steel with 0.20 to 0.30% carbon. Steel 2330 is a nickel steel containing about 3% nickel and 0.25 to 0.35% carbon. Steel 71660 is a tungsten steel having 16% tungsten, and 0.50 to 0.70% carbon.

*Tables 1 and 2 give approximate physical properties of the various types of steel, while Tables 3 through 11 set forth the official SAE Standard Specifications as they appear in the 1943 SAE Handbook. Table 12 is a classification of American Tool Steels, according to type analysis.

The following pages give the numbers and qualities of the more commonly used steels. Some of the steels described are not included in the SAE tables. However, such steels are still found in the shops, and information concerning them should be helpful.

**ALLOY STEELS WITH CARBON. SAE 1010 and SAE 1015.** These steels have high ductility and are suitable for all types of cold bending and forming operations. In the very severe procedures, however, they must be annealed after cold drawing, to prevent cracking during the forming operation.

Because of their extreme softness, these steels do not machine well, and they are particularly difficult to drill and thread.

In addition to bending and forming applications, these grades may also be used for the manufacture of casehardened parts. However, SAE 1010 is not largely used for this purpose because of the difficulty in machining.

**SAE 1020.** This steel machines better than the lower carbon SAE steels; it may also be subjected to bending and forming operations where the stresses encountered are not too severe.

This grade is also widely used for parts which are to be casehardened, and where the quality requirements in the part treated are not too high.

**SAE 1022, Formerly X1020\*.** This steel is designed to replace SAE 1020, wherever the higher manganese content is not objectionable. It machines faster and gives a better surface finish than SAE 1020; it also shows higher physical properties.

SAE 1022 is also suitable for carburized parts, carburizing 15 to 25% faster than SAE 1020 and producing a heat-treated case of great hardness and uniformity. For very thin sections, or where straightening operations are required after casehardening, SAE 1020 may prove more satisfactory.

**SAE 1025.** This steel is a mild carbon cold-rolled steel. It is used extensively for aircraft purposes. Materials can be purchased cold-rolled to accurate dimensions. This steel has an ultimate tensile strength of 55,000 lbs. per sq. in.; a yield strength of 36,000 lbs. per sq. in.; an elongation of 22%.

**SAE 1030 and SAE 1035.** These steels, in both the heat-treated and the not heat-treated condition, are used for general purposes where considerable toughness is desired. They have fair machining qualities, combined with considerable strength and ductility. Forgings of small or medium size which do not require great strength can be made from these steels. Both are used frequently for shafting where strength greater than that of regular commercial shafting is required. SAE 1035, for instance, is extensively used for oil field equipment, being known as rig iron shafting.

---

\* The 1943 SAE Handbook gives new numbers for steels formerly identified by a prefix, such as X1020 which is now *1022* or X1065 which is now *1066*. In Tables 3, 4, 6, 7, and 8, both the old and the new numbers are given for such steels.

# STEEL IN AIRCRAFT

### Table 1—Approximate Physical Properties of Various Steels (Cold Drawn)

| SAE No. | Tensile Strength (Lbs.) | Yield Point (Lbs.) | Elongation (%) | Reduction of Area (%) | Brinell |
|---|---|---|---|---|---|
| **Carbon** | | | | | |
| 1120 | 70/85,000 | 60/75,000 | 15/25 | 45/55 | 137/166 |
| 1010 | 60/70,000 | 55/65,000 | 20/30 | 50/55 | 137/149 |
| 1020 | 70/85,000 | 60/70,000 | 15/25 | 45/55 | 149/170 |
| X1020 (1020-90) | 75/90,000 | 65/75,000 | 15/25 | 50/55 | 170/187 |
| 1025 | 80/90,000 | 60/70,000 | 10/20 | 45/55 | 146/170 |
| 1035 | 95/110,000 | 75/90,000 | 10/20 | 40/55 | 170/202 |
| 1045 | 90/115,000 | 80/100,000 | 5/15 | 30/45 | 183/228 |
| 1050 | 85/110,000 | 75/85,000 | 14/20 | 35/45 | 202/228 |
| 1095 | 130/145,000 | 115/125,000 | 4.5/10 | 25/35 | 228/269 |
| X1314-X1315 | 75/85,000 | 55/65,000 | 15/25 | 45/55 | 143/170 |
| X1335 | 85/95,000 | 70/80,000 | 15/25 | 40/50 | 187/207 |
| **Nickel** | | | | | |
| 2315 | 80/100,000 | 60/80,000 | 20/30 | 55/65 | 170/196 |
| 2320 | 85/100,000 | 70/90,000 | 15/25 | 50/60 | 179/217 |
| 2330 Annealed | 95/110,000 | 85/100,000 | 10/20 | 40/55 | 182/212 |
| 2335 Annealed | 100/115,000 | 90/100,000 | 10/20 | 40/50 | 187/235 |
| 2512 | 100/120,000 | 85/100,000 | 10/20 | 35/50 | 187/228 |
| **Chrome Nickel** | | | | | |
| 3115 | 85/100,000 | 70/80,000 | 20/30 | 50/60 | 179/196 |
| 3120 | 85/105,000 | 80/95,000 | 15/25 | 40/50 | 187/212 |
| 3130 Annealed | 90/105,000 | 85/95,000 | 15/25 | 50/60 | 183/228 |
| 3135 Annealed | 90/105,000 | 80/95,000 | 15/25 | 45/65 | 183/228 |
| 3140 Annealed | 95/110,000 | 80/100,000 | 15/25 | 40/55 | 196/228 |
| **Nickel Molybdenum** | | | | | |
| 4615 | 90/105,000 | 85/100,000 | 12/20 | 45/60 | 183/212 |
| **Chrome Molybdenum** | | | | | |
| 4130° | 80/200,000 | | | | |
| 4140 Annealed | 100/115,000 | 90/100,000 | 15/25 | 40/55 | 196/228 |
| **Chrome Vanadium** | | | | | |
| 6120 | 100/115,000 | 80/100,000 | 12/20 | 45/60 | 187/228 |
| 6130 Annealed | 95/110,000 | 85/100,000 | 10/20 | 45/55 | 179/207 |
| 6140 Annealed | 100/115,000 | 90/105,000 | 10/20 | 40/55 | 196/228 |
| 6150 Annealed | 100/115,000 | 90/105,000 | 10/20 | 35/45 | 183/228 |
| **Chromium** | | | | | |
| 5120 | 90/105,000 | 80/95,000 | 20/25 | 55/65 | 170/196 |
| 5140 Annealed | 95/110,000 | 85/100,000 | 10/20 | 25/40 | 187/228 |
| 5150 Annealed | 100/115,000 | 90/100,000 | 10/20 | 25/35 | 187/228 |
| 52100 Annealed | 90/105,000 | 70/85,000 | 18/25 | 50/60 | 183/212 |

° For information on physical properties of SAE 4130, see Table 2.

**SAE 1045.** This steel is used for general purposes requiring greater strength and hardness than is obtainable with SAE 1030 and SAE 1035. It has fair machining properties in the as-rolled, cold-drawn condition, but its machining properties can be greatly improved by normalizing before cold drawing. It is extensively used for axles and shafts, both in the heat-treated and cold-drawn or turned and polished condition.

This steel, obtainable as cold-drawn wire, is used for tie rods.

AIRCRAFT SHEET METAL WORK

Table 2—Physical Properties of SAE 1025, SAE 4130, and SAE 2330

| Type | Heat Treatment | | Ultimate Tensile | Yield* Point | Bearing | Shear | Modulus of Elasticity |
|---|---|---|---|---|---|---|---|
| | Quench in Oil | Draw | Pounds per Square Inch | | | | |
| SAE 1025 Steel: Sheet—57-136-3, 47S17; Bar—57-107-9, 46S22; Tube—57-180-1, 49T1 | ...... | ...... | 55,000 | 36,000 | 90,000 | 35,000 | 28,000,000 |
| Alloy Steel Sheet and Bar, not Heat-Treated, Dia. or Thickness Equal to 1½" or less | ...... | ...... | 65,000 | 36,000 | 110,000 | 40,000 | 29,000,000 |
| Alloy Steel Bar, not Heat-Treated, Dia. or Thickness over 1½" | ...... | ...... | 55,000 | 36,000 | 90,000 | 35,000 | 29,000,000 |
| SAE 4130 Steel Tube | ...... | ...... | 90,000 | 60,000 | 140,000 | 55,000 | 29,000,000 |
| SAE 4130 Steel Tube. Welded, or Alloy Steel Welded After Heat Treatment | ...... | ...... | 80,000 | 60,000 | 125,000 | 50,000 | 29,000,000 |
| SAE 4130 Chrome-Molybdenum Steel, Heat-Treated: Sheet—57-136-8, 47S14; Bar—57-107-19, 46S23; Tube—57-180-2, 44T18 | 1625° F. | 1150° F. 970 770 600 | 125,000 150,000 180,000 200,000 | 100,000 130,000 145,000 155,000 | 175,000 190,000 200,000 220,000 | 75,000 90,000 105,000 115,000 | 29,000,000 29,000,000 29,000,000 29,000,000 |
| SAE 2330 Nickel Steel, Heat-Treated: Bar—57-107-17, 46S21 | 1500° F. | 1000 950 830 750 | 125,000 150,000 180,000 200,000 | 100,000 130,000 145,000 155,000 | 175,000 190,000 200,000 220,000 | 75,000 90,000 105,000 115,000 | 29,000,000 29,000,000 29,000,000 29,000,000 |

* Corresponds to unit set of .002" per inch. These values are for compression only, for use in column formulas. Block compression may be regarded as equal to the tensile strength.

The unit stresses in joints heat-treated after welding shall be 80% of the properties of the material in its heat-treated condition.

In welded structures where more than six members converge, a margin of safety of 50% must be maintained unless the joint is properly reinforced. (For the Army, approval of the reinforcement must be obtained.) A tube continuous through the joint shall be regarded as two members.

# STEEL IN AIRCRAFT

Table 3—Chemical Compositions of Carbon Steels°

| SAE Number | Nominal Chemical Ranges | | | |
|---|---|---|---|---|
| | C | Mn | P Max. | S Max. |
| 1008 | 0.10 max. | 0.30–0.50 | 0.040 | 0.050 |
| 1010 | 0.08–0.13 | 0.30–0.50 | 0.040 | 0.050 |
| 1015 | 0.13–0.18 | 0.30–0.50 | 0.040 | 0.050 |
| (X1015) 1016 | 0.13–0.18 | 0.60–0.90 | 0.040 | 0.050 |
| 1020 | 0.18–0.23 | 0.30–0.50 | 0.040 | 0.050 |
| (X1020) 1022 | 0.18–0.23 | 0.70–1.00 | 0.040 | 0.050 |
| 1024 | 0.20–0.26 | 1.35–1.65 | 0.040 | 0.050 |
| 1025 | 0.22–0.28 | 0.30–0.50 | 0.040 | 0.050 |
| 1030 | 0.28–0.34 | 0.60–0.90 | 0.040 | 0.050 |
| 1035 | 0.32–0.38 | 0.60–0.90 | 0.040 | 0.050 |
| 1036 | 0.32–0.39 | 1.20–1.50 | 0.040 | 0.050 |
| 1040 | 0.37–0.44 | 0.60–0.90 | 0.040 | 0.050 |
| 1045 | 0.43–0.50 | 0.60–0.90 | 0.040 | 0.050 |
| 1050 | 0.48–0.55 | 0.60–0.90 | 0.040 | 0.050 |
| 1052 | 0.47–0.55 | 1.20–1.50 | 0.040 | 0.050 |
| 1055 | 0.50–0.60 | 0.60–0.90 | 0.040 | 0.050 |
| 1060 | 0.55–0.65 | 0.60–0.90 | 0.040 | 0.050 |
| (X1065) 1066 | 0.60–0.71 | 0.80–1.10 | 0.040 | 0.050 |
| 1070 | 0.65–0.75 | 0.70–1.00 | 0.040 | 0.050 |
| 1080 | 0.75–0.88 | 0.60–0.90 | 0.040 | 0.050 |
| 1085 | 0.80–0.93 | 0.70–1.00 | 0.040 | 0.050 |
| 1095 | 0.90–1.05 | 0.30–0.50 | 0.040 | 0.050 |

° When silicon is specified in standard basic open hearth steels, silicon may be ordered only as 0.10% maximum; 0.10 to 0.20%; 0.15 to 0.30%. In the case of many grades of basic open hearth steel, special practice is necessary in order to comply with a specification including silicon. Acid bessemer steel is not furnished with specified silicon content.

*Courtesy of Society of Automotive Engineers, Inc., Twenty Nine West Thirty Ninth Street, New York.*

**SAE 1095.** This high carbon steel is used quite extensively by the airplane sheet metal worker. It is known as *drill rod* in the annealed round stock and as *spring steel* annealed in the sheet or strip. As spring steel wire, heat-treated, it is referred to as *music wire*. Music wire has a variation in tensile strength of 350,000 lbs. to 225,000 lbs. per sq. in. in the sizes from .005 to .180 in diameter.

**SAE 1120 (Open-Hearth Screw Stock).** Open-hearth screw stock was developed with the aim of combining the superior machining qualities of Bessemer Screw Steel (SAE 1112) with the advantageous physical properties of open-hearth steels. It succeeded in this aim to a certain degree. It will machine about 80% as fast as SAE 1112 and possesses physical properties closely approaching those of SAE 1020. However, it is slightly more brittle, and therefore not so well suited to forming operations, as the latter grade.

## AIRCRAFT SHEET METAL WORK

**ALLOY STEELS WITH 3½% NICKEL (SAE 2300 Series). SAE 2315 and SAE 2420.** These are essentially carburizing steels and their full value is realized only when they are so treated. The presence of the nickel lowers the critical temperature, permitting lower carburizing and quenching temperatures with consequent lessening of the tendency toward distortion.

When properly carburized, these steels, particularly of the SAE 2315 grade, produce parts which possess not only extreme surface hardness but also great toughness.

Table 4—Chemical Compositions of Free Cutting Steels°

| SAE Number | NOMINAL CHEMICAL RANGES | | | |
|---|---|---|---|---|
| | C | Mn | P Max. | S Max. |
| *Bessemer* | | | | |
| 1111 | 0.08–0.13 | 0.60–0.90 | 0.09–0.13 | 0.10–0.15 |
| 1112 | 0.08–0.13 | 0.60–0.90 | 0.09–0.13 | 0.16–0.23 |
| (X1112) 1113 | 0.08–0.13 | 0.60–0.90 | 0.09–0.13 | 0.24–0.33 |
| *Open Hearth* | | | | |
| 1115 | 0.13–0.18 | 0.70–1.00 | 0.045 max. | 0.10–0.15 |
| (X1314) 1117 | 0.14–0.20 | 1.00–1.30 | 0.045 max. | 0.08–0.13 |
| (X1315) 1118 | 0.14–0.20 | 1.30–1.60 | 0.045 max. | 0.08–0.13 |
| (X1330) 1132 | 0.27–0.34 | 1.35–1.65 | 0.045 max. | 0.08–0.13 |
| (X1335) 1137 | 0.32–0.39 | 1.35–1.65 | 0.045 max. | 0.08–0.13 |
| (X1340) 1141 | 0.37–0.45 | 1.35–1.65 | 0.045 max. | 0.08–0.13 |
| 1145 | 0.42–0.49 | 0.70–1.00 | 0.045 max. | 0.04–0.07 |

° When silicon is specified in standard basic open hearth steels, silicon may be ordered only as 0.10% maximum; 0.10 to 0.20%; 0.15 to 0.30%. In the case of many grades of basic open hearth steel, special practice is necessary in order to comply with a specification including silicon. Acid bessemer steel is not furnished with specified silicon content.

*Courtesy of Society of Automotive Engineers, Inc., Twenty Nine West Thirty Ninth Street, New York.*

Table 5—Chemical Compositions of Manganese Steels

| SAE Number | NOMINAL CHEMICAL RANGES | | | | |
|---|---|---|---|---|---|
| | C | Mn | P Max. | S Max. | Si |
| 1320 | 0.18–0.23 | 1.60–1.90 | 0.040 | 0.040 | 0.20–0.35 |
| 1330 | 0.28–0.33 | 1.60–1.90 | 0.040 | 0.040 | 0.20–0.35 |
| 1335 | 0.33–0.38 | 1.60–1.90 | 0.040 | 0.040 | 0.20–0.35 |
| 1340 | 0.38–0.43 | 1.60–1.90 | 0.040 | 0.040 | 0.20–0.35 |

*Courtesy of Society of Automotive Engineers, Inc., Twenty Nine West Thirty Ninth Street, New York.*

## STEEL IN AIRCRAFT

Grain growth in nickel steels is slow, and this, in some cases, permits the omission of the high-temperature, or core, quench, especially if carburizing temperatures have been fairly low. Recommended quenching temperature is 1500° F. for core refinement, and 1400° F. for the case.

SAE 2315 is widely used in the automotive industry for pivot pins, steering ball joints, king bolts, ball rods, camshafts, valve tappets, pin stock in silent chains, gears, pinions, and other parts requiring extreme surface hardness and freedom from distortion.

### Table 6—Chemical Compositions of Nickel Steels

| SAE Number | Nominal Chemical Ranges | | | | | |
|---|---|---|---|---|---|---|
| | C | Mn | P Max. | S Max. | Si | Ni |
| (2315) 2317 | 0.15–0.20 | 0.40–0.60 | 0.040 | 0.040 | 0.20–0.35 | 3.25–3.75 |
| 2330 | 0.28–0.30 | 0.60–0.80 | 0.040 | 0.040 | 0.20–0.35 | 3.25–3.75 |
| 2340 | 0.38–0.43 | 0.70–0.90 | 0.040 | 0.040 | 0.20–0.35 | 3.25–3.75 |
| 2345 | 0.43–0.48 | 0.70–0.90 | 0.040 | 0.040 | 0.20–0.35 | 3.25–3.75 |
| 2515 | 0.12–0.17 | 0.40–0.60 | 0.040 | 0.040 | 0.20–0.35 | 4.75–5.25 |

*Courtesy of Society of Automotive Engineers, Inc., Twenty Nine West Thirty Ninth Street, New York.*

### Table 7—Chemical Compositions of Nickel-Chromium Steels

| SAE Number | Nominal Chemical Ranges | | | | | | |
|---|---|---|---|---|---|---|---|
| | C | Mn | P Max. | S Max. | Si | Ni | Cr |
| 3115 | 0.13–0.18 | 0.40–0.60 | 0.040 | 0.040 | 0.20–0.35 | 1.10–1.40 | 0.55–0.75 |
| 3120 | 0.17–0.22 | 0.60–0.80 | 0.040 | 0.040 | 0.20–0.35 | 1.10–1.40 | 0.55–0.75 |
| 3130 | 0.28–0.33 | 0.60–0.80 | 0.040 | 0.040 | 0.20–0.35 | 1.10–1.40 | 0.55–0.75 |
| 3135 | 0.33–0.38 | 0.60–0.80 | 0.040 | 0.040 | 0.20–0.35 | 1.10–1.40 | 0.55–0.75 |
| 3140 | 0.38–0.43 | 0.70–0.90 | 0.040 | 0.040 | 0.20–0.35 | 1.10–1.40 | 0.55–0.75 |
| (X3140) 3141 | 0.38–0.43 | 0.70–0.90 | 0.040 | 0.040 | 0.20–0.35 | 1.10–1.40 | 0.70–0.90 |
| 3145 | 0.43–0.48 | 0.70–0.90 | 0.040 | 0.040 | 0.20–0.35 | 1.10–1.40 | 0.70–0.90 |
| 3150 | 0.48–0.53 | 0.70–0.90 | 0.040 | 0.040 | 0.20–0.35 | 1.10–1.40 | 0.70–0.90 |
| 3240 | 0.38–0.45 | 0.40–0.60 | 0.040 | 0.040 | 0.20–0.35 | 1.65–2.00 | 0.90–1.20 |
| 3312° 3310 | 0.08 0.13 | 0.15 0.60 | 0.025 | 0.025 | 0.20 0.35 | 3.25 3.75 | 1.40 1.75 |

° Electric furnace steel.
*Courtesy of Society of Automotive Engineers, Inc., Twenty Nine West Thirty Ninth Street, New York.*

**SAE 2330 and SAE 2335.** These are heat-treatable steels recommended where high stresses are encountered, the presence of the nickel providing increased strength without sacrifice of ductility. The presence of the nickel also makes lower quenching temperatures possible, with consequent freedom from distortion. SAE 2330 should be used for parts of large cross section where core hardness and strength are needed. This steel ordinarily is machined in the unannealed condition, but annealing will improve its ma-

Table 8—Chemical Compositions of Molybdenum Steels

| SAE Number | NOMINAL CHEMICAL RANGES | | | | | | | |
|---|---|---|---|---|---|---|---|---|
| | C | Mn | P Max. | S Max. | Si | Ni | Cr | Mo |
| 4023 | 0.20–0.25 | 0.70–0.90 | 0.040 | 0.040 | 0.20–0.35 | ...... | ...... | 0.20–0.30 |
| 4027 | 0.25–0.30 | 0.70–0.90 | 0.040 | 0.040 | 0.20–0.35 | ...... | ...... | 0.20–0.30 |
| 4032 | 0.30–0.35 | 0.70–0.90 | 0.040 | 0.040 | 0.20–0.35 | ...... | ...... | 0.20–0.30 |
| 4037 | 0.35–0.40 | 0.75–1.00 | 0.040 | 0.040 | 0.20–0.35 | ...... | ...... | 0.20–0.30 |
| 4042 | 0.40–0.45 | 0.75–1.00 | 0.040 | 0.040 | 0.20–0.35 | ...... | ...... | 0.20–0.30 |
| 4047 | 0.45–0.50 | 0.75–1.00 | 0.040 | 0.040 | 0.20–0.35 | ...... | ...... | 0.20–0.30 |
| 4063 | 0.60–0.67 | 0.75–1.00 | 0.040 | 0.040 | 0.20–0.35 | ...... | ...... | 0.20–0.30 |
| 4068 | 0.64–0.72 | 0.75–1.00 | 0.040 | 0.040 | 0.20–0.35 | ...... | ...... | 0.20–0.30 |
| 4119 | 0.17–0.22 | 0.70–0.90 | 0.040 | 0.040 | 0.20–0.35 | ...... | 0.40–0.60 | 0.20–0.30 |
| 4125 | 0.23–0.28 | 0.70–0.90 | 0.040 | 0.040 | 0.20–0.35 | ...... | 0.40–0.60 | 0.20–0.30 |
| (X4130) 4130 | 0.28–0.33 | 0.40–0.60 | 0.040 | 0.040 | 0.20–0.35 | ...... | 0.80–1.10 | 0.15–0.25 |
| 4137 | 0.35–0.40 | 0.70–0.90 | 0.040 | 0.040 | 0.20–0.35 | ...... | 0.80–1.10 | 0.15–0.25 |
| 4140 | 0.38–0.43 | 0.75–1.00 | 0.040 | 0.040 | 0.20–0.35 | ...... | 0.80–1.10 | 0.15–0.25 |
| 4145 | 0.43–0.48 | 0.75–1.00 | 0.040 | 0.040 | 0.20–0.35 | ...... | 0.80–1.10 | 0.15–0.25 |
| 4150 | 0.46–0.53 | 0.75–1.00 | 0.040 | 0.040 | 0.20–0.35 | ...... | 0.80–1.10 | 0.15–0.25 |
| 4320 | 0.17–0.22 | 0.45–0.65 | 0.040 | 0.040 | 0.20–0.35 | 1.65–2.00 | 0.40–0.60 | 0.20–0.30 |
| (X4340) 4340 | 0.38–0.43 | 0.60–0.80 | 0.040 | 0.040 | 0.20–0.35 | 1.65–2.00 | 0.70–0.90 | 0.20–0.30 |
| 4615 | 0.13–0.18 | 0.45–0.65 | 0.040 | 0.040 | 0.20–0.35 | 1.65–2.00 | ...... | 0.20–0.30 |
| 4620 | 0.17–0.22 | 0.45–0.65 | 0.040 | 0.040 | 0.20–0.35 | 1.65–2.00 | ...... | 0.20–0.30 |
| 4640 | 0.38–0.43 | 0.60–0.80 | 0.040 | 0.040 | 0.20–0.35 | 1.65–2.00 | ...... | 0.20–0.30 |
| 4815 | 0.13–0.18 | 0.40–0.60 | 0.040 | 0.040 | 0.20–0.35 | 3.25–3.75 | ...... | 0.20–0.30 |
| 4820 | 0.18–0.23 | 0.50–0.70 | 0.040 | 0.040 | 0.20–0.35 | 3.25–3.75 | ...... | 0.20–0.30 |

*Courtesy of Society of Automotive Engineers, Inc., Twenty Nine West Thirty Ninth Street, New York.*

Table 9—Chemical Compositions of Chromium Steels

| SAE Number | NOMINAL CHEMICAL RANGES | | | | | |
|---|---|---|---|---|---|---|
| | C | Mn | P Max. | S Max. | Si | Cr |
| 5120 | 0.17–0.22 | 0.70–0.90 | 0.040 | 0.040 | 0.20–0.35 | 0.70–0.90 |
| 5140 | 0.38–0.43 | 0.70–0.90 | 0.040 | 0.040 | 0.20–0.35 | 0.70–0.90 |
| 5150 | 0.48–0.55 | 0.70–0.90 | 0.040 | 0.040 | 0.20–0.35 | 0.70–0.90 |
| 52100* | 0.95–1.10 | 0.30–0.50 | 0.025 | 0.025 | 0.20–0.35 | 1.20–1.50 |

* Electric furnace steel.

*Courtesy of Society of Automotive Engineers, Inc., Twenty Nine West Thirty Ninth Street, New York.*

## STEEL IN AIRCRAFT

Table 10—Chemical Compositions of Chromium-Vanadium Steel

| SAE Number | NOMINAL CHEMICAL RANGES | | | | | | |
|---|---|---|---|---|---|---|---|
| | C | Mn | P Max. | S Max. | Si | Cr | V Min. |
| 6150 | 0.48–0.55 | 0.65–0.90 | 0.040 | 0.040 | 0.20–0.35 | 0.80–1.10 | 0.15 |

*Courtesy of Society of Automotive Engineers, Inc., Twenty Nine West Thirty Ninth Street, New York.*

Table 11—Chemical Compositions of Silicon-Manganese Steel

| SAE Number | NOMINAL CHEMICAL RANGES | | | | |
|---|---|---|---|---|---|
| | C | Mn | P Max. | S Max. | Si |
| 9260 | 0.55–0.65 | 0.70–0.90 | 0.040 | 0.040 | 1.80–2.20 |

*Courtesy of Society of Automotive Engineers, Inc., Twenty Nine West Thirty Ninth Street, New York.*

chinability. It is used to produce bushings and for the trunnions for machine gun mountings, where wear-resisting surface and shock-resisting properties are needed in combination.

In the heat-treated condition, these steels are largely used for such parts as connecting rod bolts, crankcase bolts, engine bolts, studs, keys, propeller shafts, counterweight screws, and similar units.

Aircraft bolts are heat-treated to 125,000 lbs. per sq. in. ultimate strength. Table 13 shows the rated strengths of steel aircraft bolts. This material can be bent 180° with a radius of half its thickness.

**CHROME-NICKEL (1½% Nickel) STEELS (SAE 3100 Series). SAE 3115 and SAE 3120.** These are carburizing steels, used interchangeably with SAE 2315 and SAE 2320. The combination of nickel and chrome in these alloys brings out the best properties of each element in the carburizing process. Such steels carburize more rapidly than the nickel steels and, when properly treated, produce an exceptionally hard case with a tough core and a minimum of distortion. Recommended quenching temperature is 1550° F. for core refinement and 1425° F. for the case.

However, the higher carbon grade, SAE 3120, is not recommended for thin sections, because of the danger of cracking in carburizing.

SAE 3115 machines satisfactorily without annealing. SAE 3120 is machined usually in the unannealed state, but its machinability is improved by annealing.

These steels generally are used in engine construction, for gear pins, piston pins, cam rings, push rod ends, and rollers. It is discussed here merely to give the sheet metal worker knowledge on quenching temperature.

**SAE 3130 and SAE 3135.** These are heat-treatable steels in which the physical qualities of both alloying elements are combined to give a desirable degree of toughness and resistance to shock. They are especially recommended for parts requiring high physical properties at elevated temperatures.

**Table 12—Classification of American Tool Steels According to Type Analysis**

| Content of Principal Alloys | | | | Type | Major Characteristics Affecting Utility | | | | | | |
|---|---|---|---|---|---|---|---|---|---|---|---|
| Carbon | Tungsten | Chromium | Others | | Working Hardness | Wear resistance | Toughness | Warpage | Hot hardness | Quench medium | Depth of hardening |
| 0.60/1.40 | .... | .... | .... | Plain carbon tool steel | ° | Low | ° | High | Low | Water | Low |
| 0.60/1.40 | .... | .... | 0.15/0.40 V. | Carbon-vanadium | ° | Low | ° | High | Low | Water | Low |
| 0.50/1.40 | .... | 0.30/0.90 | None, or 0.15/0.25 V. | Chromium-vanadium, or Low chromium | ° | Low | ° | High | Low | Water | Med. |
| 1.00/1.30 | 1.0/2.0 | 0.50/1.50 | 0.15/0.30 V. | High chromium, Low tungsten | C-62/C-65 | Med. | Med. | Low | Low | Oil | Med. |
| 0.80/1.00 | 0.4/0.6† | 0.40/0.60† | 1.00/1.75 Mn. | Manganese oil hardening | C-58/C-62 | Low | Med. | Low | Low | Oil | Med. |
| 0.45/0.60 | 1.0/2.0 | 1.0/1.5 | 0.15/0.30 V. | Tungsten alloy chisel | C-50/C-55 | Med. | High | Low | Med. | Oil | Med. |
| 1.10/1.40 | 3.0/5.0 | .... | .... | Finishing tool steel | C-63/C-66 | Med. | Low | High | Med. | Water | Deep |
| 0.80/1.00 | .... | 3.0/4.0 | .... | Chromium hot work | C-45/C-50 | Med. | Med. | Low | Med. | Air or Oil | Deep |
| 0.25/0.45 | 8.0/15.0 | 2.5/3.5 | .... | Tungsten hot work | C-45/C-55 | Med. | Med. | Low | High | Oil | Deep |
| 0.25/0.50 | 3.0/7.0 | 3.0/7.0 | .... | Chromium-tungsten hot work | C-43/C-50 | Med. | Med. | Low | High | Oil | Deep |
| 0.50/0.65 | .... | .... | 1.5/2.5 Si, 0.75/1.0 Mn. | Silicon-manganese tool steel | C-50/C-57 | Med. | Med. | Med. | Med. | Water Air or Oil | Med. |
| 1.50/2.25 | .... | 10/15 | .... | High carbon, High chrom. | C-58/C-66 | High | Low | Low | High | Oil | Deep |
| 0.50/0.80 | 14/20 | 3.0/4.5 | 0.75/2.25 V. | High-speed steel | C-62/C-66 | High | Low | Low | High | Oil | Deep |
| 0.60/0.80 | 14/20 | 3.0/4.5 | 0.75/2.25 V. | Cobalt high speed | C-64/C-66 | High | Low | Low | High | Oil | Deep |
| .... | .... | .... | 2.5/12.0 Co. | | | | | | | | |

° Usual working hardness and toughness depend on carbon content.
† With Manganese on high side these may be absent.
Reprinted from "Tool Steels," by James P. Gill.

# STEEL IN AIRCRAFT

### Table 13—Rated Strength of Steel Aircraft Bolts

| Size of Bolt | Body Dia. | Area of Body | Root Dia. | Area at Root of Thread | ° Tension (Root) Lbs. | † Army | | ‡ Navy and Commercial | |
|---|---|---|---|---|---|---|---|---|---|
| | | | | | | Single Shear (Body) Lbs. | Bending (Body) Inch Lbs. | Single Shear (Body) Lbs. | Bending (Body) Inch Lbs. |
| 10–32 | .189 | .0283 | .1494 | .0174 | 1740 | 1840 | .... | 2123 | .... |
| 1/4–28 | .249 | .0491 | .2036 | .0326 | 3260 | 3192 | 153 | 3683 | 192 |
| 5/16–24 | .3115 | .0767 | .2589 | .0527 | 5270 | 4986 | 300 | 5753 | 375 |
| 3/8–24 | .374 | .1105 | .3209 | .0809 | 8090 | 7183 | 518 | 8288 | 647 |
| 7/16–20 | .4365 | .1503 | .3725 | .1087 | 10870 | 9770 | 822 | 11270 | 1028 |
| 1/2–20 | .499 | .1964 | .4350 | .1486 | 14860 | 12760 | 1227 | 14730 | 1534 |
| 9/16–18 | .5615 | .2485 | .4908 | .1891 | 18910 | 16150 | 1747 | 18640 | 2184 |
| 5/8–18 | .624 | .3068 | .5528 | .2400 | 24000 | 19940 | 2397 | 23010 | 2996 |
| 3/4–16 | .749 | .4418 | .6688 | .3513 | 35130 | 33130 | 5177 | 33130 | 5177 |
| 7/8–14 | .874 | .6013 | .7822 | .4805 | 48050 | 45100 | 8221 | 45100 | 8221 |
| 1–14 | .999 | .7854 | .9072 | .6464 | 64640 | 58900 | 12272 | 58900 | 12272 |

° Based upon 100,000 lbs./sq. in. tensile strength, and the use of castle nuts. For use on all airplanes. Hexagon head bolts only, not clevis bolts, shall be used for critical tension loads. Bolts employing shear nuts or AN STD. Fine Thread Nut Plates are rated at only 50% of these values.

† Based upon 100,000 lbs./sq. in. tensile strength except for sizes ¾ diameter and larger which are based upon 125,000 lbs./sq. in.

‡ Based upon 125,000 lbs./sq. in. tensile strength.

All bolts are actually heat-treated to 125,000 lbs./sq. in.

The threaded portion of a bolt must not be used to take shear load when the shear load on the bolt is greater than 25% of the bolt's rated shear strength. When the shear load exceeds this value, not more than ¼ of the thickness of the fitting shall bear on the threads. For structural bolts in Navy airplanes, the amount of thread in bearing shall not exceed one thread.

Heavy sections of these steels may be quenched in water; thin sections should be quenched in oil. Recommended quenching temperature is 1500° F.

These steels may be used interchangeably with the 3½% nickel steels of similar carbon content, SAE 2330 and SAE 2335, and are frequently used to replace the latter materials where lower cost and better machining properties are desired. Annealing of both will aid in obtaining the most satisfactory machining properties.

**MOLYBDENUM STEELS. SAE 4130 and SAE 4140 Chrome-Molybdenum.** Chrome-molybdenum, called *chrom-molly* in the shops, is the steel alloy which has been generally adopted in aircraft construction for nearly all parts made of sheet steel and steel tubing. It is used for all welded assemblies, for sheet fabrication fittings, for landing gear axle, and supports in fuselages.

The welding qualities of this steel are excellent; it welds readily either with oxyacetylene flame or by electric arc welding. Many sheet metal fittings are fabricated from it because of its excellent forming characteristics.

The combination of chrome and molybdenum in these alloys produces a steel with deep penetration and with high physical properties at high drawing temperatures. The presence of the molybdenum allows a wider heat-treating range permitting increased production and giving greater assurance of satisfactory results. Recommended quenching temperature is 1550° F. In heavy sections, SAE 4130 may be water quenched.

These grades compare favorably with SAE 3130 and SAE 3140 in physical properties and machining qualities; their cost is also in line.

While SAE 4130 may be machined in the unannealed state, annealing is better for machining. SAE 4140 must be annealed for machining.

These steels are not used for casehardening.

The Army and the Navy, it should be noted, have their own designations for the SAE standard steels. The following example shows the SAE, the Army, and the Navy designation for SAE 4130.

|  | Commercial | Army | Navy |
|---|---|---|---|
| Bar | 4130 | AC–57–107–19 | 46S23 |
| Sheet | 4130 | AC–57–136–8 | 47S14 |
| Tube | 4130 | AC–57–180 | 44T18 |

**NICKEL-MOLYBDENUM STEEL. SAE 4615.** This is a carburizing grade capable of producing file hardness in oil quenching, thus eliminating distortion and simplifying production problems. It likewise produces a tough core with a single oil quench, and is recommended for carburized parts requiring good core ductility and a deep, exceptionally hard case.

Recommended quenching temperature is 1525° F. to refine the core and 1425° F. to refine the case.

Nickel-molybdenum steels are not furnished with high carbon content for heat-treating. The carbon must be added before heat-treating.

**CHROME-VANADIUM STEELS (SAE 6100 Series). SAE 6115 and SAE 6120.** These are carburizing steels and may be used interchangeably with SAE 3115 and SAE 3120. The presence of the vanadium as an alloying agent gives added resistance to shock and retards grain growth during the carburizing operation, providing a tough core and fine case with a single quench. This avoids the separate carburizing process with its higher temperature, which is ordinarily required for core refinement; thus the danger of distortion is lessened. In other words, in ordinary casehardening, the carburizing temperature is higher than the critical range, so, after a part is cooled from carburizing, it is necessary to reheat to just above the critical range; then it is quenched to refine the core. However, the presence of vanadium in this alloy makes it possible to carburize and heat-treat the metal with one heat.

Recommended quenching temperatures are 1575° F. to refine the core or 1450° F. to refine the case.

SAE 6115 machines satisfactorily without annealing. SAE 6120 usually is machined in unannealed state; annealing improves its machinability.

**SAE 6130 and SAE 6135.** These are heat-treatable grades, particularly recommended for use where resistance to shock is requisite, the addition of vanadium serving to accentuate the toughening effect of the chromium.

These grades do not suffer as quickly under excessive temperatures as some of the other alloys, and are therefore admirably suited for parts produced in small lots, particularly where the most modern heat-treating facilities are not available. Recommended quenching temperature is 1550° F. Annealing of both grades is desirable to obtain most satisfactory machining properties.

The physical properties of this steel as used in the manufacture of propeller hubs are: ultimate tensile strength, 135,000 lbs. per sq. in.; yield strength, 115,000 lbs. per sq. in.

**SAE 6145.** Like SAE 6130 and SAE 6135, this is a heat-treatable steel recommended for highly stressed parts. The combination of chrome and vanadium produces a clean, sound, fine-grain steel which machines to a fine finish and minimizes distortion in quenching. It is especially recommended for use where heat-treating is not under careful control. Annealing is necessary for satisfactory machining.

**SAE 6150.** This steel is used in making many coil springs in airplanes because of its high strength and fatigue properties. It can be heat-treated to an ultimate tensile strength of 200,000 lbs. per sq. in.; yield strength, 150,000 lbs. per sq. in.

**SAE 6195.** This is a steel from which ball bearings, roller bearings, and races are made.

**STAINLESS STEELS.** This class of steels, called 18–8 because it contains approximately 18% chromium and 8% nickel, is next to aluminum alloys in quantity used in aircraft, other than in engines and landing gear. It is used in making collector rings, exhaust pipes, firewalls, fillets in heated areas, and diaphragms. In some cases, entire airplanes have been made from this material.

Like copper or pure aluminum, 18–8 steel cannot be hardened by heat treatment. It is hardened by cold working, and is put into its softest state by quenching from high temperatures in cold water. This is a desirable factor, since it involves a minimum of time and effort when annealing is necessary during cold working in drop-hammer or press work.

Stainless steels are made in over thirty different alloys, each having its specific use. The most important of these 18–8 steels in the aircraft industry is the type which is corrosion-resistant. Because of the high percentage of chromium in stainless steel it resists oxidation or scaling even at temperatures up to 1650° F.

While steel weighs approximately three times as much as aluminum alloys, this unfavorable weight comparison is somewhat offset by the fact that certain fabrication units are unnecessary with aluminum alloys—such as rivets, gussets, bolts, etc. It is possible, then, that welded stainless steel, with its greater physical properties, will become the No. 1 metal in the construction of the larger aircraft.

Stainless steel parts must be formed to very close tolerances before jigging or assembling, since ordinarily spot welding is used in fabricating thin sheets, and unsightly buckling, tin-panning, or oil-cans would result if the parts had to be forced together. Whether riveting or spot welding is used in fastening thin skin to heavier gage structure, the skin should fit closely over all contours and should be drawn tightly over the structure. In addition, the riveting or spot welding should be done by the divisional method, that is, the first attachments must be located at the extremes and center of the surface to be fastened, with subsequent spots at centers of the unfastened areas, progressively throughout the surface.

**HEAT TREATMENT OF STEELS.** Heat treatment usually is a combination of operations involving heating and cooling of an alloy in the solid state. The act of cooling is called *quenching*, and is done either by immersion in water, oil, or in air at room temperature. Metals are of crystalline structure and when heated to high temperatures (but below the melting point), this structure assumes forms and grain sizes which vary with the temperatures.

Certain desirable properties are obtained from heating and cooling some steels under given controlled temperatures. For instance, when metal is cooled rapidly from a high temperature, it has a tendency to retain at normal temperatures the grain characteristics it had at the elevated temperature; the physical properties are thus changed. Temperatures at which such changes take place are called the *critical temperatures,* and the carbon content governs the exact temperature at which these changes take place. The grain size is smallest and strongest just above the critical range.

**Hardening.** Hardening is the act of both heating and quenching certain iron-base (ferrous) metals. First the alloy is heated uniformly to a temperature slightly in excess of the upper critical temperature ranges; it is then quenched in oil or water, which is agitated. This produces a fine grain, maximum hardness, and maximum tensile strength, but it also produces a condition that is too brittle for many purposes.

# STEEL IN AIRCRAFT

**Drawing.** Drawing, which is also called *tempering*, is a reheating procedure which brings the steel to the final state of heat treatment and which relieves the strains present in hardened steels, decreases their brittleness, and restores their ductility.

Table 14 shows temperatures for hardening and for drawing, as judged by colors for handwork. When the furnace is used, the temperature is read from a pyrometer.

**Table 14—Temperatures and Corresponding Colors in Heat-Treating**

| Degrees F. | High Temperature Colors | Degrees F. | Colors for Tempering |
|---|---|---|---|
| 752  | Red, Visible in Dark     | 430 | Very Pale Yellow |
| 885  | Red, Visible in Twilight | 440 | Light Yellow |
| 975  | Red, Visible in Daylight | 450 | Pale Straw-Yellow |
| 1077 | Red, Visible in Sunlight | 460 | Straw-Yellow |
| 1292 | Dark Red                 | 470 | Deep Straw-Yellow |
| 1472 | Dull Cherry-Red          | 480 | Dark Yellow |
| 1652 | Cherry-Red               | 490 | Yellow-Brown |
| 1832 | Bright Cherry-Red        | 500 | Brown-Yellow |
| 2012 | Orange-Red               | 510 | Spotted Red-Brown |
| 2192 | Orange-Yellow            | 520 | Brown-Purple |
| 2372 | Yellow-White             | 530 | Light Purple |
| 2552 | White Welding Heat       | 540 | Full Purple |
| 2732 | Brilliant White          | 550 | Dark Purple |
| 2912 | Dazzling White (Bluish)  | 560 | Full Blue |
| .... | ......................   | 570 | Dark Blue |

**Annealing.** Annealing is usually referred to as a heat treatment to induce softness. However, the purpose of annealing may be to release gases from metal, to remove stresses or workhardening effects set up in forming, to refine the crystalline structure, or to alter ductility, toughness, or other physical properties. Full annealing is heating the alloy to the upper critical temperature range and holding it at that temperature not less than one hour for each inch of section of the heaviest portion of the object being treated. Cooling usually is accomplished by turning off the furnace heat and allowing the metal to cool slowly with the cooling of the furnace.

**Normalizing.** This process has for its purpose the relieving of crystal formation promoted under the strain of cold working, welding, and like procedures. For example, in the manufacture of collector rings or exhaust pipes, the drop-hammer forming and the welding operation cause stresses which must be relieved by normalizing to reduce the possibility of fatigue failures which appear in the form of cracks.

The procedure consists in heating the parts above the critical temperature range and soaking at this heat for a time, after which the parts are removed and allowed to cool at room temperature. Steels become harder and stronger when normalized than when annealed, because of the more rapid cooling of the still air.

**Case.** Case is that outside portion of an iron-base alloy which has been casehardened, or, of which the carbon content has been substantially increased.

**Casehardening.** As the name implies, casehardening is the term used where a thin hard case is produced on the surface of soft steel, to make it resistant to wear. The ability of iron to absorb carbon at a high temperature but below the melting point makes casehardening possible.

First, the metal is impregnated on the outer surface with carbon, by immersing the part in a molten bath of a carbon-producing substance. (There are several carburizing liquids or compounds sold under various trade names and advertised in trade magazines.) The depth of penetration, or of the case, is governed by the length of this cooking time. Then, the article is removed from the compound or bath and plunged into cold water.

**Carburizing.** Carburizing is a casehardening process of adding carbon to an iron-base alloy by heating the steel to a temperature below its melting point while it is in contact with carbonaceous substances such as wood charcoal, bone, coke, and similar materials.

**HEAT TREATMENT OF SAE 4130 AND SAE 2330.** See Table 2 for the various tempers and physical properties of these materials.

**Hardening SAE 4130 Parts.** SAE 4130 parts are placed in the furnace, at a temperature of not over 1100° F. Then the furnace temperature is gradually raised to 1575 or 1600° F., depending upon the thickness of the metal. (The lower temperature is used for parts less than 1/4" thick). The parts should be held at temperature at least 15 minutes to insure uniform heating. They are then quenched in oil at room temperature (70°F.).

**Drawing SAE 4130 Parts.** The furnace should not be at a temperature, when SAE 4130 parts are inserted, above that shown

# STEEL IN AIRCRAFT

### Table 15—Metal Gage Standard °

This table shows the standard wire gages and the names of major commodities for which each is used. To determine the gage used for any commodity, notice the letter in parentheses opposite the commodity named and find the gage column below bearing the same letter in parentheses.

| | | | |
|---|---|---|---|
| Alum. and Alum. alloy except tubing (B) | Copper sheets (B) | Spring steel (A) | |
| Alum. and Alum. alloy tubing (A) | Copper tubing (A) | Steel plates (C) | |
| Brass tubing ⅜ O.D. and larger (A) | Copper wire (B) | Steel sheets (C) | |
| Brass tubing smaller than ⅜ O.D. (B) | Corr. Res. Steel & Rustless Iron (C) | Steel tubing, seamless and welded (A) | |
| Brass sheets (B) | Iron wire (D) | Steel wire (D) | |
| Brass wire (B) | Monel Metal sheets (C) | Exceptions are:— | |
| | Music wire (E) | Music wire (E) | |
| | Phos. Bronze sheet (B) | Armature binding wire (B) | |
| | Phos. Bronze strip (B) | Flat wire (A) | |
| | Phos. Bronze wire (B) | Strip steel (A) | |

| No. of Wire Gage | (A) Birmingham or Stubs | (B) American or Brown & Sharpe | (C) U.S. Standard | (D) Washburn & Moen | (E) Music Wire |
|---|---|---|---|---|---|
| 0 | .340 | .325 | .3125 | .307 | .009 |
| 1 | .300 | .289 | .2812 | .283 | .010 |
| 2 | .284 | .258 | .2656 | .263 | .011 |
| 3 | .259 | .229 | .2500 | .244 | .012 |
| 4 | .238 | .204 | .2344 | .225 | .013 |
| 5 | .220 | .182 | .2187 | .207 | .014 |
| 6 | .203 | .162 | .2031 | .192 | .016 |
| 7 | .180 | .144 | .1875 | .177 | .018 |
| 8 | .165 | .128 | .1719 | .162 | .020 |
| 9 | .148 | .114 | .1562 | .148 | .022 |
| 10 | .134 | .102 | .1406 | .135 | .024 |
| 11 | .120 | .091 | .1250 | .120 | .026 |
| 12 | .109 | .081 | .1094 | .105 | .029 |
| 13 | .095 | .072 | .0937 | .092 | .031 |
| 14 | .083 | .064 | .0781 | .080 | .033 |
| 15 | .072 | .057 | .0703 | .072 | .035 |
| 16 | .065 | .051 | .0625 | .063 | .037 |
| 17 | .058 | .045 | .0562 | .054 | .039 |
| 18 | .049 | .040 | .0500 | .047 | .041 |
| 19 | .042 | .036 | .0437 | .041 | .043 |
| 20 | .035 | .032 | .0375 | .035 | .045 |
| 21 | .032 | .028 | .0344 | .032 | .047 |
| 22 | .028 | .025 | .0312 | .028 | .049 |
| 23 | .025 | .023 | .0281 | .025 | .051 |
| 24 | .022 | .020 | .0250 | .023 | .055 |
| 25 | .020 | .018 | .0219 | .020 | .059 |
| 26 | .018 | .016 | .0187 | .018 | .063 |
| 27 | .016 | .014 | .0172 | .017 | .067 |
| 28 | .014 | .0125 | .0156 | .016 | .071 |
| 29 | .013 | .011 | .0141 | .015 | .075 |
| 30 | .012 | .010 | .0125 | .014 | .080 |

° This table was compiled from an average of the major distributors' catalogs and may not agree in some cases with data from certain small concerns.

for drawing (tempering) in Table 2. Hold the parts at the required temperature for 30 minutes to an hour, depending on metal thickness. Remove from furnace and allow to cool in still air.

**Hardening and Drawing SAE 2330 Parts.** The same procedure is used for heat-treating SAE 2330, except that such parts are raised to a temperature of 1430 to 1500° F. within a period of 45 minutes, and held at that temperature for 30 minutes, then quenched in oil. The parts are then reinserted in the furnace at the proper drawing temperature (as shown in Table 2), and held at that temperature for the proper period, depending upon size and mass of parts. They are then cooled in still air.

**METAL GAGE STANDARD.** In the aircraft industry the steels, steel alloys, and iron, like the aluminum alloys, are usually referred to in thousandths of an inch. Sometimes, however, it is necessary to speak in terms of gages of metal. In such a case, the conversion table, Table 15, will be helpful.

### QUESTIONS

1. In SAE numbering of steels, what do the three units of each number signify?
2. Analyze, for example, SAE 2350, SAE 3250, and SAE 6150.
3. Describe *music wire*.
4. What steel is used in connection with machine gun mountings?
5. What qualities does the presence of nickel impart to an alloy?
6. Combining chrome with the nickel speeds up what process?
7. What qualities are derived from the presence of vanadium as alloying agent?
8. How does the Army specify SAE 4130, sheet form?
9. How does the Navy call for SAE 4130 in bar form?
10. What class of steel is known as 18–8, and why?
11. What do we mean by the divisional method in spot welding or riveting?
12. Why is the process of hardening followed by drawing or tempering?
13. Name four reasons for annealing.
14. How does normalizing differ from annealing?
15. What has been done to steel which is referred to as *casehardened?*

# Index

## A

Abbreviations for blueprints....37-39
Abutting parts ....................191
AC standard parts................ 42
Accessories for jig...............268
Accuracy ........................ 93
Adapter ........................ 11
Adjacent parts lines.............. 51
Adjusting rivet gun.............. 26
Aircraft, aluminum in............336
**Aircraft** construction, steel in.353-370
    drawings, types of............. 55
    nut, AN castle..............40, 44
    nut, AN plain...............40, 44
    snips ........................ 13
    tubing, welding of............232
Airplane part, blueprint for......101
**Alclad** products .................335
    sheet, bearing strength of......179
    sheet, shear strength of........179
**Alcoa** aluminum and its alloys....321
    commodities .................321
**Allowance, bend** ............112, 115
    formula .....................116
**Alloy, aluminum,** see *Aluminum alloys*
    rivet heads of.........133, 134, 138
    riveting of ...................176
    rivets of ................133, 138
**Alloy** manufacturing problems....319
    steels with carbon.............354
    steels with nickel..............358
Alloying Elements ..............320
**Alloys,** Alcoa aluminum..........321
    aluminum ...................318
    aluminum, annealing ..........339
    aluminum, choice of..........332
    aluminum, heat treatment of...342
    aluminum, properties of........181
    aluminum sand-casting .......320
    annealed ....................288
    cast, resistance ...............330
    copper, melting point of........200
    hot forming ..................346
    machining ...................346
    magnesium ..............349, 351
    for rivets ....................154

**Alloys**—*continued*
    sand-casting ................323
    wrought .....................322
    wrought aluminum.320, 326, 328, 330
    wrought aluminum, forms......334
    wrought, choosing ............333
    zinc, melting point of..........200
Alternate position lines.......... 51
**Aluminum** in aircraft............336
    Alcoa ........................321
    butt welding ................213
    chemical properties of..........327
    compared with steel...........317
    electric resistance welding of...210
    electric spot welding of........211
    forming .....................343
    furnace brazing of.............201
    mechanical properties ........325
    physical properties of..........321
    torch welding of..............208
Aluminum and its alloys.........317
Aluminum and related metals.317-352
**Aluminum alloy** bolts............274
    nuts .........................275
    parts, heat treatment for.......276
    rivet heads ...........133, 134, 138
    rivets ...................133, 138
    structures ...................274
**Aluminum alloys** ...............318
    annealing ...................339
    choice of ....................332
    designation of ................ 39
    heat treatment of.............342
    hot forming .................346
    machining ...................346
    properties of .................181
    protection of .................287
    repair methods for............276
    riveting of ...................176
    selection for replacement parts.275
    tensile strength ..............318
    wrought .............326, 328, 330
**Aluminum** castings .............338
    fabricating practices...........343
    forgings .....................336
    rivets, shear strength of........180
    sand-casting alloys ...........320

# INDEX

**Aluminum**—*continued*
  sheet .........................337
  tubing .......................338
  welding ..................207, 209
  wire, rod, and bar..............338
**AN** bolts ......................40, 42
  bolts, choice of length.........305
  clevis bolts .................40, 43
  flat-head pins ................. 41
  identification numbers ........ 63
  rivet code ..................... 42
  standard parts, method of identifying .................... 64
  standard rivet forms............ 41
  standard screws ............... 41
**Angle**, cutting, for cold chisel...291
  making ........................292
Angle bracket ....................293
Angular measurements .......... 46
Annealed alloys ..................288
**Annealing** .....................339
  steel .........................367
Anodic treatment and priming...286
Arc welding .....................214
Areas of inertia..................156
Army standard parts..........40, 42
Assembling nut plate with rivet..292
**Assembly** ......................271
  heater box .................70, 71
  pulley bracket ................299
  repairs, miscellaneous techniques, and projects....271-316
  Assembly drawings .............. 56
Attachments, disc-cutting ...... 22
Axis, neutral ............116, 117, 118

## B

Backhand welding ...............224
Ball peen hammer................ 16
Bar, aluminum ...................338
**Bars**, bucking ................19, 20
  flattening ..................19, 20
Base, removing dies from........268
Basic stresses ...................289
Bastard-cut files ................. 30
Beads, reinforcing, for bends.....111
Beakhorn stake .................. 16
**Bearing** design stresses..........158
  strength of Alclad sheet........179
  strength of aluminum alloy plates ....................135
  strength of driven rivets.......136
  strength of rivets..........176-179
  strength of sheet..........177, 178
  strength of steel...............356
  value of rivets.................155
Bell crank .......................108
**Bench** ........................... 92

**Bench**—*continued*
  machines ...................... 16
  mechanics, rules for............ 92
  stakes ........................ 16
  work .......................... 87
**Bend** allowance dimensions.......114
  allowance formula ............116
  classification .................345
  line ..................112, 113, 121
  line dimensions ...............114
  radii ......................... 69
**Bending**, terms used in..........112
  of tubes and sections..........273
**Bending** brakes ..........14, 120, 121
  stress ........................290
**Bends**, allowance for.........112, 115
  patterns for ..............111-130
Bessemer steel ...................358
Beveled corner weld with filler rod.227
Binding bend ....................111
Blades, hack-saw .............23, 24
Blanking of dies..................262
Blanking and cutting..............189
**Blind** rivet, cherry..............162
  riveting ......................152
  rivets ........................160
**Blocks**, change ...............56, 57
  forming ......................296
  title .......................56, 57
Blowhorn stake .................. 16
Blueprint reading ............37-75
**Blueprints** ...................... 90
  abbreviations for ...........37-39
  miscellaneous notes on......... 69
**Board**, contour guide............247
  molding ......................251
Bob, plumb ...................... 85
**Bolt** holes, locating..............291
  threads ......................310
**Bolts** ........................... 64
  aluminum alloy ...............274
  AN .........................40, 42
  AN clevis ..................40, 43
  correct length ................305
  drill sizes for.................. 65
  head markings on............. 43
  shear strength of..............156
  threading ....................312
  torque loads for...............306
Boss for fabricated tube..........130
Bottoming tap ...................311
Box, heater, assembly..........70, 71
Box brakes ...................... 15
**Bracket**, angle ..................293
  making ...............299, 300, 302
  pulley ........................298
Bracket clamp, reinforced loop ..................126, 127
**Brakes**, bending ..........14, 120, 121

# INDEX

**Brakes**—*continued*
  box .......................... 15
**Brazing** ....................199, 200
  furnace, of aluminum..........201
**Breaks**, long, lines indicating..... 50
  short, lines, indicating.......... 50
Broken stud, extraction of........313
**Bucking** bars ..............19, 20, 149
  dolly ......................... 19
  mouse ........................152
  rivets in tube..................152
Bucks, rivet .....................149
Bulb-angle bucking bar...........149
Bulb-section bend ...............111
Burning, lead ...................198
**Butt** weld ..................226, 231
  welding of aluminum...........213

## C

Cable connector .................104
Caliper, micrometer ............. 82
Candle mold stake............... 16
Capstrips, repair of..............277
Carbon, alloy steels with.........354
Carbon steel ....................355
Carburizing steel ................368
Case .........................368
Casehardening steel .............368
Cast alloys, resistance............330
**Casting** of male die..............255
  of plaster pattern..............249
  of punch .....................254
  of zinc die....................250
Castings, aluminum .............338
Castle aircraft nut............46, 44
Cautions for oxyacetylene welding.215
C clamps .......................296
Cecostamp ..................268, 269
**Center** line ..................... 50
  punch ........................ 84
Change block ................56, 57
Channel tabulation ..............122
**Chemical composition** of carbon
    steels ......................357
  of chromium steels.............360
  of chromium-vanadium steel...361
  of free cutting steels...........358
  of manganese steels............358
  of molybdenum steels..........360
  of nickel-chromium steels......359
  of nickel steels................359
  of silicon-manganese steel......361
**Chemical** properties of aluminum.327
  welding ......................196
Cherry blind rivet................162
Chisel, cold, cutting angle for....291
**Choice** of aluminum alloys......332
  of wrought alloys..............333

Choosing drill ................... 65
Chrome-molybdenum steel.......355
Chrome-nickel steel .........355, 361
Chrome-vanadium steel ......355, 364
**Chromium steels** .................355
  chemical composition of.......360
Chromium-vanadium steel, chemical composition ............361
Circles, dimensioning ........... 47
Circular measurement ........... 78
**Clamps** .........................257
  C .............................296
  for jig ........................268
  loop type ....................123
  reinforced loop bracket...126, 127
  for square tubing.........127, 128
Clay model, developing of...245, 247
Cleaners, tests for...............288
Cleaning, drills for..........204, 205
Cleat seam ....................307
Cleco sheet holder..............141
Clevis bolts, AN..............40, 43
**Clip** ........................... 72
  making ......................291
  reinforcing ...............125, 126
Coarse-cut files ................. 30
Code, AN rivet.................. 42
Coefficient of thermal expansion..324
Coiled strip ....................338
Cold chisel cutting angle........291
**Collar**, clinch ...................307
  with wired edge...............302
Collector ring, exhaust...........268
Colors in heat-treating..........367
Combination set ................ 80
**Commercial** forms of wrought aluminum alloys .............334
  identification numbers ......... 63
Commodities in Alcoa........... 321
Comparison of terms............ 39
Composite joint ................238
**Composition** of aluminum sand-
    casting alloys .............320
  carbon steels .................357
  chemical, of chromium steels...360
  chemical, of chromium-vanadium steel ....................361
  chemical, of manganese steels..358
  chemical, of molybdenum steels.360
  chemical, of nickel steels.......359
  chemical, of nickel-chromium
    steels ......................359
  chemical, of silicon-manganese
    steel .......................361
  of free cutting steels...........358
  of soft solder .................196
  of wrought aluminum alloys....320
Compression ....................290
**Conductivity**, electrical......322, 323

# INDEX

**Conductivity**—*continued*
  thermal .............321, 322, 323
Connector, cable ................104
Construction, aircraft, steel in.353-370
**Contour** guide board............247
  saw and file machines.......... 21
  saws, teeth ................... 23
Control, flame ..................206
Controlling of wrinkles..........264
Copper, identifying alloys containing ......................275
**Copper** alloys, melting point of...200
  rivets, shear strength of........180
**Corner weld**, beveled, with filler rod ......................227
  inside .......................231
  vertical .....................231
Corrosion, resistance to..........327
Corrosion protection ............286
Corrugation, making ............304
Corrugation bends ..............111
Cotter pins, AN..............41, 44
**Countersinks** ................66, 144
  stop .........................145
**Countersunk** holes .............. 66
  rivets .......................144
Cover plate ....................294
Cowling, formed sheet for........261
Cowling piece ..................246
Cowlings, nacelle ...............266
Cracked members, repairing..281, 283
Craftsman, sheet metal..........140
Crank, bell ....................108
Creasing stake ................. 16
**Cross section** of clinch collar.....307
  of installed grommet...........307
  of weld .....................228
Crowfoot .......................265
Cubic measurement ............. 78
Cupping, rivet ................. 12
Cuts of files................... 29
Cutting of templates............ 93
Cutting and blanking............189
**Cutting** angle for cold chisel....291
  attachment, disc ............. 22
  edges of twist drill............ 9
  internally threaded parts......312
  out parts ..................... 97
  out templates ................. 97
  plane line ................... 50
  steels, free ..................358
Cylinder, making ...............306
Cylinders ......................118

### D

**Dash numbers** ................. 58
  detailed parts ................ 59
  details on other drawings...... 60

**Dash numbers**—*continued*
  hidden or obscure parts........ 63
  left- and right-hand parts...... 61
  opposite parts ................ 58
  parts with differences......... 58
  subassemblies ................ 62
**Data**, rivet set................. 12
  snap ......................... 12
Descriptions of material......... 67
**Design**, safe bearing, stresses.....158
  of welded joints .............207
  of welds ....................217
Designation of aluminum alloy... 39
Detail drawings ................. 55
Detailed parts, dash numbers for 59
**Detailing** of left-hand parts...... 60
  of right-hand parts ........... 60
**Details** with differences.......... 60
  enlarged ..................... 51
  for other drawings ........... 60
Developing of clay model....245, 247
Development of template........ 94
Diaphragm punch ...............261
**Die**, finishing of................254
  leveling of ..................254
  male, casting of..............254
  mold of .................252, 253
  pouring of ..................252
  zinc ....................255, 256
  zinc, casting of..............250
**Dies** ..........................261
  blanking of .................263
  fastening ...................256
  forming of .................262
  large .......................262
  metals used in...............260
  progressive .................263
  removing, from base..........268
  reverse .....................259
  types of ....................260
**Differences**, dash numbers for parts with ................ 58
  details with ................. 60
Dill Lok-Skru ..............162, 163
**Dimension** lines .............47, 50
  numbers .................... 47
Dimensioning circles ............ 47
**Dimensions**, bend allowance......114
  bend line ...................114
  on drawings ................. 45
  transfer of, to templates....... 96
Disassembly before repairing....275
Disc-cutting attachment ........ 22
Dissimilar metals, insulation of..193
Distance, edge .................159
**Dividers** ....................7, 80
  spring ...................... 79
Dolly, bucking ................. 19
**Double seaming** ................307

# INDEX 375

**Double seaming**—*continued*
- stake .......................... 16

Double-cut file ................... 31
Drafting, template ............... 93
Drag .............................. 251
Draw filing ....................... 32
**Drawing** to scale............... 55
- steel ..........................367
- steel parts ...............368, 370

**Drawings**, aircraft, types of...... 55
- assembly ..................... 56
- detail ........................ 55
- dimensions on ................ 45
- installation .................. 56
- larger-scale ................. 52
- revisions to ................. 56

**Drill** gage ....................... 81
- guide .........................151
- parts .......................... 8
- sizes ......................... 72
- sizes for bolts............... 65
- sizes for machine screws...... 65

Drilled holes ..................... 66
**Drilling** .......................141
- oversize holes ...............275
- removing rivets by............184

**Drills**, choosing ................ 65
- for cleaning .............204, 205
- electric, removing rivet heads with ......................185
- grinding ....................... 8
- sharpening .................... 25
- twist ........................8, 81

**Driven rivets**, bearing strength of ..........................136
- inspection of ................176
- shear strength of.............136

Drivers, screw ..................... 7
**Drop hammer** ...............241-270
- forming with ...........244, 262
- large .........................243
- large parts formed by.........267
- setup of ......................258
- small .........................243
- tools and equipment for.......242
- tubular parts formed by.......267

Dual method of identifying AN standard parts ............ 64
Duplicating punch .............. 91
Duplicator, hole ................151
Du Pont explosive rivets.....163, 165

## E

**Edge**, preparing, for weld........228
- straight ..................... 85

Edge distance ...................159
**Edges**, cutting, of twist drill...... 9
- repairing ....................281

Effect of temperature on mechanical properties .............327
Elasticity, modulus of............325
**Electric** drill, removing rivet heads with .......................185
- resistance welding of aluminum.210
- spot welders .................211
- spot welding of aluminum......211
- welding ......................196

Electric and gas welding.........196
Electrical conductivity .....322, 323
Electrode .......................214
**Elongation** .....................328
- of wrought aluminum alloys...326

Ends, shank, of rivets...........139
Enlarged details ................. 51
Equipment, oxyacetylene .......217
Equipment and tools for drop hammer ..................242
Exhaust collector ring...........268
Expander, rivet shank...........150
Expansion, thermal .........321, 324
Explosive rivets .......162, 163, 165
Extension lines ................. 50
Extracting broken stud...........313
**Extruded** sections ..............272
- shapes .......................338

## F

**Fabricated** tube, boss for.........130
- tubes ........................128

Fabricating practices for aluminum ..........................343
Fabrication, general .......189-193
Fastening dies ..................256
Fast-hitting riveter ............ 18
Fatigue, wrought aluminum alloys...........................326
Fender pliers .................... 6
File machine and contour saw... 21
**Files** ............................ 7
- cuts of ...................... 29
- shapes of .................... 29
- standard types of............. 30

Files and their uses............29-33
**Filing**, draw .................... 32
- general purpose .............. 31
- rules for .................... 32
- technique of ................. 31

**Filler rod**, beveled corner weld with ......................227
- welding with .................225

**Fillet welds** ............228, 236, 237
- in tee joint..................230
- of tubing joint...............236

Finished parts, working tolerances for ................242
**Finishing** of die................254

**Finishing**—*continued*
  protective .................... 347
First angle orthogonal projection. 69
**Fitting** skin ................. 189-193
  skin sections ................. 190
  taper pin .................... 315
**Flame,** control of ................. 206
  oxyacetylene ................. 204
  oxyhydrogen ................. 204
Flames, welding ............ 204, 205
Flange, making ................. 293
**Flange** reinforcement ........ 282, 286
  splices .............. 171, 282, 286
  weld ......................... 222
Flanged metal, fusing, without welding rod ............... 222
Flash welding .................. 213
Flat-head pins, AN ............. 41
Flattener, spar ............... 19, 150
Flattening bars .............. 19, 20
Fluids, layout .................. 92
Flush rivet ..................... 146
Flux, leaving traces of ............ 288
Fluxes for soft soldering ........ 198
Folded seam .................... 307
**Forehand** ripple welding ........ 223
  welding ...................... 224
Forgings, aluminum ............ 336
Form, shop order ................ 89
Formed sheet for cowling ........ 261
**Forming** of aluminum ........... 343
  of dies ....................... 262
  drop-hammer ............ 244, 262
  hot .......................... 346
  of parts ...................... 97
  on punch press ................ 297
  of rivet head ................. 139
  of sheet metal parts ........... 276
  stretching and shrinking in .... 294
  techniques of ................. 264
**Forming** blocks ................ 296
  machines .................... 15
  rolls ......................... 15
**Forms** of rivets, AN standard .... 41
  of wrought aluminum alloys ... 334
**Formula,** bend allowance ........ 116
  for stretch-out ........... 113, 118
**Frames,** intermediate, splicing of .................... 174, 175
  intermediate, straightening .... 280
  repairing .................... 284
Free cutting steels, composition of ........................ 358
Friction washer ................. 107
Furnace brazing of aluminum .... 201
Fuselage, repairs to ............. 285
**Fusing** flanged metal without welding rod ................. 222
  metal with oxyacetylene torch. 220

**Fusing**—*continued*
  metal without welding rod .... 221

### G

**Gage,** drill ..................... 81
  steel wire .................... 81
  wire ......................... 369
Gage standard, metal ....... 369, 370
Gages, surface .................. 83
Gas welding .................... 196
Gas and electric welding ........ 196
General fabrication .......... 189-193
General-purpose filing ........... 31
Grain of sheet metal ............ 189
Gravity, specific ............ 319, 321
Grinding twist drills ............. 8
**Grommet,** cross section of ....... 307
  making and installing ......... 308
Grooved seam .................. 307
Guide, drill .................... 151
Guide board, contour ........... 247
**Gun, rivet** .................... 18
  adjusting .................... 26
  operating .................... 26

### H

Hack saws ..................... 23
Hack-saw blades ............ 23, 24
**Hammer,** ball peen ............. 16
  drop ..................... 241-270
  drop, forming with ........... 244
  drop, large .................. 243
  drop, large parts formed by ... 267
  drop, setup of ................ 258
  drop, small .................. 243
  drop, tools and equipment for.. 242
  drop, tubular parts formed by.. 267
  riveting ..................... 16
  setting ...................... 16
Hard soldering ................. 199
**Hardening steel** ................ 366
  parts .................... 368, 370
Hardness, wrought aluminum alloys ........................ 326
Hat-section channel bends ....... 111
Heading ....................... 146
**Heads,** bolt, markings on ........ 43
  rivet, aluminum alloy. 133, 134, 138
  rivet, forming of .............. 139
  rivet, removing .............. 167
Heat treating, temperature and tempering colors in ........ 367
**Heat treatment** for aluminum alloy parts ................... 276
  conditions for ................ 342
  precipitation ................. 342
  of rivets .................... 153

# INDEX 377

**Heat treatment**—*continued*
   solution ...................... 340
   of steels ............. 356, 366-368
Heater box assembly .......... 70, 71
Heat-treatable wrought alloys....333
Heat-treated alloys containing copper, identification of....275
Helpful hints for safety.......... 1
Hem and turned edge............307
Hemmed edge seam.............307
**Hidden** lines ..................... 50
   parts, dash numbers for........ 63
Hinge, piano ....................310
Hints for safety.................. 1
Holder, sheet, Cleco.............141
**Hole**, ream ..................... 315
   rivet ........................182
   riveting patch over............187
Hole duplicator .................151
**Holes**, bolt and rivet, locating....291
   countersunk .................. 66
   drilled ....................... 66
   oversize, drilling ..............275
   punched ...................... 66
   reamed ...................... 66
   rivet ....................140, 192
   rivnut ......................151
   in skin, patching .............279
**Hollow** mandrel stake............ 16
   rivets ........................166
Honing a reamer.................314
**Horizontal** beveled weld.........227
   butt corner weld..............227
Hot forming ....................346
Hygroscopic materials ...........288

## I

**Identification** of heat-treated alloys containing copper.....275
   of rivet material...............166
   of tools ...................... 28
**Identification numbers** ........... 58
   AN .......................... 63
   commercial ................... 63
Imperfections in rivets......147, 148
Importance of measuring to mechanic ..................... 77
Inaccessible bolt and rivet holes..291
India stone .....................314
**Inertia**, areas of..................156
   moments of ...................156
Insert plate weld............236, 237
Inside flange ...................293
Inspection of driven rivets.......176
Installation drawings ............ 56
Installing patches on stressed skin ........................186
Insulation of dissimilar metals...193

**Intermediate frames**, splicing of.. .......................174, 175
   straightening ................280
Internally threaded parts, tapping or cutting .................312
Intersections, rib, repairing .....281
Iron templates ..................266
Items, round, sections for......... 54

## J

**Jig**, accessories for..............268
   clamps for ...................268
Joggle bend ....................111
Joggles or offsets ...............192
Joining sheets ..................191
**Joints**, bolts and rivets in........287
   composite ....................238
   protection of ................287
   riveted, proportions of..........158
   riveted, strength of............154
   tee, fillet welds in..............230
   tubing, fillet weld of..........236
   welded ......................230
   welded, design of..............207

## K

Kinds of measurement........... 78

## L

Lap seam ......................307
Lapped tabs ...................300
**Large** dies .....................262
   drop hammers ...............243
   parts formed by drop hammers.267
Larger scale drawing............. 52
**Layout** of angles ................ 96
   on pattern material........... 95
   of templates ..............87-110
   of templates on sheet metal.... 98
Layout fluids ................... 92
Lead burning .................. 198
Leading edges, repairing.........281
**Left-hand parts** ................ 61
   detailed ..................... 60
**Length** of bolts..................305
   of rivets .....................139
Level, transit .................... 84
Leveling of die..................254
Line of motion .................. 51
Linear measurement ............ 78
**Lines** .......................... 47
   adjacent parts ................ 51
   alternate position ............ 51
   bend ..................112, 113, 121
   bend, dimensions .............114
   center ....................... 50

378  INDEX

**Lines**—*continued*
  cutting plane ................... 50
  dimension ..................47, 50
  extension ...................... 50
  hidden ......................... 50
  indicating long breaks......... 50
  indicating short breaks........ 50
  mold ......................112, 113
  reference, laying out.......... 96
  section ........................ 48
  section, material symbols for... 54
Lining, section ................... 54
Loads, torque, for bolts..........306
Locating inaccessible holes.......291
Long breaks, lines indicating..... 50
**Loop** bracket clamp, reinforced
  ........................126, 127
  section bend .................. 111
  type clamp ....................123
Lord shock, bracket for...........300
Lugs ..............................257

### M

**Machine** screws, drill sizes for.... 65
  screws, sizes of................ 66
  settings for spot welding.......212
**Machines,** bench ................ 16
  contour saw and file............ 21
  forming ........................ 15
Machining aluminum alloys......346
Magnesium alloys ..........349, 351
**Making** angle ....................292
  bracket .......................299
  collar with wired edge.........302
  cylinder ......................306
  flange ........................293
  piano hinge ...................310
Making and installing grommet..308
Male die, casting of..............254
Mallets ........................... 16
Manganese steels, composition of.358
Markings on bolts, head.......... 43
Material symbols for section lines 54
**Materials,** description on blueprints ........................ 67
  hygroscopic ...................288
  pattern, layout on............. 95
  pattern, template on...........102
  rivet .......................... 41
  rivet, identification of..........166
  thin, welding on...............231
Measurement of temperature..... 79
**Measurements,** angular ......... 46
  circular ....................... 78
  cubic .......................... 78
  kinds of ....................... 78
  linear ......................... 78
  square ......................... 78

**Measurements**—*continued*
  weight ......................... 78
**Measuring** ....................77-86
  importance of .................. 77
**Measuring** tapes, steel........... 79
  tools .........................79-86
Mechanic, importance of measuring to ...................... 77
Mechanical properties of aluminum ......................325, 326
Mechanics, bench, rules for...... 92
**Melting point** of copper alloys....200
  of soft solder..................196
  of zinc alloys..................200
**Members,** cracked, repairing......283
  repairing .....................281
**Metal** fusing with oxyacetylene torch .......................220
  fusing without welding rod.....221
  gage standard .............369, 370
  thickness .................114, 120
**Metals,** dissimilar, insulation of..193
  used in dies...................260
Methods of repair...168-187, 276-286
Micrometer caliper ............... 82
Miscellaneous notes on blueprints 69
Model, clay, developing of..245, 247
Modulus of elasticity............325
Moisture-proofed hygroscopic materials ........................288
Mold, die ..................252, 253
**Mold** line .................112, 113
  point ......................113, 114
Molding board ...................251
**Molybdenum steels** ..............355
  chemical composition of.......360
Moments of inertia..............156
Motion, line of.................. 51
Mouse, bucking .................152

### N

Nacelle cowlings ................266
Navy standard parts ..........40, 42
Needlecase stake ................ 16
Neoprene .......................191
Neutral axis ................116-118
**Nickel steels**................355, 358
  chemical composition of.......359
Nickel-chromium steels, chemical composition ............355, 359
Nickel-molybdenum steel....355, 364
Ninety-degree angle bend........111
Normalizing steel ...............367
Number of rivets required for splices ..................173-175
Numbering of steel, SAE........353
**Numbers,** dash .................. 58
  dimension ..................... 47

# INDEX

**Numbers**—*continued*
- identification .................. 58
- identification, AN ............ 63
- identification, commercial...... 63

**Nuts,** aluminum alloy............275
- AN castle aircraft...........40, 44
- AN plain aircraft............40, 44
- tightening ....................305

## O

- Obscure parts, dash numbers for. 63
- Offset bend .....................111
- Offsets or joggles...............192
- **Open-hearth** screw stock........357
- steel ..........................358
- **Operation** of rivet gun........... 26
- of riveter ......................184
- **Opposite** assembly, left- and right-hand parts on.............. 61
- parts, dash numbers for........ 58
- Order, shop ..................... 88
- Order form, shop................. 89
- **Orthographic projection** .......... 67
- first angle .................... 68
- third angle .................... 68
- Outline of parts................. 48
- Outside flange ..................293
- Oversize holes, drilling..........275
- **Oxyacetylene** equipment ........217
- flame ..........................204
- torch, fusing metal with.......220
- welding ..................201, 215
- welding, cautions for..........215
- welding, procedures for........215
- Oxyhydrogen flame ..............204

## P

- Paint removers .................288
- Part, structural, section of........ 53
- **Parts,** abutting ..................191
- AC standard ................... 42
- aluminum alloy, heat treatment for ..........................276
- AN standard, method of identifying ..................... 64
- cutting out .................... 97
- detailed, dash numbers for...... 59
- with differences, dash numbers for ........................ 58
- exactly opposite, dash numbers for ........................ 58
- finished, working tolerances for.242
- forming ....................... 97
- hidden or obscure.............. 63
- large, formed by drop hammers ....................267
- outline of .................... 48

**Parts**—*continued*
- replacement, selection of alloy for ..........................275
- sheet metal, forming...........276
- standard, Army .............40, 42
- standard, Navy .............40, 42
- steel, drawing .................368
- steel, hardening ...............368
- steel, hardening and drawing...370
- tubular, formed by drop hammers ......................267
- twist drill .................... 8
- **Patch,** riveting, over hole........187
- round, rivet pattern for.........183
- Patches, installing, on stressed skin ........................186
- Patching holes in skin panels....279
- **Pattern,** plaster .................251
- plaster, casting of..............249
- rivet, for round patch..........183
- **Pattern material,** layout on...... 95
- template on ...................102
- Patterns for bends...........111-130
- Penetration, weld ...............228
- Personal tool requirements....... 5
- Phantom line ................... 51
- **Physical properties** of aluminum..321
- of steel .......................355
- Piano hinge ....................310
- Piece, cowling ..................246
- **Pin,** cotter ..................41, 44
- taper, fitting .................315
- **Pins,** flat-head .................. 41
- shear strength of..............156
- Pipes ..........................118
- Pittsburgh seam ................307
- Plain aircraft nut, AN.........40, 44
- **Plaster pattern** .................251
- casting of ....................249
- **Plate,** aluminum alloy, bearing strengths of ................135
- cover .........................294
- insert, weld ..............236, 237
- Pliers, fender .................. 6
- Plug tap .......................311
- Plumb bob ..................... 85
- **Point,** melting, of copper alloys...200
- melting, of soft solder.........196
- melting, of zinc alloys.........200
- mold .....................113, 114
- Points, trammel ................ 80
- Polishes, tests for...............289
- Position lines, alternate.......... 51
- Pouring of die..................252
- Practices not acceptable........287
- Precipitation heat treatment.....342
- Preheating for welding..........206
- Preparing edge for weld.........228
- Prick punch ................... 94

Priming .......................286
**Procedure** in developing Pittsburgh
    seam ......................307
    for oxyacetylene welding.......215
Production of wrought aluminum
    alloys ....................330
Products in Alclad...............335
Progressive dies .................263
**Projection, orthographic** ..........67
    first angle ....................68
    third angle ...................68
**Projects**
    Adjusting rivet gun............26
    Bracket for Lord shock........300
    Drilling rivet heads to remove
        them .....................185
    Extracting broken stud.........313
    Fabricating tubes ........128, 129
    Fillet welds on tubing.........237
    Flange welding ...............222
    Fusing metal without rod......220
    Fusing two pieces of metal together without rod.........221
    Installing patch on stressed skin ......................186
    Making angle and assembling nut plate with rivet........292
    Making bell crank ............108
    Making beveled corner weld...227
    Making 90° bracket with lapped tabs ................300, 302
    Making butt weld ............226
    Making cable connector ......104
    Making clamp for square tubing ......................127
    Making a clip .................291
    Making collar with wired edge.302
    Making corrugation ...........304
    Making cylinder ..............306
    Making fillet weld in tee joint..228
    Making friction washer........107
    Making hat section channel....122
    Making horizontal butt corner weld .....................227
    Making insert plate weld.......236
    Making inside and outside flange ....................293
    Making loop-type clamp..123, 127
    Making piano hinge ..........310
    Making reinforced loop bracket clamp ....................126
    Making reinforcing clip........125
    Making reinforcing washer.....106
    Making vertical welds.........230
    Making and installing grommet.308
    Operating rivet gun............26
    Operation of oxyacetylene equipment ................217
    Ream hole and fit taper pin....315

**Projects**—*continued*
    Removing rivets by drilling....184
    Riveting .....................184
    Riveting patch ...............187
    Riveting with riveter ..........27
    Sharpening drills ..............25
    Tapping or cutting internally threaded parts ...........312
    Threading a bolt or stud.......312
    Tightening nuts...............305
    Tool identification ............28
    Welding aircraft tubing........232
    Welding with filler rod.........225
    Welding by rotating and position process ..............232
    Welding thin material.........231
    Welding tubing ..............232
**Properties** of aluminum alloys.....181
    of aluminum, chemical........327
    of aluminum, physical.........321
    mechanical, of aluminum......325
    of tubes for splices...233, 234, 235
    of wrought alloys.............322
Proportions of riveted joints.....158
**Protection** against corrosion......286
    of joints ....................287
Protective finishing ..............347
Protractor, steel ..............80, 81
Pulley bracket ...................298
**Punch**, casting of................254
    center ......................84
    diaphragm ..................261
    duplicating ..................94
Punch press, forming on.........297
Punched holes ...................66
Push welding ....................213
Push-button rivet gun............18

**Q**

Qualities of aluminum............317
Quantity indicated by dash numbers .................60, 61
Quenching ................287, 341
Questions, review ...............
    25, 75, 103, 183, 238, 269, 350, 370
Questions and answers..........
    ..............73, 116, 120, 179

**R**

**Radii**, bend .....................69
    for 90° cold bends.............345
    scribing ....................97
Radius............113, 114, 117, 120
Raker saw tooth..................23
Rasp-cut file ....................31
Reading, blueprint ...........37-75
Reamed holes ....................66

# INDEX 381

| | Page |
|---|---|
| **Reamers** | 313 |
| stoning | 314 |
| Reaming hole | 315 |
| Reference lines, laying out | 96 |
| Reheating alloys | 288 |
| **Reinforced** loop bracket clamp | 126, 127 |
| welds | 236, 237 |
| **Reinforcement** of flanges | 282, 286 |
| of weld | 228 |
| **Reinforcing** clip | 125, 126 |
| washer | 106 |
| Relation of views to each other | 67 |
| Removers, paint | 288 |
| **Removing** dies from base | 268 |
| rivet head | 167 |
| rivet heads with electric drill | 185 |
| rivets by drilling | 184 |
| **Repair methods** for aluminum alloys | 276 |
| for skin panels | 279 |
| using rivets | 168 |
| **Repairing** aluminum alloy structures | 275 |
| cracked frame | 284 |
| cracked members | 281, 283 |
| **Repairs** to aluminum alloy structures | 274 |
| to rudder and fuselage | 285 |
| Replacement parts, selection of aluminum alloy for | 275 |
| Replacing rivets | 166 |
| Requirements, tool, personal | 5 |
| **Resistance** of cast alloys | 330 |
| to corrosion | 327 |
| Reverse dies | 259 |
| Review questions | 25, 75, 103, 183, 238, 269, 350, 370 |
| Revisions to drawings | 56 |
| Rib intersections, repairing | 281 |
| Ribs, wing and tail surface, repair methods | 276 |
| Right-hand parts detailed | 60 |
| Ring, exhaust collector | 268 |
| Ripple welding, forehand | 223 |
| **Rivet** alloy | 154 |
| bucking in tube | 152 |
| bucks | 149 |
| code, AN | 42 |
| forms, AN standard | 41 |
| gun | 18 |
| gun, adjusting | 26 |
| gun, operating | 26 |
| gun, push-button | 18 |
| heads, aluminum alloy 133, 134, 138 | |
| heads, forming of | 139 |
| heads, removing | 167 |
| heads, removing, with electric drill | 185 |

| | Page |
|---|---|
| **Rivet**—*continued* | |
| holes | 140, 182, 192 |
| holes, locating | 291 |
| materials | 41 |
| materials, identification of | 166 |
| pattern for round patch | 183 |
| set data | 12 |
| sets | 10 |
| shank expander | 150 |
| sizes | 72 |
| snaps | 10 |
| **Riveted joints**, proportions of | 158 |
| strength of | 154 |
| **Riveters** | 16 |
| bucking bars for | 19, 20 |
| fast-hitting | 18 |
| flattening bars for | 19, 20 |
| operation of | 184 |
| riveting with | 27 |
| slow-hitting | 18, 19 |
| **Riveting** | 133-187 |
| of aluminum alloy | 176 |
| blind | 152 |
| of patch over hole | 187 |
| with riveter | 27 |
| technique for | 141, 169 |
| Riveting hammer | 16 |
| **Rivets** | 133-187 |
| aluminum alloy | 133, 138 |
| aluminum, shear strength of | 180 |
| bearing strength of | 176-179 |
| bearing value of | 155 |
| blind | 160 |
| cherry | 162 |
| cherry blind | 162 |
| copper, shear strength of | 180 |
| countersunk | 144 |
| driven, bearing strength of | 136 |
| driven, inspection of | 176 |
| driven, shear strength of | 136 |
| explosive | 162, 163, 165 |
| flush | 146 |
| heat treatment of | 153 |
| hollow | 166 |
| imperfections in | 147, 148 |
| lengths of | 139 |
| removing, by drilling | 184 |
| replacement of | 166 |
| required for splices, number of | 173, 174, 175 |
| shank ends of | 139 |
| shear strength of | 156, 176-179 |
| shear value of | 154 |
| sizes of | 42, 159, 167 |
| spacing of | 159, 167 |
| tensile value of | 155 |
| types of | 160 |
| **Rivnuts** | 161 |
| holes for | 151 |

## 382 INDEX

**Rod,** aluminum .................338
  filler, beveled corner weld with.227
  filler, welding with.............225
  welding .......................216
  welding without ..........221, 222
Rods, welding ....................206
Rolling ..........................291
Rolls, forming ................... 15
**Round** items, sections for......... 54
  patch rivet pattern for........183
**Round-head** screw ............41, 44
  stake ......................... 16
Rudder repairs ...................285
Rule, shrinkage ..................244
**Rules** for bench mechanics....... 92
  for filing ..................... 32
  steel ......................... 79

### S

SAE numbering of steel..........353
**Safe** bearing design stress........158
  bearing value of rivets.........155
  shear value of rivets..........154
  tensile value of rivets.........155
Safety hints ..................... 1
Sand-casting alloys .........320, 323
Saw and file machine, contour.... 21
**Saws,** contour, teeth of........... 23
  hack ......................... 23
Scale, drawing to................. 55
Scale welding ...............224, 225
Scraping template................247
**Screw** drivers ................... 7
  stock, steel ...................357
  threads .......................310
**Screws,** AN round-head.......41, 44
  AN standard................... 41
  machine, drill sizes for........ 65
Scribe ........................... 94
Scriber .......................... 7
Scribing radii ................... 97
**Seam,** cleat ....................307
  folded .......................307
  grooved ......................307
  hemmed edge .................307
  lap ..........................307
  Pittsburgh ...................307
  standing .....................307
Seaming, double .................307
Seams ...........................307
Second-cut files ................. 30
**Section** of clinch collar..........307
  cross, of weld................228
  of structural part............. 53
**Section** lines ................... 48
  lines, material symbols for...... 54
  lining ........................ 54
Sectional views ..............51, 53

**Sections,** bending ...............273
  extruded .....................272
  for round items............... 54
  skin, fitting ..................190
Selection of aluminum alloy for
  replacement parts .........275
**Sets,** combination ............... 80
  rivet ......................... 10
Setting hammer .................. 16
Settings, machine, for spot welding ..........................212
Setup of drop hammer...........258
**Shank** ends of rivets.............139
  expander, rivet ..............150
**Shapes,** extruded ................338
  of files ...................... 29
**Sharpening** drills ............... 25
  screw drivers ................. 7
Shear, stress in..................290
**Shear strength** of Alclad sheet...179
  of aluminum rivets............180
  of bolts ......................156
  of copper rivets...............180
  of driven rivets...............136
  of pins ......................156
  of rivets..........154, 156, 176-179
  of sheet ................177, 178
  of steel .....................356
  of wrought aluminum alloys....326
**Shears** .......................... 13
  squaring ..................... 14
**Sheet,** Alclad, bearing strength of.179
  Alclad, shear strength of......179
  aluminum ....................337
  bearing strength of.......177, 178
  formed, for cowling...........261
  shear strength of.........177, 178
**Sheet** holder, Cleco..............141
  metal craftsman ..............140
  metal, grain of...............189
  metal, laying templates on..... 98
  metal parts, forming..........276
  thickness ....................291
**Sheets,** joining .................191
  splicing of ..............170, 280
**Shop order** ..................... 88
  form ......................... 89
Shop-furnished tools ............ 14
Short breaks, lines indicating.... 50
Shrinkage rule ..................244
Silicon-manganese steel, chemical
  composition ...............361
Single-cut file ................... 31
Size of rivets....................167
**Sizes,** drill ..................... 72
  drill, for bolts................ 65
  drill, for machine screws....... 65
  machine screws .............. 66
  rivet .................42, 72, 159

# INDEX

**Sizes**—*continued*
  welding-torch tips ........204, 205
**Skin**, patching holes in ..........279
  stressed, installing patches on..186
**Skin** fitting ................189-193
  sections, fitting ................190
Slip roll forming machines....... 16
Slow-hitting riveter ..........18, 19
Small drop hammer..............243
Smooth-cut files ................ 30
Snap data, rivet................. 12
Snaps, rivet .................... 10
**Snips** ......................... 13
  left and right................. 13
**Soft solder**, composition of.......196
  melting point of...............196
**Soft soldering** .................196
  fluxes for ....................198
  solders for ...................198
Solder, soft ....................196
**Soldering**, hard ................199
  soft .........................196
  soft, fluxes for................198
  soft, solders for...............198
Solders for soft soldering........198
Solution heat treatment..........340
Spacing of rivets..........159, 167
Spar flattener ...............19, 150
**Specific gravity** ...........322, 323
  of aluminum ..................321
  of aluminum alloys............319
**Splices**, flange ........171, 282, 286
  properties of tubes for 233, 234, 235
  rivets required for....173, 174, 175
  stringer .................171, 172
**Splicing** of intermediate frames..
  ..........................174, 175
  of sheets .................170, 280
  of stringers ..............170, 286
  of tubes .......................168
**Spot welding** ...................176
  electric ......................211
  electric, of aluminum..........211
  machine settings for...........212
Spring dividers ................. 79
**Square** loop clamp..............128
  measurement .................. 78
  stake ......................... 16
  tubing, clamp for.........127, 128
  type corrugations .............304
Squaring shears ................. 14
Stainless steels .................365
**Stakes**, bench .................. 16
  tinners' ...................... 17
Standard, metal gage........369, 370
**Standard** parts, AC............. 42
  parts, Army ................40, 42
  parts, Navy ................40, 42
  types of files................. 30

Standing seam ..................307
**Steel** in aircraft construction..353-370
  annealing ....................367
  carburizing ..................368
  casehardening ................368
  chrome-vanadium .............364
  chrome-vanadium, chemical
    composition of............361
  compared with aluminum......317
  drawing ......................367
  hardening ....................366
  heat treatment of........356, 368
  nickel-molybdenum ...........364
  normalizing ..................367
  open-hearth ..................358
  physical properties ...........355
  SAE numbering ...............353
  silicon-manganese, chemical
    composition ............361
  stainless .....................365
**Steel** measuring tapes........... 79
  parts, drawing ................368
  parts, hardening ..............368
  parts, hardening and drawing..370
  protractor .................80, 81
  rules ......................... 79
  tubing, welding on.............234
  wire gage ..................... 81
**Steels**, alloy, with carbon.......354
  alloy, with nickel.............358
  chrome-nickel ................361
  chromium, chemical composi-
    tion of ..................360
  cutting, free .................358
  heat-treatment of ............366
  manganese, composition of....358
  molybdenum, chemical composi-
    tion .....................360
  nickel, chemical composition of.359
  nickel-chromium .............359
  various types and grades......355
Stiffeners, repairing ............284
Stone, India ....................314
Stoning reamers ................314
Stop countersink ...............145
**Straight** edge ................. 85
  saw tooth .................... 23
**Straightening** by rolling.........291
  stringers or intermediate frames.280
**Strength**, bearing, of Alclad sheet.179
  bearing, of aluminum alloy
    plates ...................135
  bearing, of driven rivets......136
  bearing, of rivets.........176-179
  bearing, of sheet..........177, 178
  of riveted joints..............154
  shear, of Alclad sheet.........179
  shear, of aluminum rivets......180
  shear, of bolts................156

384                                                                INDEX

**Strength**—*continued*
  shear, of copper rivets..........180
  shear, of driven rivets..........136
  shear, of pins.................156
  shear, of rivets........155, 177-179
  shear, of sheet...........177, 178
  ultimate .......................326
  of wrought aluminum alloys....328
  yield .........................326
**Stress,** bending ..................290
  in shear ......................290
Stressed skin, installing patches on. 186
Stresses, basic ....................289
  safe bearing design............158
**Stretch** out ....................120
  formula for ...............113, 118
  loop-type clamp ................124
Stringer splices ..............171, 172
Stringers, splicing of.........170, 286
  straightening .................280
Strip, coiled .....................338
Structural part, section of....... 53
Structures, aluminum alloy.......274
**Stud,** broken, extracting..........313
  threading .....................312
Subassemblies, dash numbers for.. 62
Surface gages .................... 83
Sweating .........................197
Symbols for section lines, material. 54

**T**

**Tables**
  Alcoa aluminum and its
    alloys ......................321
  Amount of protrusion for rivet
    head ........................139
  Approximate machine settings
    for seam welding aluminum
    alloys ......................212
  Approximate physical properties
    of steel (cold drawn)......355
  Approximate radii for 90° cold
    bend aluminum and alumi-
    num alloy sheet ............345
  Average coefficient of thermal
    expansion per degree Fahr-
    enheit ......................324
  Average ultimate strength for
    driven rivets ...............136
  Bearing and shear strength of
    A17S-T rivets ..............176
  Bearing and shear strength of
    17S-T rivets and sheet.....177
  Bearing and shear strength of
    24S-T rivets and Alclad
    sheet .......................179
  Bearing and shear strength of
    24S-T rivets and sheet.....178

**Tables**—*continued*
  Bearing strengths of aluminum
    alloy plates ...............135
  Bend allowances for 1°........115
  Chemical composition of carbon
    steels ......................357
  Chemical composition of chro-
    mium steels ................360
  Chemical composition of chro-
    mium-vanadium steel .....361
  Chemical composition of free-
    cutting steels..............358
  Chemical composition of man-
    ganese steels ..............358
  Chemical composition of molyb-
    denum steels ..............360
  Chemical composition of nickel
    steels ......................359
  Chemical composition of nickel-
    chromium steels ..........359
  Chemical composition of silicon-
    manganese steel ..........361
  Composition and melting point
    of soft solder..............196
  Composition and uses of com-
    mon magnesium alloys with
    symbols ....................351
  Conditions for heat-treatment of
    aluminum alloys...........342
  Drill sizes for bolts............ 65
  Drill sizes for machine
    screws ..................... 65
  Fluxes and solders for soft
    soldering ..................198
  Machine settings for spot weld-
    ing aluminum alloys......212
  Melting point of copper and zinc
    alloys ......................200
  Metal gage standard..........369
  Nominal composition of alu-
    minum sand-casting alloys..320
  Nominal composition of wrought
    aluminum alloys ..........320
  Number of rivets required for
    splices .................173-175
  Physical properties of aluminum
    alloys ......................181
  Physical properties of steel.....356
  Properties of tubes for splices
    using inside sleeves........233
  Properties of tubes for splices
    using outside sleeves...234, 235
  Radii required for 90° bend in
    terms of thickness..........345
  Safe bearing design stresses for
    aluminum alloy plates and
    shapes ......................158
  Safe bearing design stresses for
    driven rivets ..............158

# INDEX

**Tables**—*continued*
Shear strength of copper and aluminum rivets............180
Shear strength of rivets, bolts, and pins .................156
Sizes of rivet AN456...........134
Sizes of template material to be stocked ..................92
Temperatures and corresponding colors in heat-treating..367
Tip sizes, size of hole, and size of drill recommended for cleaning ..................205
Types of rivet heads...........134
Typical mechanical properties of wrought aluminum alloys..326
Typical properties of sand-casting alloys .................323
Typical properties of wrought alloys ....................322
Typical tensile properties of wrought aluminum at elevated temperatures ........328
Welding-torch tips, sizes, and recommended drills for cleaning ......................204
Tabs, lapped ....................300
Tabulation, channel .............122
**Tail** post .....................285
surface ribs, repair methods....276
**Taper** pin, fitting................315
tap ..........................311
Tapes, measuring, steel..........79
Tapping internally threaded parts.312
Taps ..........................311
**Technique** of filing.............. 31
riveting ...................141, 169
**Techniques,** forming ............264
welding .....................223
Tee joint, fillet welds in........230
Temper designation .............330
**Temperature,** effect on mechanical properties .................327
measurement of .............. 79
Temperatures and corresponding colors in heat-treating.....367
Tempering, colors for............367
Tempers ........................330
**Template,** cutting of............ 93
development of .............. 94
drafting of .................. 93
on pattern material...........102
scraping ....................247
**Templates** ..................91, 294
cutting out .................. 97
iron ........................266
layout of ................87-110
sheet metal, laying........... 98
transfer of dimensions to...... 96

**Tensile** properties of wrought aluminum alloys at elevated temperatures ............328
strength of aluminum alloys....318
strength of steel..............356
value of rivets................155
**Tension** .......................290
of wrought aluminum alloys...326
**Terms,** aircraft ................ 39
used in bending..........112, 113
**Tests** for cleaners...............288
for identifying heat-treated alloys containing copper.....275
for polishes .................289
**Thermal** conductivity... 321, 322, 323
expansion ...................321
expansion per degree Fahrenheit .....................324
treatment of wrought aluminum alloys ....................339
**Thickness** of bends and radii...345
metal ....................114, 120
plus radius ..................114
of sheets ....................291
Thin materials, welding on.....291
Third angle orthogonal projection. 68
Threaded parts, tapping........312
Threading a bolt or stud........312
**Threads** on bolts...............310
screw .......................310
Tightening nuts ................305
Tinners' stakes ................ 17
**Tips, welding-torch** .......204, 205
sizes of ................204, 205
Title block ..................56, 57
Tolerance .....................77
**Tolerances** .................... 45
working, for finished parts.....242
Tongs, wiring .................. 6
**Tool** identification ............ 28
requirements, personal ....... 5
**Tools** ........................5-28
measuring ..................79-86
personal .................... 1
shop-furnished .............. 14
Tools and equipment for drop hammer ....................242
**Teeth,** raker .................. 23
straight .................... 23
wave ....................... 23
Tooth construction of contour saws 23
**Torch,** oxyacetylene, fusing metal with .......................220
welding .....................203
Torch welding of aluminum.....208
**Torque** .......................290
loads for bolts...............306
Torsion .......................290
Traces of flux.................288

# The Aviation Collection by Sportsman's Vintage Press

www.SportsmansVintagePress.com

| | |
|---|---|
| Aircraft Construction Handbook | by Thomas A. Dickinson |
| Aircraft Sheet Metal Work | by C. A. LeMaster |
| The Aircraft Apprentice | by Leslie MacGregor |
| Aircraft Woodwork | by Col. R. H. Drake |
| Aircraft Welding | by Col. R. H. Drake |
| Aircraft Sheet Metal | by Col. R. H. Drake |
| Aircraft Engines | by Col. R. H. Drake |
| Aircraft Electrical and Hydraulic Systems, and Aircraft Instruments | by Col. R. H. Drake |
| Aircraft Engine Maintenance and Service | by Col. R. H. Drake |
| Aircraft Maintenance and Service | by Col. R. H. Drake |